Structure and Function of Chloroplasts

Edited by Martin Gibbs

Structure and Function of Chloroplasts

Edited by Martin Gibbs

With 91 Figures

Springer-Verlag Berlin · Heidelberg · New York 1971

Professor Dr. Martin Gibbs
Department of Biology, Brandeis University
Waltham, MA 02154, USA

ISBN-13: 978-3-642-88309-5 e-ISBN-13: 978-3-642-88307-1
DOI: 10.1007/978-3-642-88307-1

 Library of Congress Catalog Card Number 79-130575. Typsetting and printing: Carl Ritter & Co., Wiesbaden. Bookbinding: Großbuchbinderei Karl Hanke, Düsseldorf

Softcover reprint of the hardcover 1st edition 1971

Preface

It is now about 100 years since the chloroplast has been recognized as the site of photosynthesis in plant cells. The last 20 years have seen a striking increase in interest in the structure and function of the chloroplast. Hastened on by powerful new tools such as the electron microscope and the newer methods of isolation and analysis of chloroplasts, there is presently considerable experimental work on the properties of this organelle. In such a rapidly moving field and one which is reviewed systematically is various Annual Reviews, it is not possible to present a detailed critique of the prolific literature in a book of reasonable size. Rather the decision was made to sacrifice complete coverage of the field and to indicate general areas of investigation.

In organization, problems here dealt with, are those concerned with the electron microscopy of chloroplast structure, development and conformation, genetic control of chloroplast development, characterization of some of the major components of the chloroplast and the biochemical properties of the chloroplast including the formation of adenosine triphosphate and reduced pyridine nucleotide and the assimilation of carbon dioxide into carbohydrate with subsequent conversion to secondary products.

A historical outline on the general subject "Photosynthesis and the Chloroplast" has been included to place into proper perspective the rapid developments in the several areas covered in the book.

I am particularly indebted to Dr. Roy E. McGowan who assisted with the preparation of the Index and to Carolyn Levi and Sara Kempner for a careful reading of the finished copy.

Waltham, Massachusetts

Martin Gibbs

Contents

Historical Outline

ROBERT HILL

The Ultrastructure of Plastids

KURT MÜHLETHALER

Light-Induced Chloroplast Contraction and Movement

Frank Mayer

Plastid Inheritance and Mutations

Björn Walles

Nucleic Acids and Information Processing in Chloroplasts

Christopher L. F. Woodcock and Lawrence Bogorad

Lipids of Chloroplasts

ANDREW A. BENSON

Biochemistry of Photophosphorylation

MORDHAY AVRON

Carbohydrate Metabolism by Chloroplasts

MARTIN GIBBS

Biosynthesis by Chloroplasts

TREVOR W. GOODWIN

Contributors

M. Avron	Biochemistry Department, The Weizmann Institute of Science, Rehovoth, Israel
A. A. Benson	Scripps Institution of Oceanography, University of California, La Jolla, California, USA
L. Bogorad	Department of Biology, Harvard University, Cambridge, Massachusetts, USA
M. Gibbs	Department of Biology, Brandeis University, Waltham, Massachusetts, USA
T. W. Goodwin	Department of Biochemistry, University of Liverpool, Liverpool, England
R. Hill	Department of Biochemistry, University of Cambridge, Cambridge, England
F. Mayer	Institut für Mikrobiologie, Erlangen, Germany
K. Mühlethaler	Institut für Allgemeine Botanik, Zürich, Switzerland
B. Walles	Department of Forest Genetics, The Royal College of Forestry, Stockholm, Sweden
C. L. F. Woodcock	Department of Biology, Harvard University, Cambridge, Massachusetts, USA

Historical Outline

Robert Hill

Photosynthesis and the Chloroplast

The candle would not burn again and the mouse could not breathe until the sprig of mint had restored the air. By the most skilful experiments with gases Joseph Priestley in 1771 first found that the plants could reverse the effects of combustion and respiration. He surely was to be considered the discoverer of the process we now call photosynthesis. Stephen Hales in 1727 had already suggested that plants derived part of their nourishment from the air and that light was concerned. The necessity for light in the process was first proved by Jan Ingenhousz who had originally followed the initial experiments of Priestley.

Photosynthesis, as a definable process, emerged at the time of the development of chemistry into an exact science. While Priestley would be describing oxygen as "dephlogistigated air" Antoine Laurent Lavoisier was carrying out experiments on oxidation and finally disposing of the phlogiston theory. Yet, nearly 50 years had to pass until Julius Robert Mayer, who was a surgeon, boldly enunciated in 1842 the theory of conservation of energy and its significance in biology. This could then show more exactly how the false and much denigrated phlogiston theory, so confusing to the early investigators, had owed its popularity to satisfying the need to describe effects of chemical change in terms of a separate material principle. Robert Mayer has often been much criticized for his publication said to be based on insufficient evidence. James Prescott Joule published his measurements of the mechanical equivalent of heat in 1843. The perfectionist had been anticipated which led to hard feelings, yet science in general seems to benefit from these widely differing types of personality. A theory of energy conservation had doubtless been developing in many minds for some time previously. The possibility of its application to living organisms indicated a notable advance in scientific knowledge. It seems remarkable how relatively recently the part played by light in the growth of plants has had a philosophic significance. The fact that plants do not grow properly in darkness must have been known from the earliest times; this would be a practical consideration in the sphere of the artisans not in that of the philosophers. Ingenhousz and Mayer probably had contributed the most to the appreciation of light as a condition for the assimilation of carbon by green plants. It is thus easy to realise how the very great plant physiologist, Julius von Sachs (1832 to 1897), in his "History of Botany" was led rather to give credit to Ingenhousz for the discovery than to acknowledge that of Priestley.

In Tübingen in 1837 Hugo von Mohl described his observations of the chloroplasts as discrete green bodies in the microscopic examination of plant cells. This was followed by many observations of the presence of starch in the chloroplasts of various kinds of plants. Sachs demonstrated convincingly that the production of starch in the chloroplasts of a leaf depended on light. This starch, he considered, was the first visible product of photosynthesis. SACHS was then able to complete an equation for the assimilation of carbon by green plants; carbon dioxide and water give, in light, oxygen and starch. The work of the many investigators who had contributed to this statement has been ably described by EUGENE RABINOWITCH in his "Photosynthesis and Related Processes" (1945) and by JAMES RIDDICK PARTINGTON in "A Short History of Chemistry" (1957).

The statement of photosynthesis due to SACHS could suggest that the chloroplasts, individually considered were responsible for the overall process of photosynthesis. It was known that starch could disappear from a leaf on a plant in darkness. Then JOSEF BÖHM in 1883 discovered that the starch could reappear if the leaves were floated on sugar solutions in the dark. Starch, then might well be a secondary product. The evidence would be destroyed for the chloroplast having the monopoly of the whole photosynthetic process. The opponents of SACHS could then have an apparent advantage. NATHAN PRINGSHEIM in 1882 had even suggested that chlorophyll by absorbing light protected what might be a colourless photosynthetic system in the leaf from damage. However in spite of the difficulties, such as the unequal intensities given by a prism in forming the spectrum, it came to be generally accepted that the production of starch in a leaf was induced more powerfully by the regions of greater absorption by the chlorophyll.

The function of a chloroplast in photosynthesis was most beautifully shown by the experiments of THEODOR WILHELM ENGELMANN in 1894. By the use of motile bacteria he showed in a microscopic examination of the alga *Spirogyra*, that oxygen was produced by part of a chloroplast only in a position corresponding with the illumination. This classical experiment was performed on a living cell. The movement of the bacteria towards a higher concentration is an extremely sensitive test for oxygen, it operates in the region of 10^{-6} M, and it can be used on a very small scale.

GOTTLIEB HABERLANDT was able to show the production of oxygen from a single chloroplast from a moss. The method used was to cut the leaf in a sucrose solution and then free and presumably undamaged chloroplasts could be observed separated from the cellular tissue. ENGELMANN had once stated categorically that any damage to the structure of the chloroplast resulted in a complete and irreversible stop to photosynthesis. While, in a sense, this statement may still remain in truth, at that time, it seemed to put the whole problem out of the range of current biochemical methods. This impasse was definitely sequestered by HANS MOLISCH; he was able to show production of oxygen with preparations from dried leaves. The method used was to observe the luminescence of a bacterium. This was shown by MARTINUS WILLEM BEIJERINK to be an exceedingly sensitive test for oxygen; it is more sensitive than that with motile bacteria and was used by direct visual observation. MOLISCH showed that the oxygen was produced in light and required both the insoluble residue and the soluble aqueous part when the leaves were powdered and suspended in water. In the presence of water the activity was lost on heating or by digestion with trypsin. The amounts of oxygen were so small that significant measurement was impossible. It was the physiological approach only which for many years advanced the knowledge of vegetable assimilation, the only way to obtain measurable production of oxygen was to provide the plant with carbon dioxide in the light. The quotient CO_2/O_2 was found to be nearly unity. Chlorophyll and light seemed to be the only observable conditions for a plant to carry out the assimilation of carbon dioxide. From the physiological studies on the living cells two component processes in the response of the plant to varying external conditions could be inferred. This detailed analysis first carried out so successfully in 1905 by FREDERICK FROST BLACKMAN separated a photochemical process with a temperature coefficient of nearly unity and a dark process which resembled that due to an enzyme; having a high temperature coefficient. The photosynthesis was measured by accurate gas analysis in samples of air of known CO_2 and O_2 content passing over a leaf in a chamber which could be illuminated. RICHARD WILLSTÄTTER and ARTHUR STOLL extended the work of BLACKMAN. They also made quantitative estimations of the chloroplast pigments and a series of studies published in 1913 by JULIUS SPRINGER on the chemical nature of chlorophyll itself. The degradation products of the chlorophylls they found to be almost innumerable. Thus, in the absence of the micro methods for chemical analysis now available they considered it essential to work with large amounts of material. Their neglect of the beautiful method of MIKHAIL TSWETT for separation of pigments with chromatography could be thought to account for the long interval before the modern technique was developed by ARCHER JOHN PORTER MARTIN and RICHARD LAURENCE MILLINGTON SYNGE. The presence of magnesium and the mode of combination in the chlorophyll molecule suggested to WILLSTÄTTER and STOLL the possibility of a direct reaction with CO_2. It was to be considered that the photochemical reaction was the rearrangement of a CO_2 chlorophyll compound with water in light to give a peroxide function. This could be broken down in the enzyme process to give oxygen and a reduced product from CO_2. The view that in photosynthesis the oxygen actually derives directly from CO_2, carbonic acid or a carboxylic acid seemed to be the simplest interpretation of the original physiological studies. No support for this, however, could be obtained from any of the experiments with chlorophyll itself.

At the time of these early physiological methods the beginnings may be traced of completely different approaches in experiment and in theory to the problem of carbon assimilation. It seems possible that one significant starting point for new technique was the discovery by JOHN SCOTT HALDANE in 1898 that a solution of oxyhaemoglobin from blood would part with the whole of its oxygen when ferricyanide was added. The amount of oxygen liberated could be accurately determined by measuring the pressure change in a closed vessel. This lead to a further development of manometric methods for physiological problems by HALDANE and JOSEPH BARCROFT. During the first decade of this century OTTO HEINRICH WARBURG seems to have realised the enormous advantage manometry would have if it could be used for measuring photosynthesis and developed his now famous method. By the use of microbiological techniques this could achieve a high degree of standardisation of the living material. Also the effects of substances acting as poisons or inhibitors could be analysed; WARBURG also determined the effect of intermittant illumination. This approach was greatly developed later by HANS GAFFRON and by ROBERT EMERSON in particular. EMERSON with WILLIAM ARCHIBALD ARNOLD and later with RUTH CHALMERS provided the basic experimental results which now govern much of the present theoretical interpretation in photosynthesis.

But a still more far reaching change in outlook was to come. ALBERT JAN KLUYVER and CORNELIUS BERNARDUS VAN NIEL developed the conception of many biological processes, including photosynthesis, being represented in terms of hydrogen transport. With bacteria were found forms which could reduce CO_2 in the dark utilising the energy derived from oxidation of components of the environment. Again, in the photosynthetic bacteria VAN NIEL proved that the reduction of CO_2 did not result in oxygen production, but in the simultaneous oxidation of a constituent in the medium for growth. Thus while a photosynthetic bacterium might reduce CO_2 in presence of H_2S and liberate sulphur the green plant could obtain the reduction with H_2O and liberate oxygen. This generalisation by VAN NEIL in terms of hydrogen transfer was to lead to a completely different conception of photosynthesis with the reduction of CO_2 being a process completely separate from the production of oxygen. After the late 1930s it might be said that the way had been indicated for a biochemical approach to the study of photosynthesis. During this decade the writer was fortunate enough in finding a measurable production of oxygen from chloroplasts isolated in bulk. The oxygen was measured spectroscopically using myohaemoglobin. From the experiment it was concluded that the chloroplast preparation, when illuminated, produced oxygen together with a stoichiometric reduction of an added hydrogen acceptor and that carbon dioxide was not actually involved. This confirmed the general theory developed by VAN NIEL as it applied to a green plant. The simplest interpretation of the experiments was that both oxygen and the hydrogen (or reducing equivalents) were to be regarded as being derived from the water.

It was just at this time that the use of tracer elements in biology was being developed. With the very labile isotope ^{11}C ($^1/_2$ life 21 min) SAMUEL RUBEN, MARTIN DAVID KAMEN and WILLIAM ZEV HASSID could show conclusively that the process of photosynthesis was not a straightforward photo reduction of carbon dioxide. This was followed by the discovery of RUBEN and KAMEN in 1940 of the isotope ^{14}C. With this isotope becoming generally available in 1945 MELVIN CALVIN and colleagues were able to trace in less than 10 years the whole path of carbon in a green plant from carbon dioxide to carbohydrate. This achievement owed a great part of its success to the development of techniques in chromatography and to the rapid advances of enzyme biochemistry. EFRAIM RACKER was subsequently able to show that the conversion of carbon dioxide to a carbohydrate derivative could occur in darkness with enzymes from a leaf, along the lines indicated by CALVIN. Reducing power and phosphate hydrolysis energy were made available to replace the action of light. The chloroplast would have to yield, in the light, reduced coenzyme II and adenosine triphosphate. This would involve reactions with water and inorganic phosphate. Chloroplast preparations were indeed shown to have these required properties when illuminated. This was due to the discovery by ANTHONY SAN PIETRO and HELGA M. LANG of a factor inducing the reduction of coenzyme II (variously referred to as TPN and NADP) in light and to the discovery by DANIEL ARNON, MARY BELLE ALLEN and FREDERICK ROBERT WHATLEY of photosynthetic phosphorylation. At the same time the corresponding process of photophosphorylation was discovered by ALBERT FRENKEL in a cell free preparation from a photosynthetic bacterium.

Biochemistry of photosynthesis then could be divided into two separate aspects, the photochemical H-transfer and the fate of the carbon dioxide. How then could a chloroplast accommodate both activities if indeed it could perform the complete process?

The isolation of chloroplasts in bulk as a subcellular constituent of cytoplasm showing relevant activity seems to have been the first of its type. It was followed about 10 years later by the isolation of mitochondria from animal tissues. As with the green plants sucrose was at first used for the osmotic control of media, and the original methods of isolation seem to have developed independently. At the present time the structures and activities of the chloroplast and the mitochondrion are frequently compared. The original work of DAVID KEILIN first showed the function of the cytochrome system in respiration. This involved a sequence of H- (or electron) transfers finally resulting in the reduction of molecular oxygen directly to water. There was no indication that hydrogen peroxide played any part. In the early experiments with chloroplasts the production of oxygen seemed to resemble the reverse of the respiratory system. The characteristic activities were in each system attached to an insoluble fraction. The chloroplast, indeed was found to have its characteristic cytochrome components. The presence of heme compounds in chloroplast preparations was first observed by SAM GRANICK.

The mitochondrial cytochrome system reduces the oxygen molecule to water. The four reducing equivalents can originate from carbohydrate resulting in liberation of the molecule of carbon dioxide. In photosynthesis the carbon dioxide is taken up. The illuminated chloroplast may be said to form the oxygen molecule by oxidising water. The four reducing equivalents are made available by the photochemical process for the reduction of the carbon dioxide to carbohydrate again. The essence of Priestley's discovery relating animal respiration and green plants may be represented by activities of two cytoplasmic organelles. By gradual improvements in the techniques of isolation of chloroplasts DAVID ALAN WALKER was eventually able to show how to obtain high rates of photosynthesis with this cell free system. Furthermore it has been found by MARTIN GIBBS that the pattern of CO_2 fixation in chloroplasts isolated from the cells could be identical with that obtaining in the whole leaf. Thus it came about that the original conclusions of ARNON and colleagues from their pioneer results on carbon fixation with isolated chloroplasts were vindicated. The time-honoured contention by SACHS that the chloroplasts were the sites of the complete process of photosynthesis still may be maintained.

The study of what is termed the "respiratory chain" in mitochondria was initiated by the direct spectroscopic observations which KEILIN made on behaviour of the cytochrome components in living cells. The chloroplast can equally be said to possess a "photosynthetic chain". BRITTON CHANCE initiated the more recent great development of spectroscopic methods for kinetic studies. They are now used in great variety and may be said to form an essential link between the physiological and the biochemical methods. The first application of these methods in photosynthesis was due to LOUIS NICO MARIE DUYSENS. He discovered characteristic kinetic changes in absorption due to bacteriochlorophyll in living bacteria. The studies of transient illumination effects by LAWRENCE ROGERS BLINKS with red algae and the enhancement of far red light discovered by EMERSON and RUTH CHALMERS have indicated the requirement for two photochemical systems for the production of oxygen and reduction of carbon dioxide. The photosynthetic chain which operates for the production of oxygen is now considered to consist of two parts. These would have distinct molecular composition both as regards the absorption of light and the hydrogen (or electron) carrier system.

From the recent direct biochemical and physiological studies of photosynthesis a certain kind of molecular structure of the chloroplasts may be inferred. The best interpretation of structure possible using the light microscope seems to have been given by Harold Wager by his observations recorded in 1905. The development of the electron microscope since 1940 and its application in a variety of ways to chloroplasts and chromatophores continues to give an increasingly detailed picture of structure in physical terms. The question as to whether the problem of representing the structure and function of the chloroplast from the two approaches is to be solved in unified or in complimentary terms remains to be decided in the future.

The Ultrastructure of Plastids

Kurt Mühlethaler

A. Introduction

According to light microscopic observations the plastids represent a heterogenous group of cell organelles which vary in their morphological and physiological behavior. Schimper (1885) distinguished three main groups which he termed according to their color *leucoplasts* (white), *chloroplasts* (green) and *chromoplasts* (yellow). In addition to these three common types a number of other plastids have been described. The red and brown plastids in algae have been termed *rhodo-* and *phaeoplasts* (Kylin, 1910) or the blue ones, *cyanoplasts*. The colorless plastids are widely distributed in young plant cells. These proplastids are often specialized to store reserve material such as starch, protein or lipids. They are then termed *amyloplasts*, *proteinoplasts* and *lipo-* or *elaioplasts*, respectively. These types are not clearly defined from each other. Intermediate stages are very frequent and difficult to classify in one or the other group. As first shown by Meyer (1883) and Schimper (1885) a plastid may subsequently pass through different types during its ontogenic development. According to Schimper (1885) a plastid may differentiate from a leucoplast into a chloroplast and end its development in the chromoplast stage. It was thought also that a reverse transformation could be possible. Some observations indicated that during differentiation one stage may be left out. Newer evidence collected with the electron microscope indicates that the regular differentiation of a plastid begins at the proplastid stage (v. Wettstein, 1958; Mühlethaler and Frey-Wyssling 1959). Thus the proplastids would represent the juvenile stage, the chloroplasts the organelle with full photosynthetic activity and the chromoplasts as degenerated end-forms. The specialized plastids, such as amyloplasts, proteinoplasts and elaioplasts may be grouped along these general developmental stages. The relationships between these types are indicated in Scheme 1:

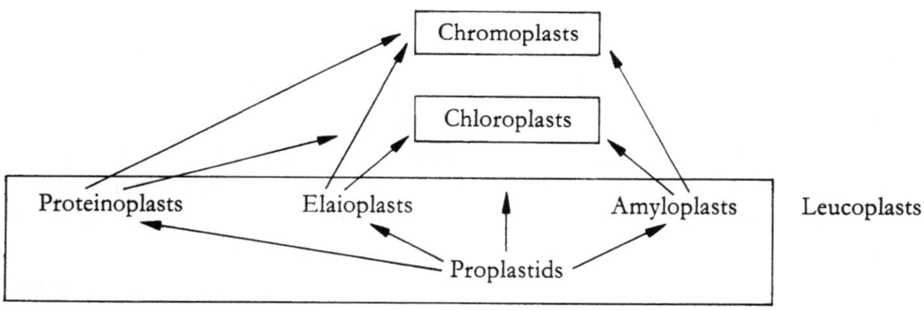

Scheme 1

To what extent the general monotropic development may be reversed is still controversial. SCHÖTZ and SENSER (1961) described a reversible conversion from the chromoplast stage into a chloroplast in petals of the *Oenothera* mutant "Weissherz". This is in contradiction to the results obtained by FREY-WYSSLING and KREUTZER [1958 (1, 2), and FREY-WYSSLING *et al.* (1955), in *Capsicum annuum* and *Ranunculus repens*. Such retrogressive changes which we would call dedifferentiation probably are quite exceptional. As a rule the differentiation is monotropic, as indicated in Scheme 1.

B. Proplastids

1. The Prolamellar Body

The early stages of plastids are amoeboid and have a size of about 0.4 to 0.9 μ in diameter. The envelope of a plastid is formed by a unit membrane and its stroma may contain a number of vesicles and lamellae (Fig. 1). These internal structures have their origin in the envelope of the plastid. Under proper illumination the inner membrane of the envelope forms sac-like flat compressed vesicles which have been termed "thylakoids" by MENKE (1961). If the leaves are kept in the dark, this lamellar system is not formed. Instead, a number of such vesicles bud off and aggregate to form a so called prolamellar body (HODGE *et al.*, 1956). This vesiculous body was observed originally by LEYON in 1953, who described it as "dense core". Subsequently other terms were introduced such as "primary granum" (STRUGGER and PERNER, 1956), "plastid center" (v. WETTSTEIN, 1957), "vesicular body" [GEROLA *et al.*, 1960 (1)], "vesicular center" (EILAM and KLEIN, 1962) and "Heitz-Leyon crystal" [MENKE, 1962 (1)]. The term "prolamellar body" has been used most frequently and will be adopted in this review. The early electron micrographs published by LEYON (1953), HEITZ (1954) and PERNER (1956) suggested this structure to be a three dimensional array of beaded strands. WETTSTEIN (1958) proposed a model of tubules arranged in a cubic lattice. The beads thus would represent points where six tubules meet. The ultrastructural arrangements of these prolamellar bodies were clarified by GRANICK (1961), GUNNING [1965 (1)], WEHRMEYER (1965) and GUNNING and JAGOE (1967). As found by GRANICK (1961) the body in *Avena* proplastids is made up of a number of tubules running in the three major axes of a cubic lattice. Where three tubules meet and fuse at the corners of the cubical "unit cells" they are swollen and their membrane surfaces are smoothly confluent (Fig. 2). The membrane surface is continuous throughout the prolamellar body. It forms the boundary between an enclosed space and the plastid stroma which penetrates the lattice between the tubules. Approximately three quarters of the body consists of plastid stroma. In sections stained with uranyl acetate GUNNING [1965 (1)] observed ribosome-like particles, centrally located between the "unit cells". According to GUNNING [1965 (1)] and GUNNING and JAGOE (1967) the ultrastructural arrangement of the tubules may be influenced by a field centered on and generated by a lattice of ribosomes, which provides a template on which membranes either come to lie or are actually formed. The size of these particles is only about two-thirds of the ribosomes seen in the cytoplasm. In addition to this cubic lattice membrane system observed in proplastids of *Avena* coleoptiles, other types of prolamellar bodies have been described.

WEHRMEYER (1965) found concentric and non-concentric types in etiolated bean leaves. In the former the basic units form a system of tetrahedrally branched tubules joined to five—and six—membered rings. The center of the lattice has the shape of a pentagonal dodecahedron. The non-concentric prolamellar bodies follow either the crystal lattice of zincblende or wurzite or a combination of both. It is now agreed

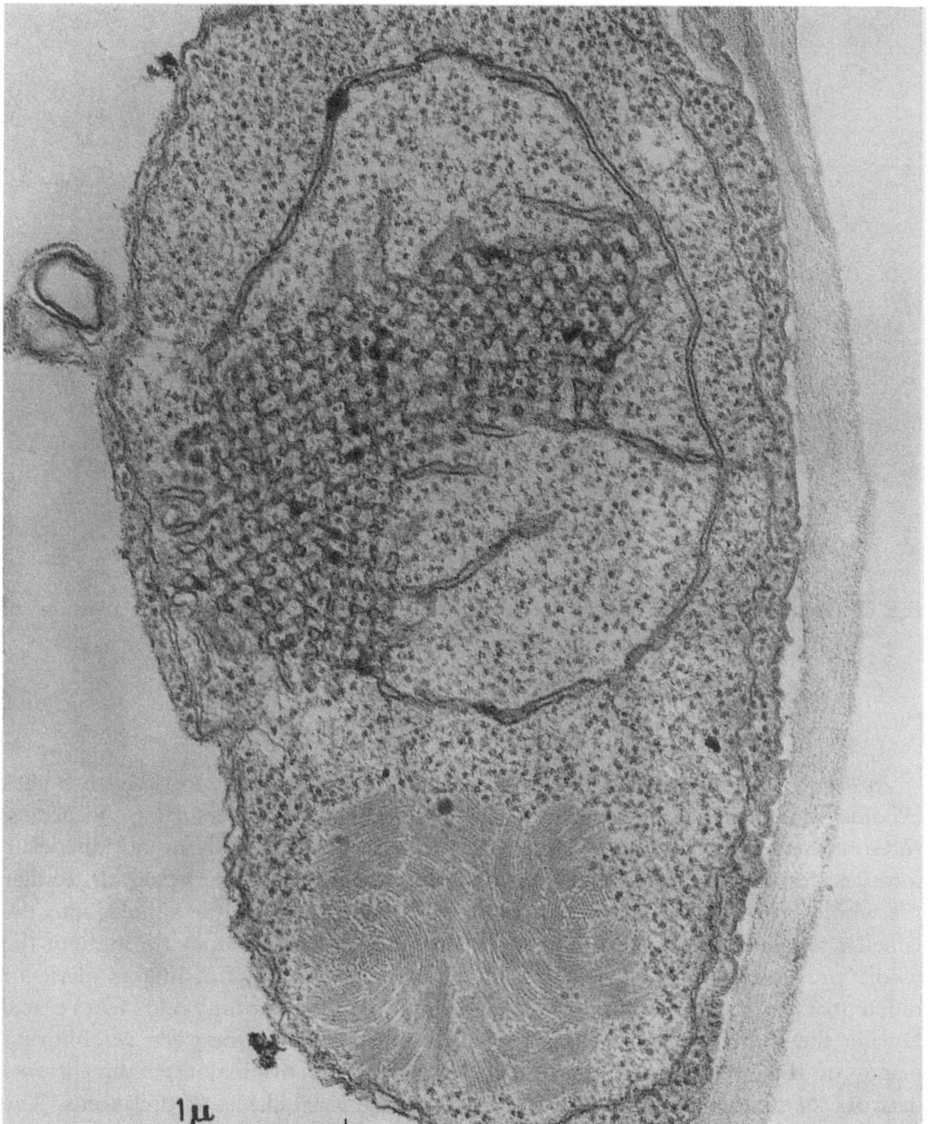

Fig. 1. Etiolated proplastid (Etioplast) in a leaf of *Avena sativa*, showing prolamellar body, stroma center, ribosomes and a few thylakoids (Courtesy Dr. B. E. S. GUNNING)

(GUNNING, personal communication) that cubic lattices are not very common. The majority of prolamellar bodies are undoubtedly based on tetrahedrally branched tubules arranged in one or the other of the lattice-types described by WEHRMEYER (1965).

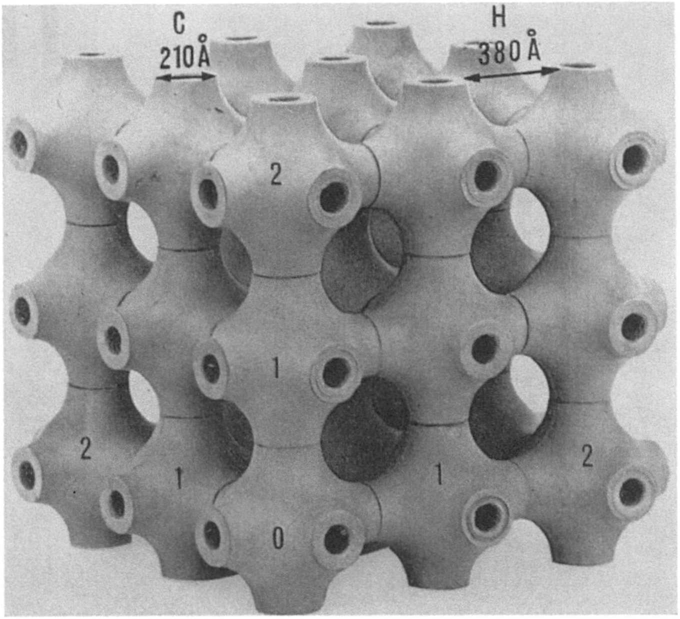

Fig. 2. Photograph of a model of the prolamellar body membrane system, consisting of 27 six-armed "nodal units" (GUNNING and JAGOE, 1967)

2. Thylakoid Formation

If leaves are illuminated after a prolonged etiolation the prolamellar bodies become disorganized and a number of thylakoids are formed on its surface. According to GUNNING and JAGOE (1967) the tubules of the prolamellar body do not dissociate into vesicles which disperse and ultimately fuse to form the new thylakoids. Rather the whole process is accomplished by a direct transition from the tubules into the lamellar thylakoids. This also implies that the surface is continuous throughout the whole process of light-induced growth of lamellae. In addition, in this phase of differentiation the conversion of protochlorophyllide into chlorophyllide takes place. During the formation of the thylakoid the connecting arms between neighboring planes of the "nodal units" are pinched off and the original three-dimensional network of tubules is transformed into two-dimensional sheets of thylakoids. The lattice spaces containing the ribosomal-like particles survive a certain time as perforations in the sheets of membrane. GUNNING and JAGOE (1967) observed that the pores disappear after 10 h of illumination. This may be correlated in time with the end of the lag period. If the leaves are etiolated again after a period of illumination a new prolamellar body will be formed. According to v. WETTSTEIN and KAHN (1960) this is done within 2 to 6 h after the bean leaf was returned to darkness. In *Avena* seedlings, GUNNING and JAGOE (1967) observed a new crystalline lattice after 45 min in darkness. The situation becomes complex after prolonged exposures to light because then the plastids contain both thylakoids and prolamellar bodies. The thylakoids formed earlier are preserved in the state they reached before the leaves were returned to darkness.

3. Constituents of the Proplastids

In addition to the prolamellar body a number of other structural details are frequently seen in proplastids. They comprise osmiophilic globules, aggregations of fine fibrils termed the "stroma center", areas similar in appearance to the bacterial nucleoplasm and large numbers of ribosomes (Fig. 3).

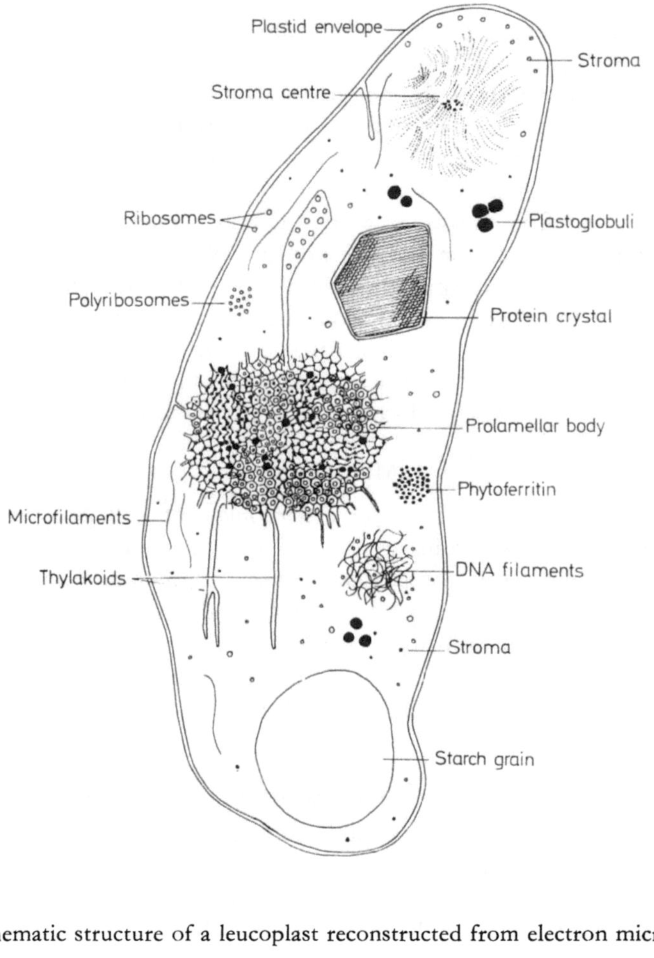

Fig. 3. Schematic structure of a leucoplast reconstructed from electron micrographs

4. Ribosomes

Ribosomal fractions were isolated from plastids of different plants (LYTTLETON, 1962; BRAWERMAN, 1963; MURAKAMI, 1962; JACOBSON *et al.*, 1963). These elements incorporate labelled amino acids into protein as the cytoplasmic ribosomes do (EISENSTADT and BRAWERMAN, 1963; BAMJI and JAGENDORF, 1966; EISENSTADT, 1967).

The ribosomes are held together by a RNA strand forming polyribosomes (GIERER, 1963; PENMAN *et al.*, 1963; SLAYTER *et al.*, 1963). The shape and orientation of these ribosomes were analysed statistically by BROWN and GUNNING (1966). As opposed to illuminated plastids, where clusters of ribosomes were observed, a random particle distribution was found in etiolated proplastids. This finding supports the biochemical evidence that light stimulates protein synthesis in plastids (KUPKE, 1962; RHODES and YEMM, 1963; MARGULIES, 1964).

5. Nucleic Acid Fibrils

Regions within the stroma with the same structure as the bacterial nucleoplasm have been described at all stages of plastid differentiation. They appear as areas of lower density and contain a network of fine fibrils, each 25 to 30 Å in diameter and of indefinite length (RIS and PLAUT, 1962; KISLEV *et al.*, 1965). These strands can be removed by treatment with deoxyribonuclease, which is confirmatory evidence that they represent DNA fibrils. First evidence for the presence of DNA in chloroplasts was obtained by the Feulgen staining method (CHIBA, 1951; METZNER, 1952; SPIEKERMAN, 1957). These results were controversial because other investigators were not able to confirm a positive reaction (LITTAU, 1958; GIVNER and MORIBER, 1964). Additional evidence for the presence of DNA in plastids came from autoradiographic methods (BRACHET, 1958; GIFFORD, 1960; WOLLGIEHN and MOTHES, 1963; BELL *et al.*, 1966). More recently the attempts to demonstrate the presence of DNA by direct chemical analysis of isolated chloroplasts were successful (BALTUS and BRACHET, 1963). All this evidence proves that the plastids contain a small amount (about 0.01%) of DNA which is located in the stroma in the form of fine fibrils.

6. Microfilaments

Occasionally it is possible to observe microfilaments, 50 to 60 Å in diameter, running lengthwise in the stroma near the plastid envelope. NEWCOMB (1967) was able to trace individual fibrils for a distance of several microns. Similar strands were described by WOHLFARTH-BOTTERMANN (1964) in the slime mold *Physarum* and by NAGAI and REBHUHN (1966) in the streaming cytoplasm of *Nitella*. It is suggestive to postulate that these filaments are the contractile elements of the cytoplasm. Since the plastids are able to contract and expand it seems possible that these elements have this function.

7. Plastoglobuli

All types of plastids contain osmiophilic globules which vary in size and number. The function of these so-called plastoglobuli (LICHTENTHALER and SPREY, 1966) is not fully understood, but it is generally assumed that they represent lipid storage granules. In proplastids they can be observed throughout the stroma as well as in the prolamellar body. In chloroplasts their number and size increases with age. This lipophanerosis is very pronounced in degenerating chloroplasts. Studies on the structure and composition of plastoglobuli have been published by MURAKAMI and TAKAMIYA (1962), BAILEY and WHYBORN (1963), GREENWOOD *et al.* (1963), LICHTENTHALER (1964, 1966, 1967), LICHTENTHALER and SPREY (1966), SPREY and LICHTENTHALER (1966), LICHTENTHALER and PEVELING (1967). After osmium fixation the globuli appear as densely stained dark bodies without a distinct internal structure. The absorption spectra of petroleum-ether extracts indicate that the globuli contain high amounts of lipophilic chloroplast quinones such as plastoquinone, α-tocopherolquinone, α-tocoquinone and vitamin K_1. The carotene content of the isolated plastoglobuli decreases with improved isolation methods which indicates that this compound is not stored in the globuli. After etiolation the plastids synthesize the lipophilic quinones at a lower rate. The formation of a number of plastoglobuli in chloroplasts indicate that a surplus amount of lipids is produced. The process of synthesis does not stop after the formation of the thylakoid system. The quinone-tocopherol content reaches its maximum shortly before the chloroplasts degenerate to chromoplasts (LICHTENTHALER and SPREY, 1966). LICHTENTHALER (1967) was able to demonstrate that thylakoid-free proplastids in etiolated seedlings of *Hordeum vulgare* contain numerous plastoglobuli. These appear rather sporadically in the chloroplasts of illuminated plants. This is additional evidence that these structures represent reservoirs for lamellar lipids which are accumulated in the dark prior to thylakoid formation.

8. Stromacenter

By means of fixation with glutaraldehyde a new structural constituent was observed by GUNNING [1965 (1)] in *Avena sativa* proplastids (Fig. 1). It was termed "stroma center" and consists of a distinct region having the appearance of an aggregated mass of thin fibrils. In transverse sections each fibril is some 85 Å in diameter. Their length is uncertain, because the fibrils pass out of the plane of the section. However fibrils measuring up to 1200 Å have been seen. The whole aggregate is usually up to 1 μ in diameter. A limiting membrane is absent and the stroma does not penetrate into this mass of bundled fibrils. Lipid, polysaccharide or nucleic acid staining reactions are negative supporting the view that this body contains protein only. Stroma centers have been noted in proplastids prior to the formation of any extensive membrane system. They exist in both, light grown and etiolated plants. With the formation of the thylakoid system the stroma center degenerates and a large number of osmiophilic globules appear in its place. With high resolution, GUNNING (1967) was able to show that the fibrils are composed of smaller units. Negatively stained fraction I protein isolated from *Avena* on the one hand and the basic unit of the fibrils on the other, show an identical morphological structure. As described by HASELKORN *et al.* (1965), fraction I protein particles appear as cubes having an edge length of about 120 Å.

After negative staining a central depression and structure of smaller units becomes apparent. Based on morphological and biochemical results the complex comprises 24 identical subunits, each having a molecular weight of 22,500. In the cube, the 24 subunits are fitted along each edge and one in the center of each of four faces. In side view three rows of three units each are packed in one face of the cube.

Particles resting on their top or bottom face where the central unit is missing appear as having a hole in the middle. The diameter of the subunits measures approximately 20 Å to 30 Å. Isolated fraction I protein has a pronounced tendency to form linear aggregates thus forming identical fibrils as seen in the stroma center. According to GUNNING (1967) stroma centers are not common structural constituents of plastids. They can be artificially produced if the proplastids are dehydrated. The available data are not sufficient to draw any conclusions about the function of these constituents. They could represent a pool for fraction I protein or the center of their synthesis. Enzymatically this protein has been characterized as ribulose-1,5-diphosphate carboxylase (TROWN, 1965). Since no ribosomes or polyribosomes are found within the stroma center it seems likely that it represents deposits of fraction I protein.

9. Phytoferritin

Besides the ribosomes, smaller particles having a diameter of about 55 Å can be noted in thin sections of proplastids. Often they are in clusters but they may also be spread uniformly throughout the stroma. Their localization, structure and distribution was studied by HYDE *et al.* (1963) in proplastids of pea embryo cells. They have been found to occur in epicotyls, root meristems and cotyledons. Their properties are identical with animal ferritin. Isolated phytoferritin particles have a diameter of about 105 Å and an iron: protein ratio of 0.17. It seems that these phytoferritin particles occur widely in angiosperm seeds and seedlings. The particles have been noted by SITTE (1961) in *Pisum* root meristems and by SCHNEPF (1962) in *Passiflora* nectary gland cells. HYDE *et al.* (1963) found them in proplastids of maize leaves. During differentiation the particles disappear and the entire molecule is disassembled. This indicates that phytoferritin represents an iron source for the development of the photosynthetic apparatus.

C. Amyloplasts, Proteinoplasts and Elaioplasts

It must be assumed that proplastids are able to produce polysaccharides, lipids and proteins. In special regions of the plant the plastids may produce one of these substances in abundance. In case the polysaccharide product is stimulated, starch grains are formed and the plastid is characterized as an *amyloplast*. The number and size of starch grains in an amyloplast may be variable. In the plastids of some tissues, such as in seeds and embryonic tissues, protein crystals may occur. In this event one speaks of *proteinoplasts*. According to KÜSTER (1951) protein crystals are abundant in embryo sacs of *Lilium sp.* and in the milk sap of *Cecropia peltata*. The third type, the *lipo-* or *elaioplast* stores lipid material in the form of globules of various sizes as described earlier. They are frequently observed in diatoms and in cells of higher plants such as in young leaves of *Vanilla* and in some orchids. In addition to these three plastid types, organelles have been observed which represent intermediate stages, storing starch as well as lipids or protein. They may be classified under the general term *leucoplast*.

As discussed earlier, the lipids are deposited in the stroma as "plastoglobuli". The starch grains are also found in the same area. A membrane has not been observed around these inclusions. On the other hand, storage protein is deposited within membrane-bounded sacs. NEWCOMB (1967) observed the membranes to be formed from the inner sheet of the plastid envelope. This means that they are identical with the thylakoid membranes. In contrast to etiolated leaf proplastids, in which the organization of the thylakoid system is strongly affected by light, the formation of sac like invaginations in root tip cells of *Phaseolus vulgaris* occurs also in the dark. This fact indicates that this "tubular complex" appears to be analogous to the highly organized prolamellar body. In these root cells the plastids are obviously specialized for protein storage. Under these conditions they do not react to light as those found in leaf cells. It might be that root plastids do not contain or synthesize protochlorophyllides which are sensitive to light. In etiolated leaf proplastids the protochlorophyllides are converted to chlorophyllides during the light-induced thylakoid formation. The protein which is stored in the tubular complex is likely to have its origin in the stroma. According to NEWCOMB (1967) ribosomes are more abundant in plastids with large protein bodies than in those devoid of storage protein. The protein crystals are variable in size and seem to be composed of smaller spherical units having a diameter of about 85 Å. In aleuron bodies the globular subunits are 150 to 160 Å in diameter (PERNER, 1965). In the early stages, the reserve protein is deposited in an amorphous state. Crystallization begins later and due to this new arrangement the bounding membranes of the sacs become deformed, conforming to the crystal outlines.

D. Chloroplasts

1. Architecture of the Lamellar System

In the light microscope a chloroplast can be differentiated into a clear structureless stroma which contains a number of small green discs. The latter have been termed "grana" and were thought to contain exclusively the pigments with photosynthetic activity. Investigations with the help of the electron microscope have shown the grana to be composed of a stack of lamellae. As mentioned before, these lamellar units have been termed thylakoids by MENKE (1961). In shape and distribution these thylakoids may differ considerably from one another (Fig. 4). Some are restricted to the granum area whereas others may cross the whole chloroplast stroma. The small lamellae in the granum area are called grana-thylakoids. Those having a larger extension are generally referred to as stroma-thylakoids. Both types are alternating at random in the lamellar system. In order to describe this system WEIER *et al.* (1965, 1966) have suggested the following terminology: As shown in Fig. 5 the contact area of two neighboring thylakoids form a *partition*. In the closely stacked thylakoids the individual membranes cannot be resolved and appear as dense bands in cross section. Both ends of the stacked thylakoids are in contact with the stroma and form the *margin*. In case the thylakoid crosses an intergrana region the membrane may form *frets*. The central space of the thylakoid is referred to as *loculus*. An *end-granal membrane* is the outermost thylakoid membrane in a granum stack. The space between two adjacent lamellae would represent an *interthylakoid*—or *interdisc space*. Some confusion is caused in this terminology because many authors do not use the terms in their original meaning.

Due to light microscopic observations it was generally concluded that the photosynthetic pigments are located solely within the grana. As a consequence the processes of photosynthesis were believed to take place in this area only. The fact that the thylakoid membrane appears somewhat thicker in the partitions led some authors to believe the chemical composition to be different in the grana region than in the stroma layers. It has been shown by LINTILHAC and PARK (1966) that the chloro-

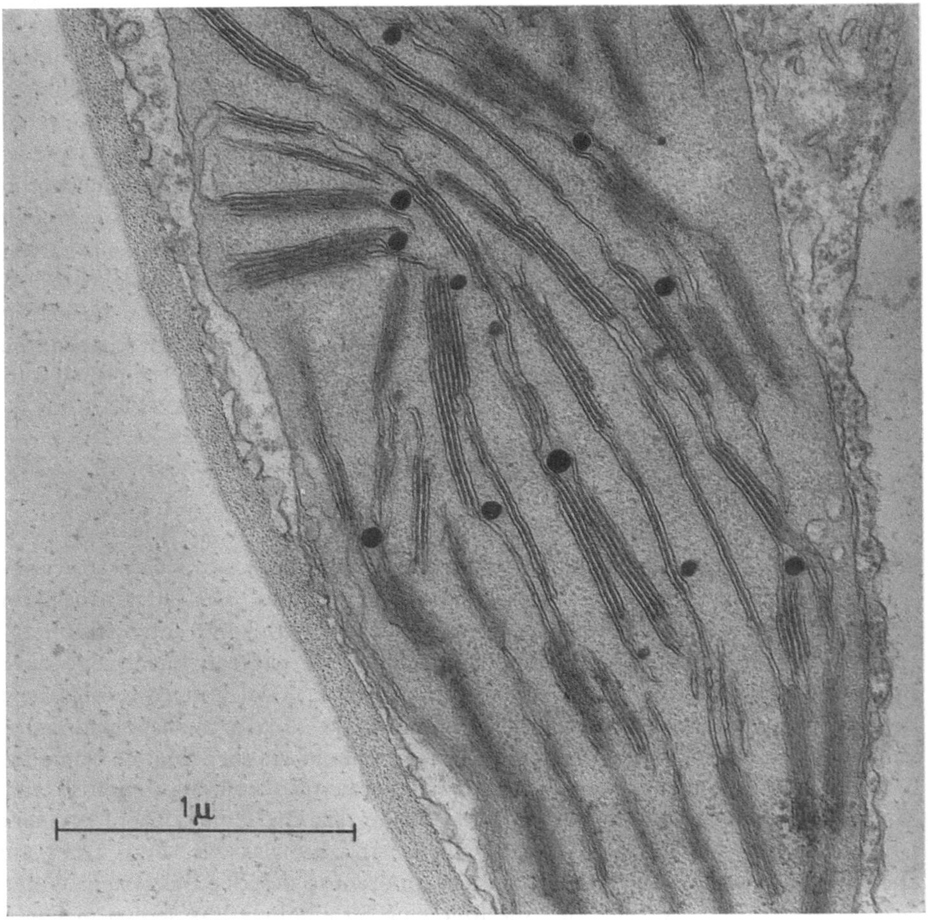

Fig. 4. Chloroplast in an *Avena* leaf, showing the thylakoid system and a few plastoglobuli

phyll content is the same throughout the whole lamellar system. In ultraviolet light, isolated grana- and stroma thylakoids exhibit the same fluorescence which means that they contain the pigment in the same concentration. The higher number of thylakoids in a granum region may be responsible for the more intensive green color.

The relationship between the stroma and grana system has been described by numerous authors: STEINMANN (1952), MÜHLETHALER and FREY-WYSSLING (1959),

v. Wettstein (1959), Gerola *et al.* [1960 (2)], Menke (1960), Mühlethaler (1960), Sitte (1962), Wehrmeyer [1963 (1, 2), 1964]), Weier *et al.* (1963, 1965), Heslop-Harrison (1963), Schötz (1965). Each stroma thylakoid is depicted as pursuing an independent course through several grana. From their work Weier *et al.* (1965) concluded that the thylakoid is fenestrated dividing the lamella into compartments. Newer investigations by Sitte (1962), Wehrmeyer [1963 (2), 1964)], Schötz (1965), Diers and Schötz (1966) have shown that the stroma thylakoids are not punctured by pores of different size but form large uninterrupted areas extending through several grana. There are some indications that a "fret" system is formed in plastids which have been etiolated previously. During the formation of the lamellae the intertubular spaces of the prolamellar body are preserved for a short period of time. These pores disappear during differentiation. The fret system is probably formed only after etiolation and may represent a transition stage.

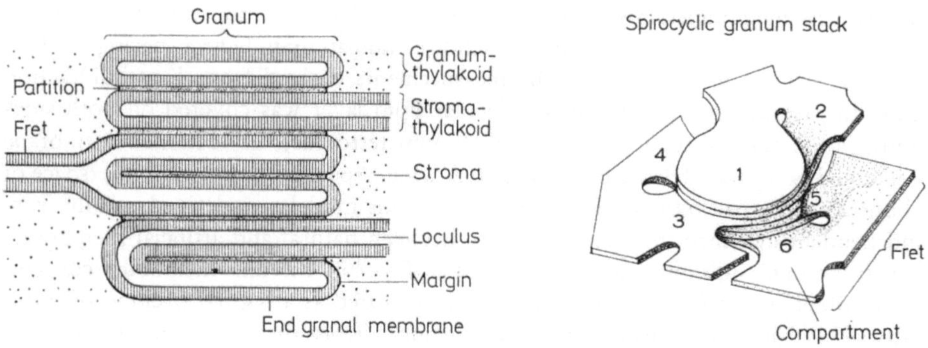

Fig. 5. Models of the arrangement of stroma and grana thylakoids. Granum formation by spirocyclic growth according to Wehrmeyer (1964). The nomenclature used has been proposed by Weier *et al.* (1965)

The arrangement and complexity of the lamellar system is best shown in the model proposed by Wehrmeyer (1964) (Fig. 5). The first thylakoids expand by marginal growth and form a basic system of fenestrated or continuous lamellae. At the margin, or by lateral evagination, the smaller grana discs are formed and stacked in a spirocyclic arrangement (Fig. 5). The formation and variability of the thylakoid system can be influenced by genetic factors (Döbel and Hagemann, 1963), the spectral composition of light (Fasse-Franzisket, 1955) or its intensity (Abel, 1965; Wild, 1958). Electron micrographs of sectioned chloroplasts are difficult to interpret because components which are continuous with each other at different levels may appear as independent structural units. In these cases serial sections or the study of whole isolated fragments may be helpful.

2. Ultrastructure of the Thylakoid Membrane

The recent improvement of resolution in modern electron microscopes does not automatically imply considerable gains in the knowledge of ultrastructure. Artefacts arising by fixation, embedding and sectioning procedures may change the original molecular architecture of a cell structure to a considerable extent. As we approach molecular dimensions with the high resolution microscope smallest changes of the colloidal state such as coagulation or crystallization effects, uncoiling or displacement of macromolecules may yield a structure considerably different from the living condition. The effect of fixation can be readily visualized if one part of the chloroplast sediment is preserved with osmium tetroxide and the other with permanganate. After LUFT (1956) introduced permanganate as a fixation agent a consistent three layered pattern appeared in the membrane of different cell organelles. This led ROBERTSON (1960) to postulate his unit-membrane concept. The comparison of X-ray diffraction measurements on nerve myelin (FINEAN, 1955) and electron microscopy led to a model which corresponded roughly to the one proposed earlier by GORTER and GRENDEL (1925) and DANIELLI and DAVSON (1935). The unit membrane as seen in the electron microscope has an average thickness of 75 Å. It was believed to contain a central bimolecular leaflet of lipids, covered on both sides with protein layers. Strong support for this unit membrane concept came from studies made on artificial membranes. STOECKENIUS (1959) was able to demonstrate that artificial myelin figures from the phospholipid fraction of human brain have a spacing of 40 Å. This value is in agreement with the measurements derived from X-ray diffraction data. In preparations where the bimolecular lipid layer was covered on both sides with globin and subsequently fixed in osmium tetroxide, the sectioned membrane contained two parallel dense lines 25 to 50 Å wide, separated by a lighter interspace of about 25 Å (STOECKENIUS, 1959, 1962). Since this model was derived mainly from independent evidence the close similarity between natural and artificial membranes constituted an additional strong argument in favor of the unit membrane concept.

In older schemes the lipid layer has been pictured as a regular packing of linear fatty acid chains. This arrangement was deduced from the birefringence effects of nerve myelin and chloroplasts. Their optical anisotropy was ascribed to the lipids. When the lipids are removed they produce striking birefringent myelin forms (WEBER, 1933; MENKE, 1938; FREY-WYSSLING, 1937). As first shown by MENKE (1938), the chloroplasts of *Closteria* exhibit a negative-textured birefringence and a positive intrinsic birefringence. The latter remains unchanged when the platelet effect is suspended by the infiltration of glycerol. Therefore it must be ascribed to the optically anisotropic lipids. Extracted lipids produce myelin forms which are optically positive with reference to the radius of the tubes. Such a high birefringence can only be explained if the lipid molecules are packed in a regular array and with their longitudinal axes oriented specifically with respect to the lamellar plains. In chloroplasts the positive birefringence disappears if the lipids are extracted. Based on X-ray diffraction studies on model lipid systems LUZZATI *et al.* (1962, 1966) concluded that the chains are in a "liquid" state. This interpretation is based on the existence of a broad and diffuse X-ray band in the region of 4.5 Å, which is also found in liquid paraffins. This interpretation has been questioned by SEGERMAN (1965), who objected that the diffuse reflection in a power diagram is not proof of a highly random arrangement of chains, but may correspond to a rather ordered chain packing associated with substantial fluctuations within the chains. According to PETHICA (1967) the balance of evidence is against the idea of randomly coiled-chains of lipid molecules in membranes. It appears that the chains are rather straight and will exhibit a high degree of ordered packing. At the same time packing disorders may occur, allowing the chains to rotate freely. This would imply that the lipids present in biological membranes are more likely to be in a semi-solid state than in a random liquid arrangement. According to the view of PETHICA (1967) a transformation of sub-structure will depend to a high degree on the packing configurations of the mixed lipid hydrocarbon moieties. Biologists oppose the idea that structures in which metabolic activities take place are in an unordered "liquid"-like state. In spite of the fact that we do not know at the moment what functions have to be related to the lipids, their different chemical structure and constant proportions in different cell organelles seem to favor the idea that their presence is necessary for many important reactions in the thylakoid membranes. A semi-solid packing of chains may contribute to the stabilization of the sheet structure. It also would allow to arrange a variety of lipid molecules into a structured complex.

The conventional methods do not yield much information concerning the structural arrangement within the protein layers. In most schemes the molecules are pictured in a random, straight conformation.

The tripartite unit membrane model has been questioned for chloroplast lamellae by PARK and PON (1961). In repeating an early experiment by FREY-WYSSLING and STEINMANN (1953) they confirmed the finding that isolated granum discs consisted of flat vesicles with a globular substructure. Based on these results they postulated a new model composed of two layers of globular units, 100 Å in thickness and 200 Å in diameter and surrounded by a 30 Å thick osmiophilic envelope. These 200 Å entities with their attached membranes were termed *quantasomes* (PARK and PON, 1963). PARK and BIGGINS (1964) observed these units in chloroplast fractions with high Hill activity and expected them to be the structural components of the photochemical-acting system. Thus it could be assumed that a single quantasome would represent the smallest unit which will perform the light reaction of photosynthesis. It would be the morphological expression of the photosynthetic unit as formulated by EMERSON and ARNOLD (1933). In a recent experiment HOWELL and MOUDRIANAKIS [1967 (1)] used an electron-dense tetrazolium salt (p-iodonitrotetrazolium violet) as a Hill oxidant to determine whether "quantasomes" were the sites of photoreduction. The reduced tetrazolium salt yields an insoluble formazane. Electron microscopic studies revealed that the photoreduced stain on the thylakoids was not confined to quantasomes: the entire membrane appeared to participate uniformly in the deposition of the stain. Preparations devoid of all resolvable membrane-bound particles are still fully Hill active. According to the experiments carried out by HOWELL and MOUDRIANAKIS [1967 (1)] photoreduction is a total membrane phenomenon rather than a function of a resolvable membrane particle. The nature and composition of these membrane-bound particles were also studied by HOWELL and MOUDRIANAKIS [1967 (2)]. They demonstrated that the particles can be separated from the thylakoids with ethylenediaminetetraacetate (pH 8.0). These isolated particles were purified from other chloroplast phosphorylation activities and characterized in terms of their sedimentation coefficient, substructure and enzymatic activities. After negative staining they measure approximately 100 Å square and have a sedimentation coefficient of about 13S. The particles do not participate in photoreduction reactions but show a Ca^{2+} dependent ATPase activity (VAMBUTAS and RACKER, 1965). For this reason HOWELL and MOUDRIANAKIS [1967 (2)] recommend the term "*13S photophosphorylase*" instead of the term "quantasome". The arrangement of these particles can best be seen with the freeze-etch technique (Moor *et al.*, 1961) (Fig. 6). With this procedure large areas of the thylakoid membrane are disclosed in fractioned chloroplasts allowing a detailed study of the size, structure and arrangement of these units (Mühlethaler *et al.*, 1965). In most chloroplasts they are distributed at random but in some cases also regular arrays have been noted. They appear as cubic units with a diameter of 100 to 120 Å. In preparations of lamellar fragments from ultra-

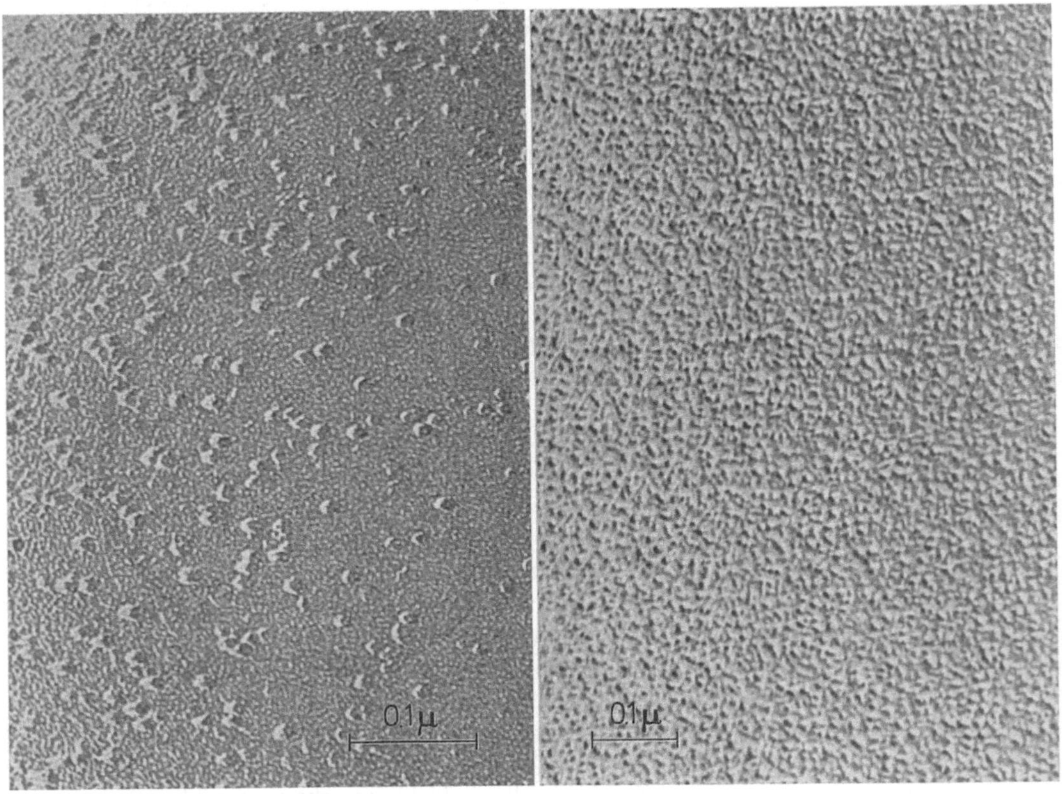

Fig. 6. Outer and inner faces of a thylakoid membrane after freeze-etching *(Spinacia oleracea)*

sonically ruptured chloroplasts they can be shown to be present with the negative staining method (Fig. 7). It was noted that the particles were generally composed of at least 4 subunits. As a typical feature it was observed that they show a central depression. Similar units were found by Haselkorn *et al.* (1965) in isolated fraction I protein. Due to its biochemical activities its presence can be expected in a thylakoid. As described by Howell and Moudrianakis [1967 (2)], the structure of "13S phosphorylase" is very similar to that of ribulose-1,5-diphosphate carboxylase. Both have a cuboidal shape, the same subunit structure and a similar central depression, when viewed from the surface (Fig. 7). The Ca^{2+} dependent ATPase and ribulose

1,5-diphosphate carboxylase particles differ only in their size: The ribulose 1,5-diphosphate carboxylase particle measures about 120 Å along one edge, the Ca^{2+} dependent ATPase about 100 Å. In freeze-etched preparations thylakoids may be observed where these ATPase particles have been torn from the surface. Where they were attached, small perforations with a diameter of about 100 Å become apparent. This indicates the particles to be partly embedded in the supporting layer.

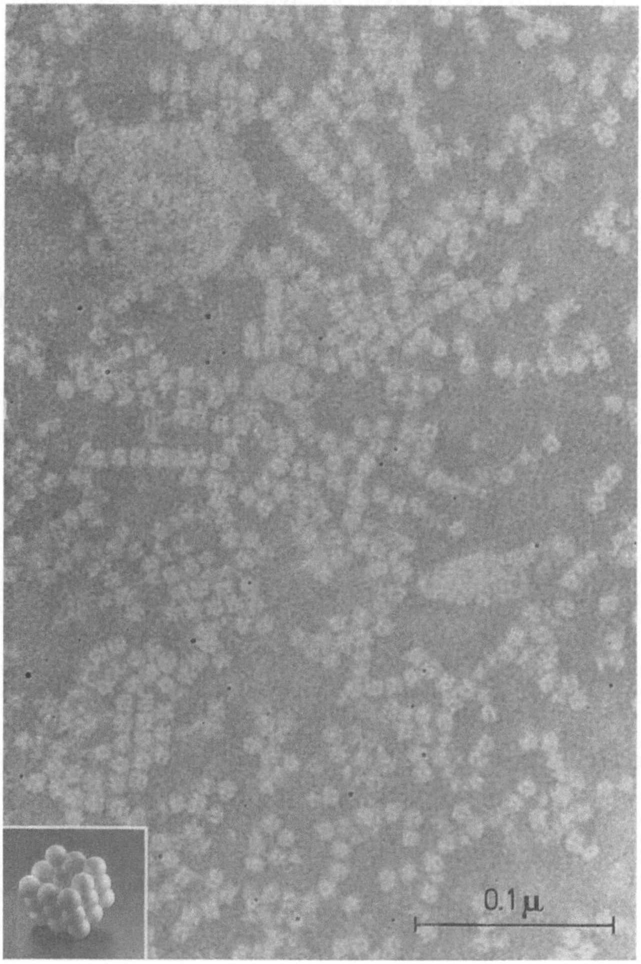

Fig. 7. Isolated ATPase particles from spinach chloroplasts. Inset: a possible model of the arrangement of subunits (Photograph E. WEHRLI)

This basic sheet has a fine granular structure and is obviously composed of smaller units with an approximate diameter of about 40 Å. Such a substructure was postulated by KREUTZ and MENKE (1962) on the basis of their low angle X-ray diffraction patterns. A single thylakoid membrane would be composed of a compact lipid layer of about 35 Å in width and an additional globular protein sheet with a substructure periodicity of 36 Å (Fig. 8). Additional support for a particulate composition of the protein layer came from biochemical studies. As shown by CRIDDLE (1966), isolated membranes can be disintegrated by detergents. The determination of the molecular weight did not yield a consistent particle size. Depending on the mode of preparation the chlorophyll-protein ratio of the solubilized preparations varied over a weight range of 40. A first method for the isolation of chloroplast lamellar protein, free of soluble proteins and small molecules has been worked out by MENKE and JORDAN (1959).

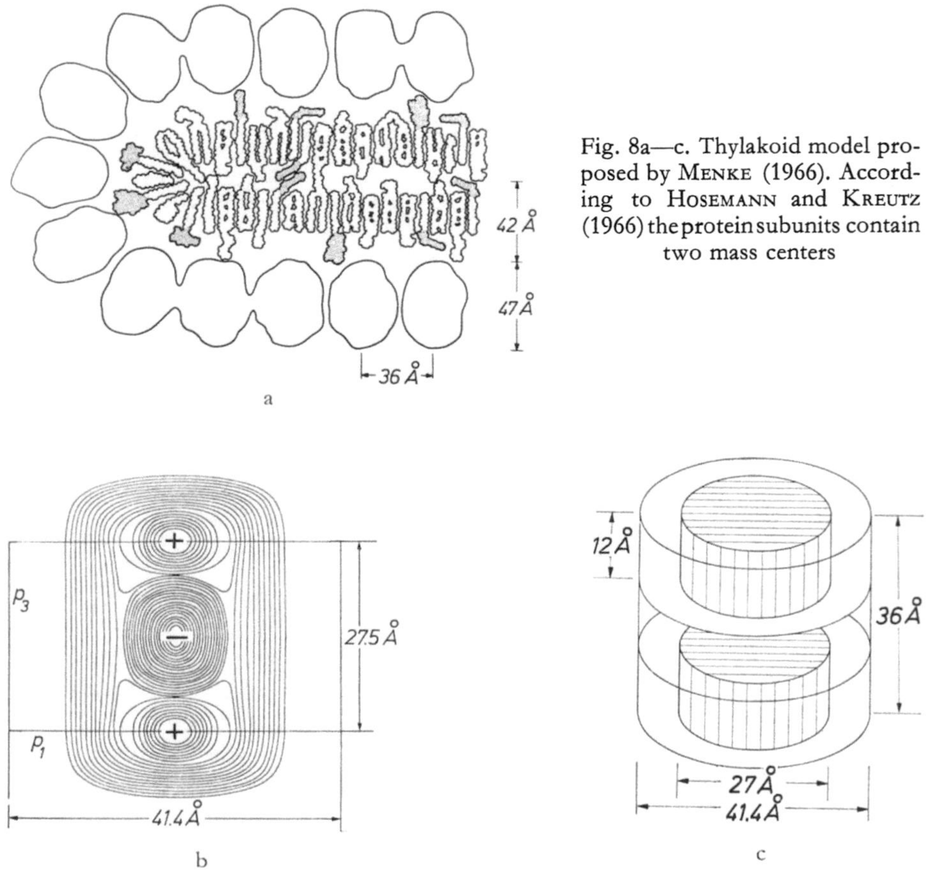

Fig. 8a—c. Thylakoid model proposed by MENKE (1966). According to HOSEMANN and KREUTZ (1966) the protein subunits contain two mass centers

According to this method, lamellar compound constitutes 40 to 54% of the total weight of the chloroplast preparation [MENKE, 1962 (2)]. The amino acid composition of this protein, analyzed by WEBER (1962) in four different chloroplast sources, was found to be quite similar. Also two mutants of *Antirrhinum majus*, defective in chlorophyll production, did not differ in the amino acid content of the lamellar structure proteins. The solubilized protein moved as a uniform 5.9S peak in the ultracentrifuge, which corresponds to an average molecular weight of 165,000. This protein was found to contain some highly bound components which participate in the light reaction of photosynthesis. According to CRIDDLE (1966) the Weber preparations represent complex mixtures of a number of different thylakoid proteins and not only pure "structural protein". These preparations representing approximately 80% of the total chloroplast protein should rather be referred to as lamellar or membrane protein. The structural protein is a specific subfraction which determines primarily the organization and mechanical stability of the membrane. This structural protein is obtained in yields amounting to about 40% of the total thylakoid protein. According to CRIDDLE (1966) it has an average molecular weight of 25,000 and a sedimentation coefficient of 2.3S. From terminal amino acid analyses he concluded that the preparation contained predominantly a single polypeptide chain. Structural protein molecules readily aggregate with themselves into insoluble polymers and form molecular complexes with chlorophyll, lipid, and surprisingly also with myoglobin. Many other membrane systems have been analyzed and it could be shown that these also contain a protein fraction corresponding to structural protein. This indicates that the same building units are used in metabolically different lamellar systems. The biochemical evidence could suggest that the thylakoid membrane is made up of a rigid sheet of structural protein to which other enzymatically active proteins are attached. Such an interpretation is in agreement with observations made on freeze-etched chloroplasts. It can be calculated that particles having a molecular weight of 25,000 correspond to spheres of about 40 Å. Similar particles can be detected in the electron microscope and by low angle X-ray diffraction. Clear evidence for a globular substructure was obtained by GIESBRECHT and DREWS (1966) in thylakoids of the bacterium *Rhodopseudomonas viridis*. The general surface structure is identical to chloroplast lamellae. In these photosynthetic membranes the subunits measure 40 to 50 Å in diameter and are arranged in a hexagonal pattern. A crystalline array of subunits of similar size was also observed in cytomembranes of *Nitrosocystis oceanus* (Fig. 9) by REMSEN *et al.* (1967). These results indicate that there is indeed a component which determines the mechanical stability of a membrane. Recently HOSEMANN and KREUTZ (1966) as well as KREUTZ and WEBER (1966) analyzed the tertiary structure of this protein. They found a "cuff-link" conformation which means that the molecule contains two mass centers (Fig. 8). Between these two centers a zone with an electron density lower than that of water was calculated. This region of the molecule could be ascribed to a hole in the structure. The main part of the protein molecule is concentrated in the top and bottom area, each having a width of 10 Å. The hydrophobic zone between these mass centers measures 16 Å and contains four protein strands which run from the top to the bottom area. The hydrophobic zone has a square shape with the same dimensions as a porphyrin ring.

This led HOSEMANN and KREUTZ (1966) to postulate that a non-phytylated chlorophyll (chlorophyllide) molecule is present in this zone, oriented in such a way that the ring planes are arranged parallel to the lamellar surface. The additional pigment molecules are deposited between the protein and lipid sheets. Sixteen of the structural protein molecules are packed in a regular lattice-like arrangement. The outer thylakoid layer would thus be composed of regularly arranged 40 Å particles in which larger protein complexes are embedded (Fig. 10).

Controversial views exist also in relation to the localization of the lipid molecules. It was generally accepted that they are packed in a bimolecular sheet between the bordering protein layers. KREUTZ and MENKE (1962) objected to this view and postulated an asymmetric structure with only two strata: an inner lipid layer covered by regularly arranged protein molecules (Fig. 8). The lipid layer would also include chlorophyll molecules oriented with their porphyrin rings towards the protein sheet.

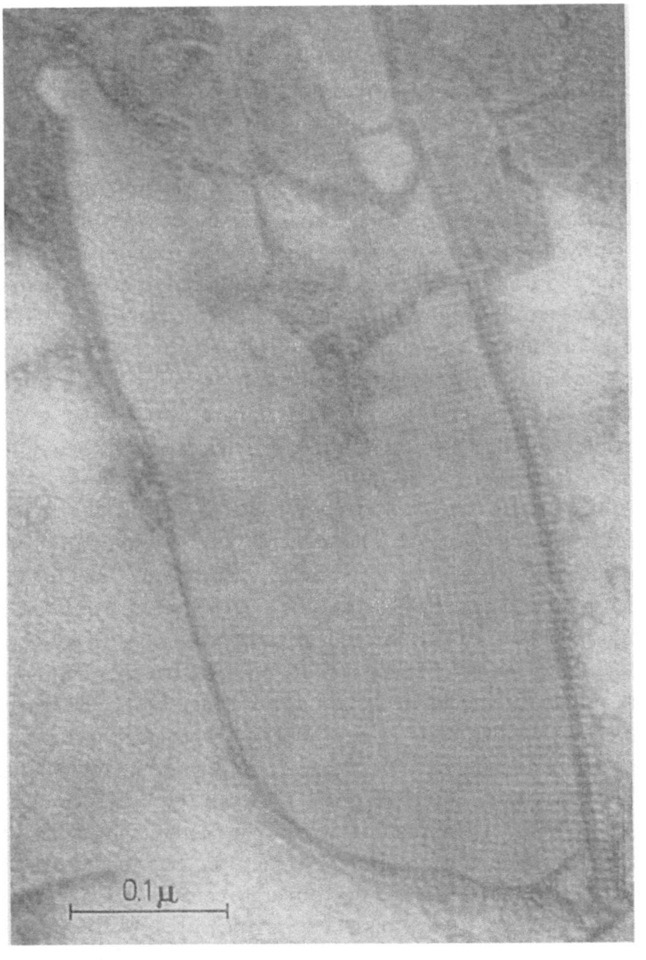

Fig. 9. A membrane fragment of the lamellar system in the bacterium *Nitrosocystis oceanus*. After negative staining a regular pattern with a periodicity of about 40 Å becomes apparent (Photograph Dr. CH. C. REMSEN)

A third concept of membrane structure has been put forward by WEIER and BENSON (1967). Their model is also based on the presence of protein subunits which, as we have described previously, has been confirmed by several investigators using different techniques. WEIER and BENSON's (1967) thylakoid model closely resembles the one first proposed by FREY-WYSSLING and STEINMANN in 1953. At this time the membrane was thought to consist of a single sheet of approximately spherical subunits each about 50 Å in diameter. The chlorophyll molecules were pictured to be in the same layer along the outer margins of the subunits. Based on birefringence studies FREY-WYSSLING and STEINMANN (1953) concluded that the porphyrin rings cannot be parallel with each other nor at right angles to the surface of the subunits. A similar arrangement has also been postulated by GOEDHEER (1957): the chlorophyll molecules would be arranged in a lipid layer which would surround the protein units. According to this view a thylakoid membrane would consist of two layers of lipoprotein molecules with the pigments occurring only on the outer surface of the subunits. In contrast to GOEDHEER's model, WEIER and BENSON (1967) postulate a single sheet of lipoprotein subunits. These particles would be lined with hydrophilic materials such as highly surface-active glycolipids (see Fig. 10 and p. 141). Chlorophyll would then be placed between adjacent protein layers and the cytochrome molecules would be packed within the protein subunits.

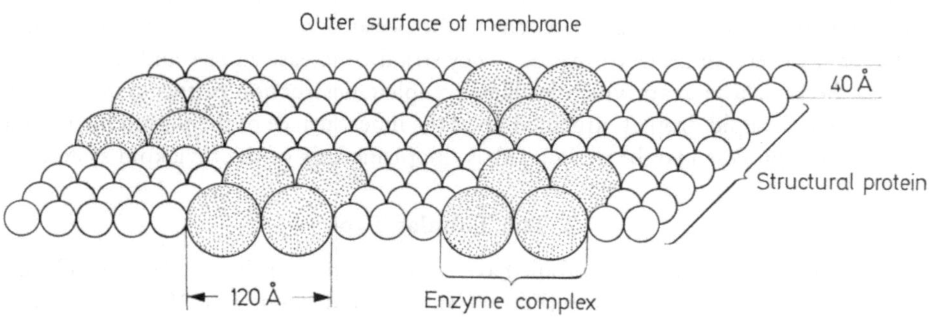

Fig. 10. Model of a thylakoid surface, composed of regularly arranged structural protein particles and enzyme complexes

The spatial relationship between protein and lipid is still controversial. Studies of cellular membranes by ultraviolet optical rotatory dispersion led to the conclusion that there must be membrane proteins, whose amino acid sequences impose tertiary and quaternary structures in which two hydrophilic peptide regions are widely separated by a hydrophobic zone. WALLACH and ZAHLER (1966) envision the two hydrophilic sections to lie at the membrane surface, connected by hydrophobic rods penetrating the membrane normal to its surface. The connecting bar would consist of helical peptide segments packed amidst the hydrocarbon residues of membrane lipids. Thus the protein-lipid association would strongly depend upon the primary structure of the hydrophobic peptide segment. Support for such an arrangement between lipids and protein come from results published by FLEISCHER *et al.* (1967) on lipid-depleted mitochondria. It was found that the cristae retain the unit membrane structure, if more than 95% of their lipid had been extracted previously. In addition the space between the two dark layers did not change, which would indicate that they are kept in place by a connecting bar.

It can be hoped that the discrepancies between the different thylakoid models may be clarified if efforts are made to combine the results obtained by various techniques of investigation.

E. Chromoplasts

In the Introduction it was mentioned that chromoplasts represent the end-forms of plastid differentiation. This stage may be reached through a metamorphosis from proplastids or chloroplasts. The classification is made according to the mode of aggregation of the carotene. In some chromoplasts this pigment is found in lipid droplets, which are termed globuli. In others they form filaments resembling myelinic tubes and a third type contains carotene crystals. As may be expected intermediate types are frequently found.

1. The Globular Type

As an example for the globular type, the chromoplasts found in the perigone of *Ranunculus repens* [Frey-Wyssling and Kreutzer, 1958 (1)] or *Aloe plicatilis* (Steffen and Walter, 1958) may be cited. In the perianth of *Ranunculus* the first globuli appear at the end of grana formation. They steadily increase in size and subsequently press the thylakoid apart which leads to complete destruction of this system. In the final stage the lipid droplets are lined up at the inner surface of the plastid membrane. The whole process may be designated as lipophanerosis of the thylakoid system.

2. The Tubular Type

Chromoplasts containing tubuli have been found in fruits of *Rosa sp.* (Grilli, 1965), fruits of *Capsicum annuum* [Frey-Wyssling and Kreutzer, 1958 (2)], *Solanum capsicastrum* (Steffen and Walter, 1958) and in the spadix appendix of *Typhonium* and *Arum* (Gerola *et al.*, 1963; Schnepf and Czygan, 1966). These tubular chromoplasts are spindle-shaped and vary considerably in size. The plastids exhibit a strong birefringence between crossed Nicol prisms. As the fruit matures the chloroplasts become paler green, starch and chlorophyll gradually disappear and the plastid assumes the shape of a spindle. During this change, tubules are formed in an increasing number and aggregate into bundles. The diameters of single tubules vary between 40 to 50 Å. Since lipase and lipid solvents do not destroy them, they must contain mainly protein. The fact that the major absorption direction of the carotenoid molecules is parallel with the longitudinal axis makes it possible to determine their arrangement in the tubules. It is most likely that the molecules are arranged in an axial direction. The different changes which occur during the transformation from the thylakoid system into the tubular structure are still unknown.

3. The Crystalline Type

These chromoplasts contain microscopic crystals of carotenoids. They are abundant in the orange-colored roots of cultivated varieties of *Daucus carota* or in *Narcissus poeticus* coronas. The crystal may vary in its shape and filamentous, polygonal spiral forms have been found (Straus, 1953, 1961). The majority of the chromoplasts belong into the polygonal class and most of them contain tetragonal, rectangular plates.

The development and ultrastructure of these plastids have been described by STEFFEN and RECK (1964) and also by FREY-WYSSLING and SCHWEGLER (1965). In the root tip of young seedlings the proplastids are filled with starch. Crystals are first formed in older roots around this starch deposit. As the carotene crystals increase in size the reserve material gradually disappears. The carotene crystallizes in tubular sheets. A fully developed "crystal" represents a multilayered tube with a lamellar periodicity of about 200 Å. With the exception of the plastid envelope which does

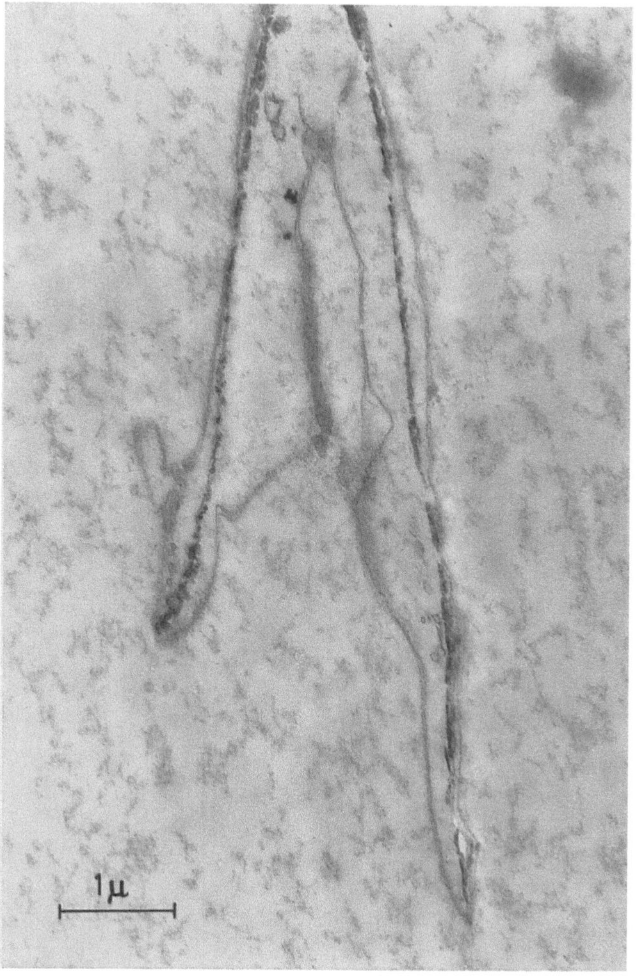

Fig. 11. Chromoplast of a carrot cell containing a tube-like crystal (Photograph F. SCHWEGLER)

not break, all the other contents disintegrate to a watery fluid (Fig. 11). The lamellation of these tubes is best seen in freeze-etched preparations where the lamellar packets split along the sheet surfaces. In a recent investigation of the corona of *Narcissus poeticus*, KUHN (1967) observed that the first sheet of carotene is formed within the thylakoid lumen or loculus.

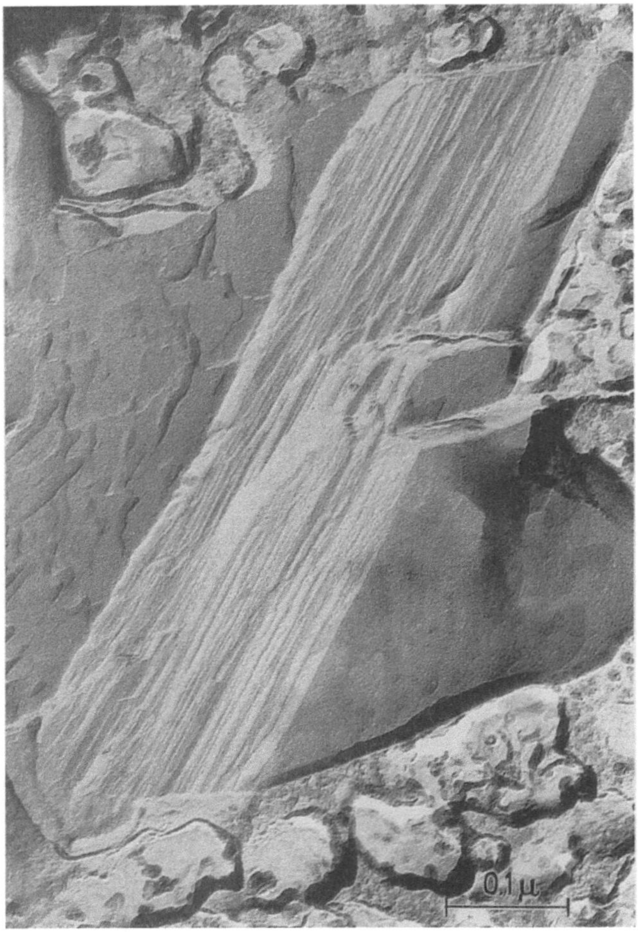

Fig. 12. Carotene crystal of the corona of *Narcissus* (Photograph H. KUHN)

According to STRAUS (1961) isolated chromoplasts contain between 19 to 56% carotene. The pigment-free portion contains 22% protein, 3.3% RNA, 21% phospholipid and 37% other ether-alcohol-soluble matter. In contrast to this analysis KUHN (1967) was not able to find any indication for the presence of lipids or proteins in the crystals. This means that the sheet-like tubules are made up of carotene only. In his objects the sheet periodicity was measured to be 50 Å (Fig. 12). Since the crystals display a striking anisotropy of absorption (dichroism), the carotene molecules must be arranged in a regular fashion (FREY-WYSSLING, 1967). The optical studies led to the conclusion that the carotene molecules are arranged parallel to the sheet plane and not perpendicular like the fatty acid chains in myelin figures.

As pointed out, all three chromoplast types represent end stages of plastid differentiations and they may be characterized as senile plastids.

Acknowledgments

I wish to express my thanks to Miss RUTH RICKENBACHER for her help with the manuscript and to Miss SONJA TÜRLER for preparing the drawings.

F. Literature

ABEL, R.: Über die Beeinflussung der Chloroplastenstruktur durch Licht bei *Antirrhinum majus*. Naturwissenschaften **43**, 136—137 (1956).

BAILEY, J. L., WHYBORN, A. G.: The osmiophilic globules of chloroplasts. II. Globules of the spinach beet chloroplast. Biochim. biophys. Acta (Amst.) **78**, 163—174 (1963).

BALTUS, E., BRACHET, J.: Presence of deoxyribonucleic acid in the chloroplasts of *Acetabularia mediterranea*. Biochim. biophys. Acta (Amst.) **76**, 490—492 (1963).

BAMJI, M. S., JAGENDORF, A. T.: Amino acid incorporation by wheat chloroplasts. Plant Physiol. **41**, 764—770 (1966).

BELL, P. R., FREY-WYSSLING, A., MÜHLETHALER, K.: Evidence for the discontinuity of plastids in the sexual reproduction of a plant. J. Ultrastruct. Res. **15**, 108—121 (1966).

BRACHET, J.: New observations on biochemical interactions between nucleus and cytoplasm in *Amoeba* and *Acetabularia*. Exp. Cell Res. Suppl. **6**, 78—96 (1958).

BRAWERMAN, G.: The isolation of a specific species of ribosomes associated with chloroplast development in *Euglena gracilis*. Biochim. biophys. Acta (Amst.) **72**, 317—331 (1963).

— Nucleic acids associated with the chloroplast of *Euglena gracilis*. In: Biochemistry of chloroplasts, Vol. I, p. 301—317 (T. W. GOODWIN, Ed.). London-New York: Academic Press 1966.

BROWN, F. A. M., GUNNING, B. E. S.: Distribution of ribosome-like particles in *Avena* plastids. In: Biochemistry of chloroplasts, Vol. I, p. 365—373 (T. W. GOODWIN, Ed.). London-New York: Academic Press 1966.

CHIBA, Y.: Cytochemical studies on chloroplasts. I. Cytologic demonstration of nucleic acids in chloroplasts. Cytologia (Tokyo) **16**, 259—264 (1951).

CRIDDLE, R. S.: Protein and lipoprotein organization in the chloroplast. In: Biochemistry of chloroplasts, Vol. I, p. 203—231 (T. W. GOODWIN, Ed.). London-New York: Academic Press 1966.

DANIELLI, J. F., DAVSON, H. A.: A contribution to the theory of permeability of thin films. J. cell, comp. Physiol. **5**, 495—500 (1935).

DIERS, L., SCHÖTZ, F.: Über die dreidimensionale Gestaltung des Thylakoidsystems in den Chloroplasten. Planta **70**, 322—343 (1966).

DÖBEL, P., HAGEMANN, R.: Elektronenmikroskopischer Nachweis echter Mischzellen bei *Antirrhinum majus status albomaculatus*. Biol. Zbl. **82**, 749—751 (1963).

EILAM, Y., KLEIN, S.: The effect of light intensity and sucrose feeding on the fine structure in chloroplasts and on the chlorophyll content of etiolated leaves. J. Cell Biol. **14**, 169—182 (1962).

EISENSTADT, J.: Protein synthesis in chloroplasts and chloroplast ribosomes. In: Biochemistry of chloroplasts, Vol. II, p. 341—349 (T. W. GOODWIN, Ed.) London-New York: Academic Press 1967.

— BRAWERMAN, G.: The incorporation of amino acids into protein of chloroplasts and chloroplast ribosomes of *Euglena gracilis*. Biochim. biophys. Acta (Amst.) **76**, 319—321 (1963).

Emerson, R., Arnold, W.: The photochemical reaction in photosynthesis. J. gen. Physiol. **16**, 191—205 (1933).

Fasse-Franzisket, U.: Die Teilung der Proplastiden und Chloroplasten bei *Agapanthus umbellatus l'Hérit*. Protoplasma (Wien) **45**, 194—227 (1955).

Finean, J. B.: Recent ideas on the structure of myelin. In: Biochem. problems of lipids, p. 129—131. Proceed. 2nd Intern. Conf. Univ. Ghent. London: Butterworths 1955.

Fleischer, S., Fleischer, B., Stoeckenius, W.: Fine structure of lipid-depleted mitochondria. J. Cell Biol. **32**, 193—208 (1967).

Frey-Wyssling, A.: Der Aufbau der Chlorophyllkörner. Protoplasma (Wien) **XXIX**, 279 bis 299 (1937).

— Über die Carotinkristalle in der Nebenkrone der Narzissenblüte. Anales edafol. y agrobiol (Madrid) **26**, 25—32 (1967).

— Kreutzer, E.: (1) Die submikroskopische Entwicklung der Chromoplasten in den Blüten von *Ranunculus repens* L. Planta **51**, 104—114 (1958).

— — (2) The submicroscopic development of chromoplasts in the fruit of *Capsicum annuum* L. J. Ultrastruct. Res. **1**, 397—411 (1958).

— Ruch, F., Berger, X.: Monotrope Plastiden-Metamorphose. Protoplasma (Wien) **45**, 97—114 (1955).

— Schwegler, F.: Ultrastructure of the chromoplasts in the carrot root. J. Ultrastruct. Res. **13**, 543—559 (1965).

— Steinmann, E.: Ergebnisse der Feinbau-Analyse der Chloroplasten. Vjschr. naturforsch. Ges. Zürich. **98**, 20—29 (1953).

Gerola, F. M.: Ricerche sulle infrastrutture cellulari dello spadice di Arum. G. botan. ital. **70**, 177—183 (1963).

— Cristofori, F., Dassu, G.: (1) Ricerche sullo sviluppo dei cloroplasti di pisello *(Pisum sativum)*. II. Sviluppo dei cloroplasti in piantine eziolate e loro modificazioni durante il successivo invedimento. Caryologia **13**, 179—197 (1960).

— — — (2) Ricerche sulle infrastrutture delle cellule del mesofillo di piante sane e virosate di tabacco *(Nicotiana tabacum)* A: Piante sane. Caryologia **13**, 352—366 (1960).

Gierer, A.: Function of aggregated reticulocyte ribosomes in protein synthesis. J. molec. Biol. **6**, 148—157 (1963).

Giesbrecht, P., Drews, G.: Über die Organisation und die makromolekulare Architektur der Thylakoide "lebender" Bakterien. Arch. Mikrobiol. **54**, 297—330 (1966).

Gifford, E. M., Jr.: Incorporation of ^{3}H-thymidine into shoot and root aspices of *Ceratopteris thalictroides*. Amer. J. Bot. **47**, 834—837 (1960).

Givner, J., Moriber, L. G.: An attempt to demonstrate DNA in the chloroplasts of *Euglena gracilis* using fluorescence microscopy. J. Cell Biol. **23**, 36A (1964).

Goedheer, J. C.: Optical properties and in vivo orientation of photosynthetic pigments. Thesis, University Utrecht 1957.

Gorter, E., Grendel, E.: On bimolecular layers of lipid on the chromocytes of the blood. J. exp. Med. **41**, 439—443 (1925).

Granick, S.: The chloroplast: Inheritance, structure and function. In: The cell, Vol. II, p. 489—602 (J. Brachet, A. E. Mirsky, Eds.). New York: Academic Press 1961.

Greenwood, A. D., Leech, R. M., Williams, J. P.: The osmiophilic globules of chloroplasts. I. Osmiophilic globules as a normal component of chloroplasts and their isolation and composition in *Vicia faba*. Biochim. biophys. Acta (Amst.) **78**, 148—162 (1963).

Grilli, M.: Ultrastrutture e stadi involutivi di alcuni tipi di cromoplasti. G. botan. ital. **72**, 83—92 (1965).

Gunning, B. E. S.: (1) The greening process in plastids. 1. The structure of the prolamellar body. Protoplasma (Wien) **60**, 111—130 (1965).

— (2) The fine structure of chloroplast stroma following aldehyde osmium-tetroxide fixation. J. Cell Biol. **24**, 79—93 (1965).

— Jagoe, M. P.: The prolamellar body. In: Biochemistry of chloroplasts, Vol. II, p. 655—676 (T. W. Goodwin, Ed.). London-New York: Academic Press 1967.

— Steer, M. W., Cochrane, P.: Fraction I protein and stromacentre fibrils. Proc. Roy. Microscop. Soc. **2**, 378 (1967).

Haselkorn, R., Fernández-Morán, H., Kieres, F. J., van Bruggen, E. F. J.: Electron microscopic and biochemical characterization of fraction I protein. Science **150**, 1598—1601 (1965).

Heitz, E.: Kristallgitterstruktur des Granum junger Chloroplasten von *Chlorophytum*. Exp. Cell Res. **7**, 606—608 (1954).

Heslop-Harrison, J.: Structure and morphogenesis of lamellar systems in grana-containing chloroplasts. I. Membrane structure and lamellar architecture. Planta **60**, 243—260 (1963).

Hodge, A. J., McLean, J. D., Mercer, F. V.: A possible mechanism for the morphogenesis of lamellar systems in plant cells. J. biophys. biochem. Cytol. **2**, 597—608 (1956).

Hosemann, R., Kreutz, W.: On the tertiary structure of the protein layers of chloroplasts. Naturwissenschaften **53**, 298—304 (1966).

Howell, S. H., Moudrianakis, E. N.: (1) Hill reaction site in chloroplast membranes: Non-participation of the quantasome particles in photoreduction. J. molec. Biol. **27**, 323—333 (1967).

— — (2) Function of the "Quantasome" in photosynthesis: Structure and properties of membrane-bound particle active in the dark reactions of photophosphorylation. Proc. nat. Acad. Sci. (Wash.) **58**, 1261—1269 (1967).

Hyde, B. B., Hodge, A. J., Kahn, A., Birnstiel, M. L.: Studies on phytoferritin. I. Identification and localization. J. Ultrastruct. Res. **9**, 248—258 (1963).

Jacobson, A. B., Swift, H., Bogorad, L.: Cytochemical studies concerning the occurrence and distribution of RNA in plastids of *Zea mays*. J. Cell Biol. **17**, 557—570 (1963).

Kirk, J. T. O.: Nature and function of chloroplast DNA. In: Biochemistry of chloroplasts, Vol. I., p. 319—340 (T. W. Goodwin, Ed.). London-New York: Academic Press 1966.

Kislev, N., Swift, H., Bogorad, L.: Nucleic acid of chloroplasts and mitochondria in Swiss chard. J. Cell Biol. **25**, 327—344 (1965).

Kreutz, W.: Strukturuntersuchungen an Plastiden. VI. Über die Struktur der Lipoproteinlamellen in Chloroplasten lebender Zellen. Z. Naturforsch. **19 b**, 441—446 (1964).

— Menke, W.: Strukturuntersuchungen an Plastiden. III. Röntgenographische Untersuchungen an isolierten Chloroplasten und Chloroplasten lebender Zellen. Z. Naturforsch. **17 b**, 675—683 (1962).

— Weber, P.: About the protein structure of quantasomes. Naturwissenschaften **53**, 11—14 (1966).

Kuhn, H.: Ontogenie der Chromoplasten in der Nebenkrone von *Narcissus poeticus*. Diploma Thesis, ETH 1967 (unpublished).

Kupke, D. W.: Correlation of a soluble leaf protein with chlorophyll accumulation. J. biol. Chem. **237**, 3287—3291 (1962).

Küster, E.: Die Pflanzenzelle, p. 326, 379. Jena: Gustav Fischer 1951.

Kylin, H.: Über Phykoerythrin und Phykocyan bei *Ceramium rubrum*. Hoppe-Seyler's Z. physiol. Chem. **69**, 169—239 (1910).

Leyon, H.: The structure of chloroplasts. II. The first differentiation of the chloroplast structure in *Vallota* and *Taraxacum* studied by means of electron microscopy. Exp. Cell Res. **5**, 520—529 (1953).

— The structure of chloroplasts. IV. The development and structure of the *Aspidistra* chloroplast. Exp. Cell Res. **7**, 265—273 (1954).

Lichtenthaler, H. K.: Untersuchungen über die osmiophilen Globuli der Chloroplasten. Ber. dtsch. botan. Ges. **77**, 398—402 (1964).

— Plastoglobuli und Plastidenstruktur. Ber. dtsch. botan. Ges. **79**, 82—88 (1966).

— Beziehungen zwischen Zusammensetzung und Struktur der Plastiden in grünen und etiolierten Keimlingen von *Hordeum vulgare*, L. Z. Pflanzenphysiol. **56**, 273—281 (1967).

— Peveling, E.: Plastoglobuli in verschiedenen Differenzierungsstadien der Plastiden bei *Allium cepa* L. Planta **72**, 1—13 (1967).

— Sprey, B.: Über die osmiophilen globulären Lipideinschlüsse der Chloroplasten. Z. Naturforsch. **21 b**, 690—697 (1966).

Lintilhac, P. M., Park, R. B.: Localization of chlorophyll in spinach chloroplast lamellae by fluorescence microscopy. J. Cell Biol. **28**, 582—585 (1966).

Littau, V. C.: A cytochemical study of the chloroplasts in some higher plants. Amer. J. Bot. **45**, 45—53 (1958).

Luft, J. H.: Permanganate—a new fixative for electron microscopy. J. biophys. biochem. Cytol. **2**, 799—801 (1956).

Luzzati, V., Husson, F.: The structure of the liquid crystalline phases of lipid-water systems. J. Cell Biol. **12**, 207—219 (1962).

— Reiss-Husson, F., Rivas, E., Gulik-Krzywieti, T.: Structure and polymorphism in lipid-water systems and their possible biological implications. In: Biological membranes: Recent progress of sciences. Ann. N.Y. Acad. Sci. **137**, 409—414 (1966).

Lyttleton, J. W.: Isolation of ribosomes from spinach chloroplasts. Exp. Cell Res. **26**, 312—317 (1962).

Margulies, M. M.: Effect of chloramphenicol on light-dependent synthesis of proteins and enzymes of leaves and chloroplasts of *Phaseolus vulgaris*. Plant Physiol. **39**, 579—585 (1964).

Menke, W.: Über den Feinbau der Chloroplasten. Kolloid Z. **85**, 256—259 (1938).

— Einige Beobachtungen zur Entwicklungsgeschichte der Plastiden von *Elodea canadensis*. Z. Naturforsch. **15 b**, 800—804 (1960).

— Über die Chloroplasten von *Anthoceros punctatus*. Z. Naturforsch. **16 b**, 334—336 (1961).

— (1) Über die Struktur der Heitz-Leyonschen Kristalle. Z. Naturforsch. **17 b**, 188—190 (1962).

— (2) Structure and chemistry of plastids. Ann. Rev. Plant Physiol. **13**, 27—44 (1962).

— Versuche zur Aufklärung der molekularen Struktur der Thylakoidmembran. Z. Naturforsch. **20 b**, 801—805 (1965).

— The structure of chloroplasts. In: Biochemistry of chloroplasts, Vol. I, p. 3—18 (T. W. Goodwin, Ed.). London-New York: Academic Press 1966.

— Jordan, E.: Über das lamellare Strukturproteid der Chloroplasten von *Allium porrum*. Z. Naturforsch. **14 b**, 234—240 (1959).

Metzner, H.: Über den Nachweis von Nukleinsäuren in den Chloroplasten höherer Pflanzen. Naturwissenschaften **39**, 64—65 (1952).

Meyer, A.: Das Chlorophyllkorn in chemischer, morphologischer und biologischer Beziehung. Leipzig: Arthur Felix 1883.

Moor, H., Mühlethaler, K., Waldner, H., Frey-Wyssling, A.: A new freezing ultramicrotome. J. biophys. biochem. Cytol. **10**, 1—13 (1961).

Mühlethaler, K.: Die Struktur der Grana- und Stromalamellen in Chloroplasten. Z. wiss. Mikroskop. **64**, 445—452 (1960).

— Frey-Wyssling, A.: Entwicklung und Struktur der Proplastiden. J. biophys. biochem. Cytol. **6**, 507—512 (1959).

— Moor, H., Szarkowski, J. W.: The ultrastructure of the chloroplast lamellae. Planta **67**, 305—323 (1965).

Murakami, S.: Prolamellar body of the proplastid in barley root cells. Experientia (Basel) **18**, 168—169 (1962).

— Takamiya, A.: Osmiophilic granules of spinach chloroplasts. Fifth Intern. Congr. Electron Microscopy Philadelphia, Vol. II, p. XX-12. New York-London: Academic Press 1962.

Nagai, R., Rebhun, L. I.: Cytoplasmic microfilaments in streaming *Nitella* cells. J. Ultrastruct. Res. **14**, 571—589 (1966).

Newcomb, E. H.: Fine structure of protein-storing plastids in bean root tips. J. Cell Biol. **33**, 143—163 (1967).

Park, R. B., Biggins, J.: Quantasomes: Size and composition. Science **144**, 1009—1011 (1964).

— Pon, N. G.: Correlation of structure with function in *Spinacia oleracea* chloroplasts. J. molec. Biol. **3**, 1—10 (1961).

— Chemical composition and the substructure of lamellae isolated from *Spinacia oleracea* chloroplasts. J. molec. Biol. **6**, 105—114 (1963).

Penman, S., Scherrer, K., Becker, Y., Darnell, J. E.: Polyribosomes in normal and poliovirus-infected Hela cells and their relationship to messenger-RNA. Proc. nat. Acad. Sci. (Wash.) **49**, 654—662 (1963).

Perner, E. S.: Die ontogenetische Entwicklung der Chloroplasten von *Chlorophytum comosum*. Z. Naturforsch. **11 b**, 560—567 (1956).

— Elektronenmikroskopische Untersuchungen an Zellen von Embryonen im Zustand völliger Samenruhe. Planta **65**, 334—357 (1965).

PETHICA, B. A.: Structure and physical chemistry of membranes. Protoplasma (Wien) **63**, 145—156 (1967).

REMSEN, CH. C., VALOIS, F. W., WATSON, ST. W.: Fine structure of the cytomembranes of *Nitrosocystis oceanus*. J. Bact. **94**, 422—433 (1967).

RHODES, M. J. C., YEMM, E. W.: Development of chloroplasts and synthesis of proteins in leaves. Nature (Lund.) **200**, 1077—1080 (1963).

RIS, H., PLAUT, W.: Ultrastructure of DNA-containing areas in the chloroplast of *Chlamydomonas*. J. Cell Biol. **13**, 383—391 (1962).

ROBERTSON, J. D.: The ultrastructure of cell membranes and their derivatives. The structure and function of subcellular components. Biochem. Soc. Symposia Cambridge, Engl. **16**, 3—43 (1959).

— A molecular theory of cell membrane structure. In: Verh. 4. Intern. Kongr. Elektronenmikroskop., Bd. II, p. 159—171 (W. Bargmann, D. Peters, C. Wolpers, Eds.). Berlin-Göttingen-Heidelberg: Springer 1960.

— Origin of the unit membrane concept. Protoplasma (Wien) **63**, 218—245 (1967).

SCHIMPER, Untersuchungen über die Chlorophyllkörner und die ihnen homologen Gebilde. Jb. wiss. Botan. **16**, 1—27 (1885).

SCHNEPF, E.: Plastidenstrukturen bei *Passiflora*. Protoplasma (Wien) **54**, 310—313 (1962).

— Über die Zusammenhänge zwischen Heitz-Leyon-Kristallen und Thylakoiden. Planta **61**, 371—373 (1964).

— CZYGAN, F. C.: Feinbau und Carotinoide von Chromoplasten in Spadix-Appendix von *Typhonium* und *Arum*. Z. Pflanzenphysiol. **54**, 345—355 (1966).

SCHÖTZ, F.: Zur Frage der Vermehrung der Thylakoidschichten in den Chloroplasten. Planta **64**, 376—380 (1965).

— SENSER, F.: Reversible Plastidenumwandlung bei der Mutante "Weissherz" von *Oenothera suaveolens* DESF. Planta **57**, 235—238 (1961).

SEGERMAN, E.: Surface chemistry. In: Proc. 2nd Scand. Symp. on surface activity, p. 157—180 (V. RUNSTRÖM-REIO, Ed.). Copenhagen: Munksgaard 1965.

SITTE, P.: Zum Bau der Plastidenzentren in Wurzelproplastiden. Protoplasma (Wien) **53**, 438—442 (1961).

— Zum Chloroplasten-Feinbau bei *Elodea*. Port Acta biol. A, **6**, 269—278 (1962).

SLAYTER, H. S., WARNER, J. R., RISCH, A., HALL, C. E.: The visualization of polyribosomal structure. J. molec. Biol. **7**, 652—657 (1963).

SPIEKERMAN, R.: Cytochemische Untersuchungen zum Nachweis von Nukleinsäuren in Proplastiden. Protoplasma (Wien) **48**, 303—324 (1957).

SPREY, B., LICHTENTHALER, H. K.: Zur Frage der Beziehungen zwischen Plastoglobuli und Thylakoidgenese in Gerstenkeimlingen. Z. Naturforsch. **221 b**, 697—699 (1966).

STEFFEN, K., RECK, G.: Chromoplastenstudien. III. Die Chromoplastengenese und das Problem der Plastidenhülle bei *Daucus carota*. Planta **60**, 627—648 (1964).

— WALTER, F.: Die Chromoplasten von *Solanum capsicastrum* und ihre Genese. Planta **50**, 640—670 (1958).

STEINMANN, E.: An electron microscope study of the lamellar structure of chloroplasts. Exp. Cell Res. **3**, 367—372 (1952).

STOECKENIUS, W.: An electron microscope study of myelin figures. J. biophys. biochem. Cytol. **5**, 491—500 (1959).

— The molecular structure of lipid-water systems and cell membrane models studied with the electron microscope. In: The interpretation of ultrastructure, Vol. I, p. 349—367 (R. J. C. HARRIS, Ed.). New York-London: Academic Press 1962.

STRAUS, W.: Chromoplasts-development of crystalline forms, structure, state of pigments. Botan. Rev. **19**, 147—186 (1953).

— Studies on the chromoplasts of carrots. Protoplasma (Wien) **53**, 405—421 (1961).

STRUGGER, S., PERNER, E.: Beobachtungen zur Frage der ontogenetischen Entwicklung des somatischen Chloroplasten. Protoplasma (Wien) **46**, 711—742 (1956).

TROWN, P. W.: An improved method for the isolation of carboxy-dismutase. Probable identity with fraction-I-protein and the protein moiety of protochlorophyll holochrome. Biochemistry **4**, 908—918 (1965).

VAMBUTAS, V. K., RACKER, E.: Partial resolution of the enzymes catalyzing photophosphorylation. J. biol. Chem. **240**, 2660—2667 (1965).

WALLACH, D. F. H., ZAHLER, P. H.: Protein conformations in cellular membranes. Proc. nat. Acad. Sci. (Wash.) **56**, 1552—1559 (1966).

WEBER, I.: Myelinfiguren und Sphaerolithe aus *Spirogyra* chloroplasten. Protoplasma (Wien) **19**, 455—462 (1933).

WEBER, P.: Über lamellare Strukturproteide aus Chloroplasten verschiedener Pflanzen. Z. Naturforsch. **17 b**, 683—688 (1962).

WEHRMEYER, W.: (1) Über die Membranbildungsprozesse im Chloroplasten. I. Zur Morphogenese der Granamembranen. Planta **59**, 280—295 (1963).

(2) Qualitative und quantitative Untersuchungen über den Bau der Stromamembranen der ausdifferenzierten Chloroplasten von *Spinacia oleracea* L. Z. Naturforsch. **18 b**, 60—66 (1963).

— Zur Klärung der strukturellen Variabilität der Chloroplastengrana des Spinats in Profil und Aufsicht. Planta **62**, 272—293 (1964).

— Zur Kristallgitterstruktur der sogenannten Prolamellarkörper in Proplastiden etiolierter Bohnen. Z. Naturforsch. **20 b**, 1270—1296 (1965).

WEIER, T. E., BENSON, A. A.: The molecular organization of chloroplast membranes. Amer. J. Bot. **54**, 389—402 (1967).

— BISALPUTRA, T., HARRISON, A.: Subunits in chloroplast membranes of *Scenedesmus quadricanda*. J. Ultrastruct. Res. **15**, 38—56 (1966).

— STOCKING, C. R., BRACKER, C. E., RISLEY, E. B.: The structural relationships of the internal membrane system of in situ and isolated chloroplasts of *Hordeum vulgare*. Amer. J. Bot. **52**, 339—352 (1965).

— — THOMSON, W. W., DREVER, H.: The grana as structural units in chloroplasts of mesophyll of *Nicotiana rustica* and *Phaseolus vulgaris*. J. Ultrastruct. Res. **8**, 122—143 (1963).

v. WETTSTEIN, D.: Chlorophyll-letale und der submikroskopische Formwechsel der Plastiden. Exp. Cell Res. **12**, 427—506 (1957).

— The formation of plastid structures. The photochemical apparatus. Brookhaven Symposia in Biol. **11**, 138—159 (1958).

— Developmental changes in chloroplasts and their genetic control. In: Developmental cytology, p. 123—160 (D. RUDNICK, Ed.). New York: Ronald Press Co. 1959.

— KAHN, A.: Macromolecular physiology of plastids. In: Europ. Reg. Conf. Electron Microscopy Delft, Vol. II, 1051—1054. Delft: De Nederlandse Vereniging voor Electronenmicroscopie 1960.

WILD, A.: Experimentelle Beeinflussung des Granamusters einer abweichenden Plastidensorte von *Antirrhinum majus*. Planta **50**, 379—387 (1958).

WOHLFARTH-BOTTERMANN, K. E.: Cell structures and their significance for ameboid movement. Int. Rev. Cytol. **16**, 61—131 (1964).

WOLLGIEHN, R., MOTHES, K.: Über DNS in den Chloroplasten von *Nicotiana rustica*. Naturwissenschaften **50**, 95—96 (1963).

Light-Induced Chloroplast Contraction and Movement

Frank Mayer

A. Introduction

The first observation of chloroplast movement was made in 1850 by v. Mercklin. He found that the arrangement of chloroplasts of fern prothallia in very young cells differed from that in old ones. The relationship between light and chloroplast arrangement and shape was recognized by Böhm (1856). An extensive review by Senn comprising observations in a variety of species appeared in 1908. This work has not lost its value and is still worth referring to.

Interest in chloroplast movement has been recently stimulated by the introduction of new tools such as better-defined light sources and interference filters. In addition, studies in other scientific areas such as physics and chemistry prompted a more thorough investigation of this phenomenon.

B. Changes in Chloroplast Shape

Chloroplasts, as seen in the light microscope, can be flat or contracted. They may be situated in the cell so that they are seen from one side—profile position—or from above; then they show their face area.

As early as 1908 it was shown by Senn that light *per se* is not the only factor responsible for changes in chloroplast shape. Senn recognized that temperature, water and chemical environment are important factors which determine chloroplast shape. In 1953 Busch demonstrated that endogenous and rhythmic reactions influence chloroplast shape.

In many plants changes of chloroplast shape, i.e. damage caused by unilateral light, can be prevented by plastid movements. These movements may be interpreted as a "defense reaction". However this "escape" from damage by chloroplast movement is no longer possible, if a clinostat is employed which allows irradiation with the same intensity from different directions by rotating the plant.

Senn who investigated light dependent changes of chloroplast shape concluded that any plant, perhaps any cell, needs individual light conditions for the optimal absorption of light which is necessary for photosynthesis. The optimal intensity is lower for shade plants than for sun plants. Any alteration of these light conditions causes a deviation from the optimal physiological chloroplast state and a change of the optimal chloroplast shape, i.e., a change of the ratio of surface to volume.

Light of different spectral regions causes unequal effects. In *Hormidium flaccidum*, *Funaria* (Senn, 1908) and *Selaginella martensii* (Mayer, 1964) red light, but not blue light causes chloroplast contraction. In*Mougeotia* chloroplasts regain their flat shape in blue, but not in red light (Senn, 1908).

Upon irradiation, isolated chloroplasts have been shown to exhibit the phenomenon of reversible shrinkage, i.e., a 20 to 50% descrease of their original volume. A simultaneous increase of the face area of the plastids can be observed. The action spectrum of shrinkage and the effect of certain inhibitors indicate a connection of this process with photophosphorylation (ZURZYCKI, 1966).

By employing intact leaf cells of *Mnium undulatum* ZURZYCKI (1966) found that the effect of long-wave and short-wave light on the face area of chloroplasts is qualitatively different. He has discussed his findings in terms of two processes:

1. The enlarging of the face area which is caused by either red light or blue light of low and middle intensity. This process seems to be coupled with photophosphorylation.

2. The shrinking of the face area which is caused by strong blue light. The latter overcomes the chloroplast's tendency of flattening. The shrinking is known to be controlled only by short-wave light. It is not effected by inhibitors of photophosphorylation such as NH_4-ions and o-phenanthroline.

Definitive evidence is not available to determine, whether these effects are related to the known influence exerted by light of different wave-lengths and intensities on the viscosity of the cell plasm (VIRGIN, 1951, 1952; SEITZ, 1967).

C. Light-Induced Chloroplast Movements

Chloroplasts of various origin are able to avoid damage by high light intensities in two ways: they can change their position by either moving to a shadowed cell area or by turning their profile toward the light source. This is known as the "high-intensity response".

In unilateral light of low intensity the chloroplasts normally turn their larger surface towards the light source. This is called the "low-intensity response". Examples of these two reactions are illustrated in Fig. 1.

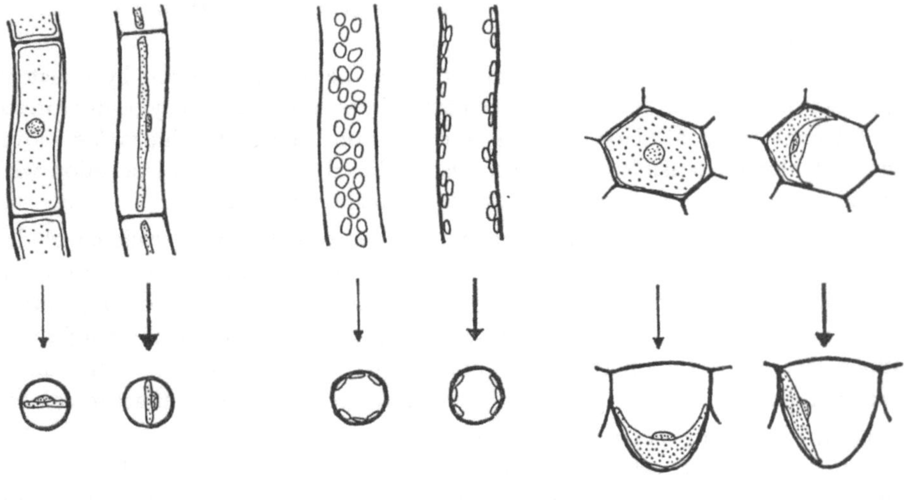

Fig. 1. Light-dependent chloroplast movements. Left: *Mougeotia;* center: *Vaucheria;* right: *Selaginella.* ↓ Low intensity response; ↓ high intensity response. (According to HAUPT, 1964)

In order to bring about the light-dependent orientation of chloroplasts it is necessary for the cell to absorb the light, to determine the light direction and to distinguish between high and low intensities. Moreover, there has to be at least one cytoplasmic system that facilitates the change of chloroplast positions, provided this is not done by the chloroplast itself. This system must be supplied at the right moment with energy from the cell's metabolism since the energy received from the light stimulus is not always sufficient for the displacement of one or more chloroplasts.

1. The Photoreceptor Problem

Investigations of the photoreceptor systems are aimed at answering specific questions such as:

> Of what nature are the compounds (photoreceptors) absorbing the light which induces chloroplast movements?
> Where are these photoreceptor molecules localized within the cell?
> Is it possible to obtain information about the orientation of the photoreceptor molecules?
> Are the primary effects of light solely on the photochemical processes?
> Are the same photoreceptors present in different plants?

In the following some experiments will be described that attempt to answer these questions.

a) Action Spectra of Chloroplast Movements

Chloroplasts are the organelles of photosynthesis. Chloroplast displacements lead to optimal conditions for photosynthesis. Therefore, it appears quite plausible that light absorbed by the chlorophylls is responsible for the induction of chloroplast movements. The absorption spectra of the different chlorophylls and of the accessory pigments are known. However most reports in the literature (for an exception see Dorscheid and Wartenberg, 1966) give action spectra for light induced chloroplast movements which differ from the chlorophyll absorption curves. Examples for action spectra of light induced chloroplast movements are shown in Fig. 2.

The examination of a variety of plants indicated that ultra violet and blue light up to about 520 nm induces chloroplast movements. The action spectra are similar to each other. Furthermore, the action spectra of chloroplast movements in both high and low light intensities roughly match each other. This suggests that the same photoreceptor is responsible for both responses. An absorption spectrum closely corresponding to the action spectra of chloroplast movements is that of riboflavin (Fig. 2).

However these data alone do not constitute unequivocal proof that riboflavin is identical to the photoreceptor. It can be envisaged that other pigments, which may be localized in the vacuole or in the cell wall, could act as "screening pigments". Clearly, such pigments could also emit fluorescent light upon UV-irradiation. This fluorescence, if absorbed by the photoreceptor pigment, would simulate an UV-absorption, while in reality the photoreceptor does not absorb at these wave lengths.

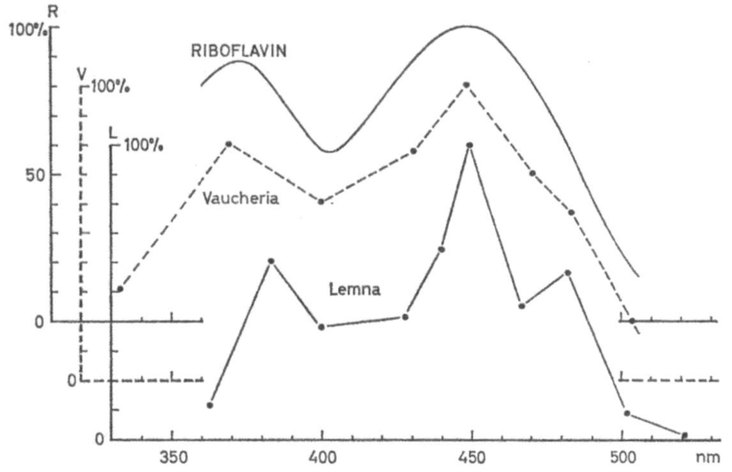

Fig. 2. Action spectra of chloroplast movements in *Vaucheria* and *Lemna* and absorption spectrum of riboflavin: the percentages of chloroplast orientation in *Vaucheria* (V) and *Lemna* (L) are plotted versus the wavelength. Ordinate R: absorption of riboflavin in percent. (According to HAUPT, 1964)

b) The Location of the Photoreceptor

In agreement with the idea of the photoreceptor being a flavin, HAUPT (1964) suggested that riboflavin may be located in the cytoplasm rather than in the chloroplast. He proposed to induce chloroplast movements by irradiating the cell without

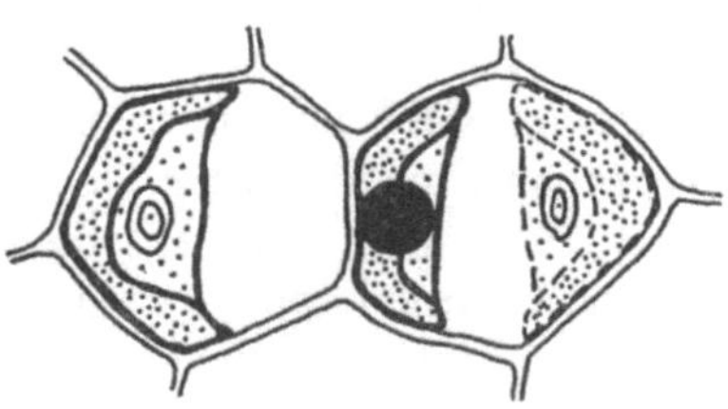

Fig. 3. Irradiation of a single *Selaginella* cell (right cell) with a light spot of 6 μ diameter (solid circle). Broken lined: chloroplast position before irradiation; solid lined: chloroplast position after irradiation; left cell: non-irradiated control

exposing the chloroplast. This was verified by FISCHER-ARNOLD (1963), who employed *Vaucheria* cells. During irradiation, the streaming of cytoplasm and chloroplasts is maintained in these cells. If only a small region of the cell is exposed to light chloroplasts accumulate at this spot. It could be observed, that chloroplasts which randomly enter the irradiated region together with the streaming cytoplasm stop their movement and remain there. This reaction, however, does not occur until 5 min after the onset of irradiation. Also, if no chloroplast passes the irradiated spot during he first 5 min, the chloroplast entering this region first will remain. From this result,

Fischer-Arnold concluded that the cytoplasm and not the plastid is the primary target of the light stimulus leading in this experiment to the accumulation of chloroplasts in the light exposed region. In agreement with this concept and with results of Bock and Haupt (1962), Mayer (1964) found, that partial irradiation of *Selaginella martensii* with blue light of low intensity causes chloroplast movement in the epidermal cells even if the chloroplast is not hit by light. Chloroplast migration into the lighted area can be observed (Fig. 3).

These results indicate that in *Vaucheria* and *Selaginella* the photoreceptor is not located within the chloroplast itself but rather in the cytoplasm. If the photoreceptor is a flavin, this finding favors the hypothesis that light perception occurs in the cytoplasm.

c) Orientation of the Photoreceptor Molecules

A photoreceptor molecule exhibits the highest light absorption if its main absorption plane is parallel to the vibration plane of the polarized light; least absorption is experienced if these two directions cross each other. An experiment concerning the induction of polarity in *Fucus* zygotes is described by Jaffe (1958). He employed linear polarized light for the irradiation of zygotes and found that the area on the surface of a zygote where germination of a rhizoid occurs is determined by the vibration plane of the polarized light. From his results Jaffe concluded that the photoreceptor molecules responsible for the germination induced in this fashion are oriented with respect to the cell surface.

Experiments concerned with this effect of polarized light on the photoreceptor pigments of chloroplast movements and of photodinesis are described by Haupt and Bock (1962), Haupt (1964), Mayer (1964), Zurzycki (1967) and by Seitz (1964), respectively. Also Fischer-Arnold's experiment with *Vaucheria* (see p. 38) is useful for a more precise localization of the photoreceptor within the cytoplasm. Under the experimental conditions the cytoplasm—with the exception of its cortical layer —continues to stream, but the chloroplasts stop moving in the irradiated cell area after a lag-time of 5 min. Therefore it can be concluded that the photoreceptor is located within the immovable layer of the cytoplasm, i.e., the cortical layer or the plasmalemma.

Similar results were obtained in *Selaginella martensii* (Mayer, 1964). Upon irradiation with plane polarized blue light of low intensity from above and perpendicularly to the leaf surface the single chloroplast of each cell becomes "band-shaped" after some time (Fig. 4). This change is brought about by the chloroplast which moves away from those cell walls located *perpendicularly* to the electric vector of the polarized light. If—under the same conditions—the light intensity is raised, the chloroplast loses its "band-shape" and eventually returns to a "band-shape", where—in contrast to the low intensity response—the flanks *parallel* to the electric vector are detached from the side walls (Fig. 4). These results suggest that there is an absorption gradient in the plane of the cell's cross section, which is induced by plane polarized light. Maximum absorption by the photoreceptors in the neighboring cortical layer occurs at the cell walls parallel to the electric vector of the polarized light while least absorption occurs at the walls that are oriented perpendicularly to the plane of vibration. Under high-intensity irradiation the chloroplast "avoids" the areas of maximal absorption; under low-intensity conditions the chloroplast is "attracted" by the irradiated area.

How is such a gradient formed and maintained during irradiation under these conditions? An explanation given by Mayer (1964) postulates that dichroic photoreceptor molecules in the cytoplasm are oriented parallel to the cell surface. This means that the main absorption plane of the molecules is parallel to the neighboring cell wall (Fig. 5).

Another conclusion was also derived from these results. If the orientation of pigment molecules would have been disturbed during the low intensity response this would have prevented the subsequent high intensity reaction. Since the cortical

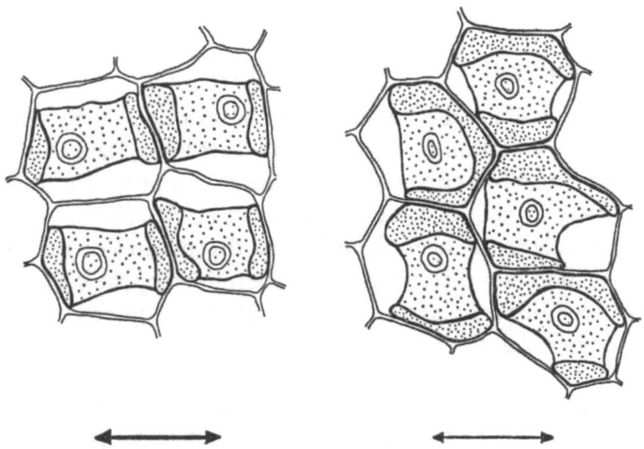

Fig. 4

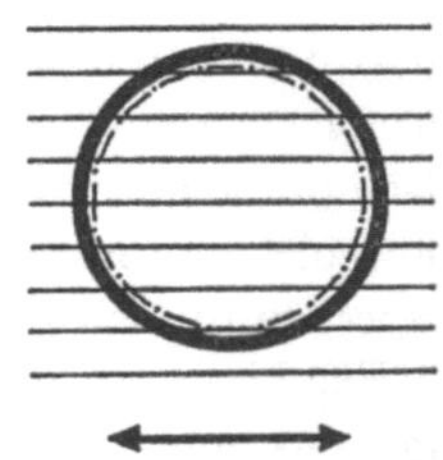

Fig. 5

Fig. 4. Movement of *Selaginella* chloroplasts under polarized light. Left: high intensity response (◄━►); right: low intensity response (⟷). The arrows indicate the vibration plane of the polarized light. (According to Mayer, 1964)

Fig. 5. Schematic drawing of a cell of *Selaginella*. Shown within the cell wall are the photoreceptor's planes of preferential light absorption [symbolized by dashes (- - - -), and dots (····)], oriented parallel to the cell surface. The superimposed grid of parallel lines indicates the plane of vibration of the incident polarized light. Maximum absorption occurs in those regions of the cell where these lines are parallel to the photoreceptor symbols. (According to Mayer, 1964)

plasm is the only cytoplasmic layer which is not displaced by the responses, it can be assumed that the photoreceptor molecules are located in this cortical layer or, maybe, in the plasmalemma. Experimental results obtained with *Funaria hygrometrica* (Zurzycki, 1967) are in complete agreement with these conclusions concerning the location and orientation of the photoreceptor molecules.

2. Experiments in ***Mougeotia spec.***, a Green Alga

Because of its clear structural organization *Mougeotia* is particularly suited for light-physiological studies. Its chloroplast movements were investigated mainly by Haupt and his co-workers.

The response of *Mougeotia* chloroplasts to low and high intensity irradiation is shown in Fig. 1. In contrast to other plants which do not react to red light by chloroplast displacement, in *Mougeotia* red light of low intensity is even more effective than blue light and leads to the face position of the single chloroplast (Haupt, 1959). It would be, therefore, conceivable that the induction of this response is mediated by light absorption in the chlorophyll pigments. This can be decided by the following simple experiment. After a short exposure to red light (about 660 nm) which provides a clearly visible low intensity reaction in the control, cells were exposed to far-red light (about 730 nm). No chloroplast movement was observed. This indicates that the photoreceptor responsible for this reaction cannot be a chlorophyll but rather the red-far red system (phytochrome system). This pigment system is known to be converted into its physiologically active form (P_{FR}) by red light. This activated form can be reversed to the red absorbing form (P_R) by far-red light. This process can be repeated. The experimental study of these reactions is simplified by the fact that the response caused by light absorption in the phytochrome system posesses the characteristic of an induction phenomenon. This means that one short irradiation is sufficient for the whole reaction to take place in the dark. It is well known that the red-far red pigment system functions as a photoreceptor system in many plant reactions induced by light (see Mohr, 1960). However, its function as a photoreceptor for the induction of chloroplast displacement is new and somewhat unexpected.

a) Temperature Experiments in Mougeotia

Mugele (1962) irradiated *Mougeotia* at different temperatures, namely at 2, 10, 20 and 30 °C with light of equal energy and quality. At all induction temperatures the cells reacted similarly if they were allowed to perform their response under favorable temperature conditions during the following dark period. This result demonstrates that the temperature at the period of induction (irradiation) does not affect the induction itself. In another series of experiments the speed of chloroplast movement as a function of temperature was determined. The results obtained for reaction temperatures of 10, 20 and 30 °C, respectively, indicated that within this range the chloroplast completes its turn faster with increasing temperature. The temperature coefficient (Q_{10}) was determined to be about 2. This is in agreement with the values obtained for chemical reactions and suggests, that the low intensity response is, therefore, limited by chemical reactions. Moreover, Mugele (1962) succeeded in separating the induction phase (photochemical process) from the response (the subsequent chemical reactions). Thus, if kept at 4 to 5°C rather than at 10°C at after irradiation a decrease in the chloroplast turning rate was observed. A weak but significant reaction was still obtained after several hours. When returned to 20 °C after low temperature treatments of varying duration some response was always observed. The extent of the response was always dependent on the duration of the low temperature, but a significant reaction was measured even 9 h after induction.

Furthermore, it could be shown that the reversal of induction by far red light was also temperature-dependent. The following experiment was conducted: plants were induced by red light at 2 °C. One sample was treated with far-red light at low temperature and subsequently warmed to 20 °C. Another sample was warmed directly after the induction and then was irradiated with far-red light. The result showed that the induction effect was reversed at 2 °C as well as at 20 °C. The reversing effect of far-red light is, therefore, temperature-independent. HAUPT (1959), by employing *Mougeotia*, could show that optimal reversion effects by far-red irradiation at 20 °C are obtained only if the far-red treatment is given within 5 min after the red light induction. These experiments indicate that dark decomposition of a red light-induced gradient is decreased by lowering the temperature. It was, therefore, understandable, that the interval between red and far-red irradiation at 4 °C could be extended to 3 h without diminishing the reversion effect of the far-red light.

According to MUGELE (1962) these results present evidence that the reaction at low temperature is stopped immediately after the production of P_{FR}, i.e., directly before the beginning of the subsequent chemical reactions. The pigment molecules, P_R and P_{FR}, therefore, do not change their distribution in the dark at low temperatures for at least several hours after induction. The question of the location of the light acceptor molecules re-arises. An answer appears to be given by experiments conducted by HAUPT and his co-workers during their extensive investigations on the connection between the pigments of the red-far red system with chloroplast movement in the *Mougeotia* cell.

b) Localization of the Red-Far Red System in Mougeotia

In order to appreciate fully the following section it should be mentioned again that for the induction of chloroplast movement it is necessary to have pigment molecules of different states of excitation in order to have an absorption gradient.

If a *Mougeotia* cell is irradiated with red plane-polarized light which vibrates parallel to the long axis of the cell no reaction will occur. If, however, the vibration plane stands vertically on the long axis the chloroplast will turn from profile to face position. A possible explanation of this phenomenon was given by HAUPT and BOCK (1962). They argue that the lack of a response upon irradiation with parallel vibrating light reflects the absence of an absorption gradient, i.e., all peripheral pigment molecules absorb equal amounts of light. Polarized light which vibrates perpendicularly to the long axis of the cell causes the chloroplast to turn. Consequently, absorption differences do exist under these condition. An explanation of these findings can be provided by the hypothesis which postulates that the photoreceptors are oriented parallel to the cell surface in the cytoplasm. As is illustrated in Fig. 6, light vibrating perpendicularly to the long axis of the cell is exclusively or mainly absorbed by pigment molecules located at the front or the rear side of the cell. Thereby, an absorption gradient is generated between the front and the rear side and the two flanks. The absorption of polarized light vibrating parallel to the long axis of the cell by the photoreceptor molecules is about equally distributed over the surface. Consequently, in this case no gradient can be generated.

Future experiments may prove the model shown in Fig. 6 too simple because the pigment molecules are visualized as rods. It is known, however, that many pigment molecules are shaped like plates. Therefore, it could be possible that light vibrating parallel to the long axis but striking the board-shaped molecules at different angles will be absorbed differently.

Fig. 7 depicts the absorption gradient in light vibrating perpendicularly to the cell axis. This gradient is tetrapolar. It is symmetric in Fig. 7a. It must be kept in mind, however, that this representation does not account for the screening effect of the chloroplast, although this effect is considerable. At a wavelength of 650 nm the intensity of the light passing the chloroplast in profile position is attenuated to about 30%. Thus the tetrapolar gradient is not strictly symmetric (Fig. 7b). Finally, if the absorption within the cell is very high, a bipolar gradient should result, as illustrated in Fig. 7c. The following experiment was

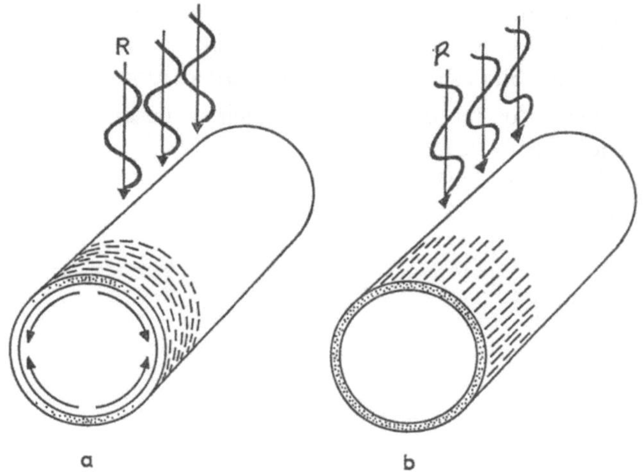

Fig. 6a and b. Absorption properties of a *Mougeotia* cell (without chloroplast). The dots (····) symbolize the absorption gradient of the cell in the cytoplasm; the gradient pattern is indicated by the 4 arrows within the cell at the left side. a Vibration plane perpendicular to the cell's long axis; only those photoreceptors parallel to the vibration plane are drawn. b Vibration plane parallel to the cell's long axis; only those photoreceptors parallel to the vibration plane are drawn. (According to HAUPT, 1960)

Fig. 7a—c. Three possibilities for the absorption gradient in a cylindrical cell (drawn without chloroplast): a tetrapolar gradient, symmetrical; b tetrapolar gradient, not symmetrical; c bipolar gradient. Dots (····) denote the region of absorption; the gradient is given by the arrows. (According to HAUPT and BOCK, 1962)

devised to decide whether the low-intensity response is affected differently by these various gradients. Cells have been irradiated from opposite sides. The results were compared with those obtained by unilateral irradiation. During equal periods of irradiation the bilaterally irradiated cell receives exactly twice the energy of the unilaterally exposed cell. The bilaterally exposed cell shows an increased response as compared to the unilaterally treated cell. The same effect is shown by cells which are unilaterally irradiated with twice the intensity. Thus, it can be concluded that the distribution of the total intensity of irradiation between front and rear sides does not influence the response. This means that in case of the low intensity response it is not decisive whether the gradient is a tetrapolar symmetric or assymmetric or a bipolar one.

The assumption that the photoreceptor molecules are located within the ectoplasm and are oriented parallel to the surface of the cell alone provides no information about the orientation of these molecules within the surface plane. Fig. 8a—c demonstrates three possibilities of surface-parallel orientation:

a) equal distribution of the molecules with their main absorption plane parallel and perpendicular to the cell axis,
b) randomly oriented molecules, and
c) orientation along an assumed screw around the cell.

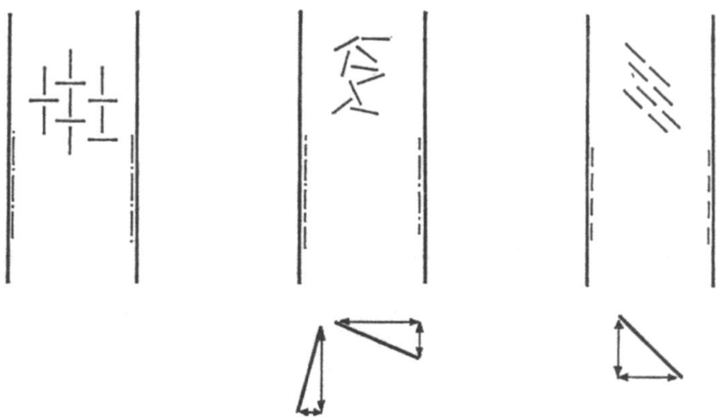

a b c

Fig. 8a—c. Different possibilities of a surface-parallel orientation of the photoreceptor molecules in a cylindrical cell (section of a *Mougeotia* cell): Dashes (- - - -) and dots (· · · ·) represent the photoreceptor's plane of preferential absorption. Sections a to c illustrate in the upper part: photoreceptors of the cell's portion facing up; in the lower part: photoreceptors located in the plane of drawing. In sections b and c, in addition, the absorption of the respective photoreceptors is shown by vectorial analysis. (According to Haupt and Bock, 1962)

Upon irradiation all three arangements may lead to the same physiological effect. All can be described by vectorial analysis as orientations along and across the long cell axis. The decision which of these possibilities is correct was aided by the clear structural organization of the *Mougeotia* cell. As mentioned above about 30% of the red light intensity is transmitted by the chloroplast in profile position (Fig. 9b). This should have an important consequence if the pigment molecules in the cortical plasma layer are oriented in an assumed screw. Bock and Haupt (1961) postulated that the pitch of the screw is 45°. If the vibration plane of red light under the conditions of irradiation shown in Fig. 9 and 10 is parallel to the direction of the thread at the front side optimum absorption occurs at the front only. If, however, the vibration plane is parallel to the thread at the rear side only this side absorbs. Since the absorption of red light within the chloroplast is 70%, a considerable decrease of

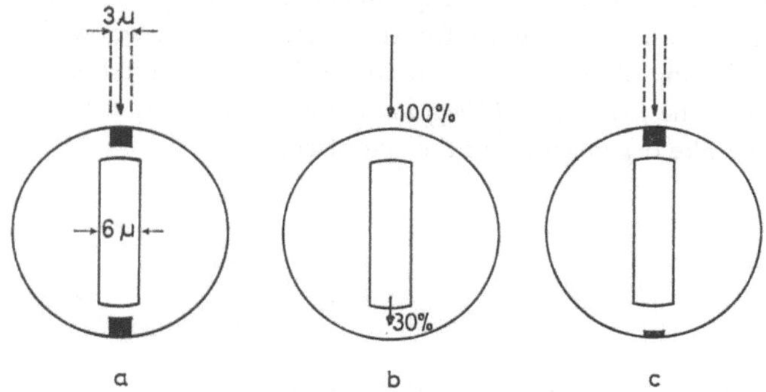

Fig. 9a—c. Absorption properties of a *Mougeotia* cell during irradiation with a light spot of 3 μ diameter. Simplified view of a cross section of the cell with the chloroplast in profile position. The square dots indicate phytochrome absorption in the ectoplasm. a without absorption; b loss of intensity during the passage of light through the cell and the chloroplast; c with loss of intensity caused by the chloroplast. (According to HAUPT and BOCK, 1962)

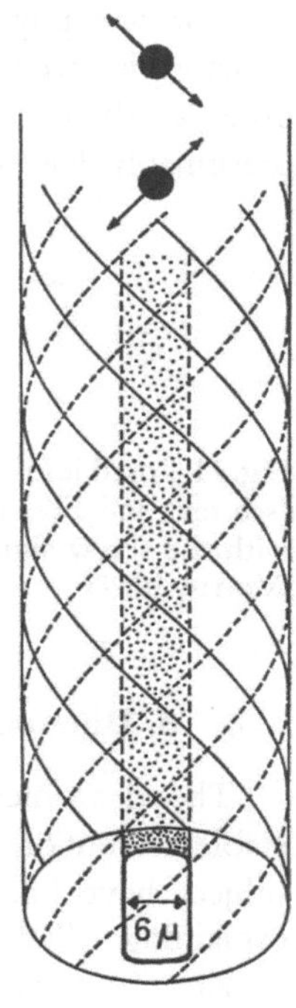

Fig. 10. Screw-like orientation of the phytochrome molecules in the cortical plasma of a *Mougeotia* cell (see Fig. 7c). The photoreceptor molecules are oriented in a way that they are able to absorb red light vibrating parallel to the screw. The solid circles represent light spots (drawn to scale), the arrows indicate the two different planes of vibration of incident light to be compared. (According to HAUPT and BOCK, 1962)

the effect is experienced in the second case. Under the following conditions the effect is roughly $^1/_3$:

1. the diameter of the light spot does not exceed the thickness of the chloroplast;
2. the "photoreceptor screw" has a pitch of exactly 45°;
3. all pigment molecules are oriented exactly parallel to this assumed screw;
4. the pigment molecules are absolutely dichroic;
5. in passing the chloroplast the light is not depolarized nor is its polarization plane turned.

Requirements 2 to 5 are most likely not fully satisfied. Therefore it has been predicted that a decrease of $^2/_3$ would not be obtained experimentally. In fact, the experimental results are in good agreement with these considerations. The ratio of the two effects was found between 2:1 and 2.5:1. Moreover, the turn of the assumed screw was shown to be identical to that of the chloroplast of *Spirogyra*, a closely related alga. Bock and Haupt (1961) interpret their result in favor of a screw-like organization of the cortical plasma layer in *Mougeotia*. They do not exclude the possibility of pitches differing from 45°, since in their experiments they always employed cells of one physiological state. An indication for the existence of a screw-like structure is given also by a finding of Haupt and Wirth (1967). These investigators observed the peculiar shape of the filament of *Mougeotia* shown in Fig. 11, which is presumably due to the screw-like organization of the cytoplasm.

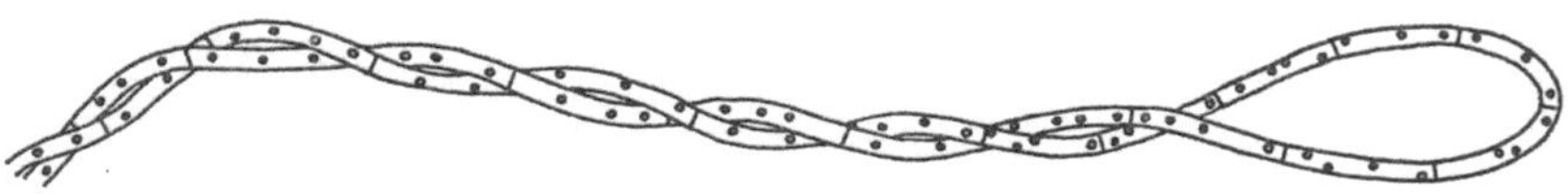

Fig. 11. Possible influence of a hypothetical screw-like organization of the *Mougeotia* cytoplasm (see text and Fig. 10) on the growth of the cell and the filament. Double helix formation with the screw sign as expected from theoretical considerations (for details see Haupt and Wirth, 1967)

3. Different Models for the Mechanism of Chloroplast Movement

Theories which explain the mechanism of chloroplast movement require a considerable amount of experimental and morphological data. A number of hypotheses on this subject have been published. However, none of these hypotheses is capable of explaining *all* the data available and, therefore, none is completely satisfactory.

Zurzycki (1962) who has been working on the problem of chloroplast movement for some time, has summarized the published hypotheses concerning the mechanism of chloroplast movement. Relatively early two alternatives had been recognized. The first implies that the chloroplast itself actively migrates in the cytoplasm while the second proposes a passive movement of the chloroplast within the streaming cytoplasm. The combination of both possibilities is also considered. Some of the hypotheses which are based on these alternatives will be described (see Zurzycki, 1962, and Kamiya, 1966).

a) The Peristromium Hypothesis

Senn (1908) thought that the peristromium — a chlorophyll-free layer of the chloroplast directly below its double membrane — is the structure responsible for movement. He supposed that chloroplast displacements are accomplished by amoeboid-like protrusions of the peristromium which attach to the ectoplasm and which by contraction pull the chloroplast into the proper orientation. These movements are induced by differences in the light intensity of different regions of the same chloroplast, and the plastid moves in the direction of the light gradient toward the optimal light conditions. This hypothesis is no longer convincing since there are many plastids without any visible protrusions. Whenever protrusions could be observed, they usually had no relation to the direction of displacement. Moreover, it is known for a number of plants that their photoreceptor is located in the cytoplasm.

b) The Surface Tension Hypothesis

PFEFFER (1904) and VOERKEL (1934) proposed that changes of differences in the surface tension on opposite sides of the chloroplast caused by a dissimilarity in light absorption are responsible for movement. They postulated that the light-induced synthesis of certain compounds alters the viscosity of the cytoplasm and the permeability of the membranes thus generating the surface tension at the boundary of cytoplasm and chloroplast.

VOERKEL suggests that the synthesis of active compounds depends on the quality of the light. Only visible and far-red light are believed to increase the amount of these compounds. This idea is based on the observation, that in *Funaria* visible and far-red light induce face position of the plastids, whereas UV-irradiation results in profile position.

While this hypothesis is quite appealing it should be taken with some caution (ZURZYCKI, 1962). The synthesis and the decomposition of these compounds should be intensity dependent, as light of the same quality given at low or high intensity may result in a low or high intensity response in various plants.

c) The Electrokinetic Hypothesis

SUESSENGUTH (1938) and HEILBRUNN and DAUGHERTY (1939) suggested that electrokinetic changes might participate in chloroplast movements. Under different light conditions the plastids may have positive or negative potentials. Potential differences between plastid and medium may be generated by photosynthetic and other light reactions (BRAUNER and BRAUNER, 1937). However, as HOFFMAN (1957) pointed out these electrokinetic forces cannot account for chloroplast movements, since their effective distance is too short.

d) The Hypothesis of Pulling Fibers

KNOLL (1908) observed that in cells of mosses, chloroplasts were displaced by contraction and extension of netlike plasmatic fibers. The plastids were thought to be attached to the net. BORESCH (1914) who also observed fiber nets noticed no frequent relation between these structures and the direction of chloroplast movement. The light-dependent movement continued after destruction of the fibers by acetone. ZURZYCKI (1957) who traced chloroplast movements by cinematographic analysis could demonstrate that the path marked by a plastid is very complex. Thus, the hypothesis of pulling fibers is not sufficient to explain all types of movement.

e) Recent Hypotheses

Chloroplasts embedded in cytoplasm droplets mechanically isolated from the Characeae frequently show a striking rotation, even several hours after preparation. Under similar conditions chloroplast chains sometimes exhibit a creeping motion. If the plastids are prevented from rotation, a streaming of the medium surrounding the chloroplast can be observed in the opposite direction. JAROSCH

(1956, 1957), who restudied this phenomenon, photographed loops of plasmatic fibers with selfmotility able to form polygons with rotating angles. This observation could possibly explain the mechanism of plastid rotation provided that some of these fibrillar structures remain fixed on the surface of chloroplast during the preparation of plasma droplets. JAROSCH envisaged that these fibrillar elements are positioned parallel to each other. Based on this idea he explained the existence of repulsion stimuli at the boundary zone between ecto- and endoplasm responsible for the streaming of cytoplasm. KAMITSUBO (1965) could demonstrate the *de novo* synthesis of fibrillar structures in the cortical plasma of centrifuged *Nitella* cells and NAGAI and REBHUN (1966) showed these fibrillar structures in the electron microscope. It should be mentioned, however, that *Nitella* did not show chloroplast movement, but only plasmic streaming. In this context studies by KAMIYA (1966) on the motility of cytoplasm are worth mentioning. Also according to KAMIYA the mechanism that accounts for the gliding of the endoplasm along the ectoplasm can be attributed to the motile micro fibers. SEITZ (1967) measured light-dependent changes of cytoplasmic streaming, viscosity and chloroplast movement in the same cell. He found a common action spectrum for each phenomenon. This fact favors the assumption that there are relations between these different reactions of the cell. These results may provide a possible answer to the question of chloroplast movement. Another aspect is concerned with ATP-requiring, reversible changes of protein structures which always accompany cytoplasmic motility. A good deal of future investigations ought to be devoted to this aspect of plastid movements.

D. Literature

BOCK, G., HAUPT, W.: Die Chloroplastendrehung bei *Mougeotia*. III. Die Frage der Lokalisierung des Hellrot-Dunkelrot-Pigmentsystems in der Zelle. Planta **57**, 518—530 (1961).

BÖHM, J. A.: Beiträge zur näheren Kenntnis des Chlorophylls. S.-B. Akad. Wiss. Wien, Math.-nat. Kl. **22**, 479ff (1856).

BORESCH, K.: Über fadenförmige Gebilde in den Zellen von Moosblättern und Chloroplastenverlagerung bei *Funaria*. Z. Botan. **6**, 97—156 (1914).

BRAUNER, L., BRAUNER, M.: Untersuchungen über den photoelektrischen Effekt in Membranen. I. Weitere Beiträge zum Problem der Lichtpermeabilitätsreaktionen. Protoplasma (Wien) **28**, 230—261 (1937).

BUSCH, G.: Über die photoperiodische Formänderung des Chloroplasten von *Selaginella serpens*. Biol. Zbl.. **72**, 598 (1953).

DORSCHEID, T., WARTENBERG, A.: Chlorophyll als Photoreceptor bei der Schwachlichtbewegung des *Mesotaenium*-Chloroplasten. Planta **70**, 187—192 (1966).

FISCHER-ARNOLD, G.: Untersuchungen über die Chloroplastenbewegung bei *Vaucheria sessilis*. Protoplasma (Wien) **56**, 495 (1963).

HAUPT, W.: Die Chloroplastendrehung bei *Mougeotia*. I. Über den quantitativen und qualitativen Lichtbedarf der Schwachlichtbewegung. Planta **53**, 484—501 (1959).

— Die Chloroplastendrehung bei *Mougeotia*. II. Die Induktion der Schwachlichtbewegung durch linear polarisiertes Licht. Planta **55**, 465—479 (1960).

— Photoreceptorprobleme der Chloroplastenbewegung. Ber. dtsch. bot. Ges. **76**, 313 — 322 (1964).

— BOCK, G.: Die Chloroplastendrehung bei *Mougeotia*. IV. Die Orientierung der Phytochrommoleküle im Cytoplasma. Planta **59**, 38—48 (1962).

— WIRTH, H.: Nachweis einer Schraubenstruktur in der *Mougeotia*-Zelle. Plant and Cell Physiol. **8**, 541—543 (1967).

HEILBRUNN, L. V., DAUGHERTY, K.: The electric charge of protoplasmic colloids. Physiol. Zoöl. **12**, 1—12 (1939).
HOFFMAN, J. G.: The life and death of cells. New York: Hannover House 1957.
JAFFE, L.: Tropistic responses of zygotes of the Fucaceae to polarized light. Exp. Cell Res. **15**, 282—299 (1958).
JAROSCH, R.: Protoplasmaströmung und Chloroplastenrotation bei Characeen. Phyton (Buenos Aires) **6**, 87—107 (1956).
— Zur Mechanik der Protoplasmafibrillenbewegung. Biochim. biophys. Acta (Amst.) **25**, 204—205 (1957).
— Die Protoplasmafibrillen der Characeen. Protoplasma (Wien) **50**, 93—108 (1958).
KAMITSUBO, E.: Motile protoplasmic fibrils (in Japanese). Kagaku to Kôkyô (Osaka) **35**, 91—95 (1965).
KAMIYA, N.: Motilität des Plasmas der lebenden Zelle. Naturw. Rundschau **7**, 270—282 (1966).
KNOLL, F.: Über netzartige Protoplasmadifferenzierungen und Chloroplastenbewegung. S.-B. Akad. Wiss. Wien, math.-nat. Kl. **117**, 1224—1241 (1908).
MAYER, F.: Lichtorientierte Chloroplasten-Verlagerungen bei *Selaginella martensii*. Z. Botan. **52**, 346—381 (1964).
v. MERCKLIN, C. E.: Beobachtungen an dem Prothallium der Farnkräuter. St. Petersburg 1850 (see SENN).
MOHR, H.: Photomorphogenetische Reaktionssysteme in Pflanzen. 1. Teil: Das reversible Hellrot-Dunkelrot-Reaktionssystem und das Blau-Dunkelrot-Reaktionssystem. Ergebn. Biol. **22**, 67—107 (1960).
MUGELE, F.: Der Einfluß der Temperatur auf die lichtinduzierte Chloroplastenbewegung. Z. Botan. **50**, 368—388 (1962).
NAGAI, R., REBHUN, L. J.: Cytoplasmic microfilaments in streaming *Nitella* cells. J. Ultrastruct. Res. **14**, 571—589 (1966).
PFEFFER, W.: Pflanzenphysiologie. II. Aufl., II. Bd. Kraftwechsel. Leipzig: W. Engelmann 1904.
SEITZ, K.: Das Wirkungsspektrum der Photodinese bei *Elodea canadensis*. Protoplasma (Wien) **58**, 621—640 (1964).
— Wirkungsspektren für die Starklichtbewegung der Chloroplasten, die Photodinese und die lichtabhängige Viskositätsänderung bei *Vallisneria spiralis*. Z. Pflanzenphysiol. **56**, 246—261 (1967).
SENN, G.: Die Gestalts- und Lageveränderungen der Pflanzen-Chromatophoren. Leipzig: W. Engelmann 1908.
SUESSENGUTH, K.: Neue Ziele der Botanik. München: J. F. Lehmann 1938.
VIRGIN, H.: The effect of light in the protoplasmic viscosity. Physiol.Plantarum **4**, 255 (1951).
— An action spectrum for the light induced changes in the viscosity of plant protoplasma. Physiol. Plantarum **5**, 575 (1952).
— Further studies on the action spectrum for light-induced changes in the protoplasmic viscosity of *Helodea densa*. Physiol. Plantarum **7**, 343 (1954).
VOERKEL, H. S.: Untersuchungen über die Phototaxis der Chloroplasten. Planta **21**, 156—205 (1934).
ZURZYCKA, A., ZURZYCKI, J.: Cinematographic studies on photoactic movements of chloroplasts. Acta Soc. Botan. Polon. **26**, 177—206 (1957).
ZURZYCKI, J.: The mechanism of the movements of plastids. Handbuch der Pflanzenphysiologie, Band 17/2, pp. 940—978. Berlin-Göttingen-Heidelberg: Springer 1962.
— The possible role of photophosphorylation in the movements of plastids. In: Currents in photosynthesis, pp. 235—242 (J. B. THOMAS, J. C. GOEDHEER, Eds.). Rotterdam: Ad. Donker Publ. 1966.
— Properties and localisation of the photoreceptor active in displacement of chloroplasts in *Funaria hygrometrica*. II. Studies with polarized light. Acta Soc. Botan. Polon. **36**, (1967).

Plastid Inheritance and Mutations

Björn Walles

A. Introduction

The structure and function of chloroplasts are controlled by nuclear and extranuclear genetic factors. Mutations in these genetic systems occur spontaneously and can also in some cases be induced by appropriate agents, *e.g.* radiation or mutagenic chemicals. All hereditary changes that influence the morphology or physiology, *i.e.* the phenotype of chloroplasts will — independent of their mode of inheritance — be referred to in this article as *chloroplast mutations* and the plants affected by them are accordingly called *chloroplast mutants*. As defined here a chloroplast mutation is not necessarily localized in the chloroplast itself.

In chloroplast mutants the amount of chlorophyll is generally more or less reduced. This phenotypic manifestation is the most conspicuous symptom of the mutation and explains the traditional term "chlorophyll mutant". In the majority of the "chlorophyll" (= chloroplast) mutants the ability to synthesize chlorophyll has not been lost. When grown in darkness, seedlings of most mutants produce protochlorophyll which they can convert into chlorophyll *a* upon illumination. This chlorophyll is photo-oxidized to various degrees, depending on the reaction norms of the individual mutations (Kandler and Schötz, 1956; Koski and Smith, 1951; Smith, Durham and Wurster, 1959). In order to resist bleaching, the chlorophyll must presumably be combined with a thylakoid protein and the extent of photo-oxidation in the different chloroplast mutants is connected with the ability of their plastids to develop a lamellar system. In addition the chlorophyll seems to require protection by carotenoids, since in all carotenoid-deficient mutants studied the chlorophyll is almost completely bleached by photo-oxidation (Anderson and Robertson, 1960; Claes, 1954; Faludi-Dániel and Láng, 1964; Habermann, 1960; Mackinney, Rick and Jenkins, 1956). Most chloroplast mutants are lethal. Variegated types are viable if they produce a sufficient amount of green photosynthesizing tissue. Some pale green types can also survive. There is considerable variation in viability between different mutants and this property is apparently not directly correlated with the content of chlorophyll (*cf.* Demerec, 1935) but is dependent on photosynthetic efficiency (*cf.* Schötz, 1956).

For practical purposes it is convenient to classify the chloroplast mutants in higher plants according to a simple system based on leaf colours. Several classification systems have been employed but they often suffer from their not being applicable to all types of higher plant or from being too elaborate to handle in practice. The scheme given in Table 1 — originally presented in 1962 at a mutation conference in Sweden — is intended to overcome these obstacles.

Table 1. *Classification of chloroplast mutants in phanerogams*
According to Å. Gustafsson, B. Walles *and* D. von Wettstein *(1962) (unpublished)*

Group I	Plants of one colour	
	1. *Albina*	White leaves
	2. *Xantha*	Yellow leaves
	3. *Viridis*	Yellow green to pale green leaves
Group II	Plants of two or more colours	
	4. *Alboxantha*	Leaf tip white, base yellow
	5. *Xanthalba*	Leaf tip yellow, base white
	6. *Alboviridis*	Leaf tip white, base green
	7. *Viridoalbina*	Leaf tip green, base white
	8. *Tigrina*	Chlorophyll deficiency in the form of transverse stripes
	9. *Striata*	Chlorophyll deficiency in the form of longitudinal stripes
	10. *Maculata*	Spotted chlorophyll deficiency
	11. *Marginata*	Chlorophyll deficiency of the leaf margin or leaf center
	12. *Costata*	Chlorophyll deficiency of the leaf veins or intercostal areas
	13. *Transformiens*	The colour of the first developed leaves differs from that of later leaves
Group III	Colour-changing plants	
	14. *Albescens*	Leaves bleach to a white—white yellow colour
	15. *Lutescens*	Leaves bleach to a yellow—pale green colour
	16. *Virescens*	Leaves are initially chlorophyll-deficient but turn green later on

N.B. Types No. 4 to 9 are restricted to grasses and other monocotyledons. Type No. 12 becomes *striata* in most monocotyledons.

B. Inheritance of Chloroplast Mutations

1. Gene Mutations

The sum of the chromosomal genes (showing Mendelian inheritance) constitutes the *genome*. In this genetic system gene mutations (changes in the sequence of the bases or loss and addition of single bases in the DNA code of the chromosomes) and deficiencies (loss of pieces of DNA) give rise to chloroplast defects. In practice we usually cannot distinguish between the two, except in a few favourable cases, *e.g.* when a chromosome aberration can be demonstrated cytologically. Most of the gene mutations which affect chloroplasts are recessive.

2. Plastome Mutations

The *plastome* (Renner, 1929, 1934) comprises all the extranuclear genetic factors which appear localized to the plastids and control their phenotype. When a wild type plastome and a mutated plastome occur together in a plant, the two plastome types segregate at the cell divisions and by this process homoplastomatic (*cf.* Michaelis,

1957) bleached and normal green cell lineages are produced. The somatic sorting-out of different plastomes follows a certain general pattern. In heteroplastomatic seedlings carrying a mutated plastome in addition to the normal plastome, the cotyledons and often the next few leaves as well, are variegated in a fine mosaic of white (or in some mutants another bleached colour such as yellow or pale green) and normal green areas.

Fig. 1. Green shoot (to the left in the background), white shoot (to the right in the background) and variegated shoots of *Nicotiana tabacum* status *albomaculatus*. By courtesy of G. ERIKSSON

The bleached and green sectors are progressively larger in the successively formed leaves and the plants develop into sectorial and periclinal chimeras, which can produce pure green and pure bleached shoots (Fig. 1).

Plastome mutations appear spontaneously with a frequency of about 0.02 to 0.06% (MALY, 1958). It has been reported that plastome mutations may be induced experimentally, in *Epilobium* by treatment with ^{35}S (MICHAELIS, 1958), in *Arabidopsis* by X-rays (RÖBBELEN, 1962) and alkylating agents (RÖBBELEN, 1965), and in *Nicotiana* by ethyl methanesulfonate (DULIEU, 1967).

a) Maternal Inheritance

In many species the mutated plastome is only transmitted to the offspring by the egg cell. CORRENS (1909) was the first geneticist to analyze the maternal heredity of variegation, of which he discovered a case in *Mirabilis jalapa*. The experimental crosses yielded the following results:

1. Flowers on green sectors of the variegated plant produce only normal green offspring;

2. Flowers on bleached sectors produce only bleached offspring;

3. Flowers on variegated shoots yield a mixed progeny of green, variegated and bleached seedlings in varying ratios, depending on the proportion between green and bleached tissue in the maternal flowers, *e.g.* flowers that have mainly green tissues yield predominantly green offspring.

In all three cases the result is independent of the colour of the father plant. This kind of maternally inherited variegation was designated by Correns as "status *albomaculatus*".

b) Biparental Inheritance

The status *paralbomaculatus* type (Correns, 1931) of variegation was first described in the classic paper by Baur (1909) on variegated forms of *Pelargonium zonale*. These forms are periclinal chimeras which can develop pure green or pure white shoots. The plastid defect is inherited from both parents in the following non-Mendelian way:

1. Flowers on green shoots when self-pollinated yield only green progeny;

2. Flowers on white or variegated shoots when self-pollinated yield only white progeny (the flowers originate from subepidermal white cell layers of the chimeric plants);

3. Green mother × white (or variegated) father yields a mixed offspring, consisting of green, variegated and white seedlings;

4. White (or variegated) mother × green father yields a mixed offspring as in crosses of type 3.

c) Gene-Induced Plastome Mutations

Some recessive genes are found to induce plastome mutations. This phenomenon is, for instance, illustrated by the action of the *iojap* (*ij*) gene in maize (Rhoades, 1943, 1946). Plants of the constitution *ij ij* are striped in green and white. The cross *Ij Ij* ♀ (green) × *ij ij* ♂ (striped) gave a green F_1 progeny, as is to be expected when the variegation is due to a recessive gene. However, the reciprocal cross, *i.e.* *ij ij* ♀ (striped) × *Ij Ij* ♂ (green) usually yielded a mixed F_1 of green, white and striped seedlings in any ratios. The case was further investigated by making reciprocal crosses *Ij ij* (striped) × *Ij Ij* (green). The offspring should here segregate 1 *Ij Ij*: 1 *Ij ij*. With variegated *Ij ij* plants as male parents the whole F_1 population was green. But with the variegated *Ij ij* plants as female parents the colour of the F_1 seedlings varied between the crosses and was apparently only determined by the constitution of the egg cells. In some crosses all progeny plants were green; in other crosses green, white and striped individuals were obtained in varying ratios and a few crosses gave an all-white progeny. Obviously the plastid defect arises in *ij ij* plants but is then inherited according to the status *albomaculatus* scheme and cannot be cured by substitution of the *ij* gene with its wild type allele.

Another interesting case of gene-induction of plastome mutations was analyzed in *Nepeta cataria* (Woods and Dubuy, 1951). In plants that were homozygous for the *m* gene several kinds of defective plastids, differing in size and pigment content as well as in internal structure, were induced. At least 15 distinct types were identified in the plant material. In a single leaf cell as many as three different types of defective plastid could be found in addition to normal chloroplasts. In this species the plastome factors are transmitted both by egg cells and by pollen. After the cross green × variegated the variegated F_1 seedlings had defective plastids of the same phenotype as that characteristic for their mutation-carrying parent.

In the "Okina-mugi" strain of barley a nuclear gene induces a plastome mutation giving rise to "white" plastids [Imai, 1936 (1)]. In such a variegated plant yellow plastids appeared as the result of a spontaneous plastome mutation. Like the wild type plastome this "yellow" plastome could be gene-induced to mutate to "white", but could also undergo spontaneous "back-mutation" to produce green plastids. The plastome causing yellow plastids is thus very unstable and liable to undergo both induced and spontaneous mutations.

Gene mutations that bring about plastome mutations can be induced with X-rays [Hagemann and Scholz, 1962; Röbbelen, 1966 (2); v. Wettstein, 1961; v. Wettstein and Eriksson, 1965].

d) Humulus japonicus

Winge (1917, 1919) analysed a maternally inherited variegation in *Humulus japonicus*. In this case there is no somatic segregation of the extranuclear genetic factors and the offspring from variegated female plants consists entirely of variegated seedlings. From this mode of inheritance Winge concluded that the mutation responsible for the variegation is not in the plastome but is localized elsewhere in the cytoplasm, *i.e.* the defect is due to a "*plasmone*" mutation. The variegated *Humulus* strain was later studied by Correns (1937) and Imai [1936 (2), 1937] who confirmed the genetic results obtained by Winge. Correns was convinced that the variegation is due to a "labile" state of the cytoplasm. In Imai's opinion the variegation is due to a mutated plastome; the variegated plants contain two kinds of plastomes: "green" and "white". These are "automutable", *i.e.* each type is by mutation repeatedly transformed into the other and this mechanism prevents the production of all-green or all-white seedlings in the progeny from variegated female plants. Tilney-Bassett (Kirk and Tilney-Bassett, 1967) holds the view that the variegated *Humulus* is homoplastomatic for a mutated plastome. This plastome is assumed to be sensitive to environmental influences and as a consequence of this, in the course of the development of the plant, produces phenotypically normal chloroplasts in some cells while other cells get pale (bleached) plastids. At the time of origin of the defect there must have been heteroplastomatic *Humulus* plants which possessed both the wild type plastome and the mutated plastome. These individuals should have given a mixed progeny of variegated and green seedlings until, after a sufficient number of generations, the segregation of the two plastome types was completed. If this idea is correct, it must be assumed that the segregation process was finished when Winge made his crosses and that he worked with a homoplastomatic material of mutant plants. Segregation into variegated and green lines was observed by Kappert (1953) in an extranuclear mutant of *Plantago*. Tilney-Bassett presents a list of several other mutants in various species that may similarly be interpreted as involving plastomes with a variable expression.

e) Oenothera

In some plant genera variegated F_1 plants are obtained after species crosses. This "hybrid variegation" is interpreted as the result of incompatibility between particular genomes and alien plastomes. Unique advantages for investigations of this and other phenomena which have a bearing upon plastid genetics are provided by the genus

Oenothera. The following account will be restricted to *Euoenothera*, the largest and genetically most extensively studied subgenus.

To understand the genetics of *Oenothera* it is necessary to consider the cytological characteristics of this genus (see review by CLELAND, 1962). The majority of the *Oenothera* species constitute a special kind of permanent heterozygotes (called "complex-heterozygotes"). Repeated translocations between their 14 chromosomes have in most forms brought about the situation that in the meiotic prophase all chromosomes form a closed ring in which paternal and maternal chromosomes alternate. This ring of chromosomes persists into metaphase. At anaphase separation of adjacent chromosomes to opposite poles produces two kinds of gametes genetically identical with the two paternal ones that formed the plant. This arrangement leads to only two linkage groups' being disclosed in genetic studies. Each such "Renner complex" has been given a name, *e.g.* the complexes that constitute *O. lamarckiana* are named *velans and gaudens*. Formation of homozygous plants is prevented by the presence of lethal genes which affect either the diploid or the haploid generations. In *O. lamarckiana* each complex carries recessive lethal factors which cause the death of the homozygous embryos *(e.g. velans.velans)*. In other species the occurrence of gametophyte lethals has the effect that each complex is transmitted only by one of the gamete types. In *O. biennis* the egg cell usually carries the *albicans* complex, whereas the pollen only transmits the *rubens* complex. *O. hookeri* is a homozygous species that forms exclusively bivalents in its meiosis. This species can be considered a complex-homozygote, possessing a single "complex" in duplicate. This "complex" is called ^{h}*hookeri* (haplo-*hookeri*) by analogy with the Renner complexes of other species.

From the cross *O. lamarckiana* × *O. hookeri* the offspring consists of plants of two different genotypes: ^{h}*hookeri.gaudens* and ^{h}*hookeri.velans*. Of these, the first combination yields green plants, whereas plants of the constitution ^{h}*hookeri.velans* are green when *O. hookeri* is mother but yellow when *O. lamarckiana* is mother (RENNER, 1924). Besides this, a few of the ^{h}*hookeri.velans* seedlings are variegated, *e.g.* bleached ^{h}*hookeri.velans* plants (with *lamarckiana* cytoplasm) can produce green leaf spots. The variegation is attributed to plastids transmitted from the pollen. It appears that the *lamarckiana* plastids cannot function normally when combined with the ^{h}*hookeri.velans* genotype, even in their own *lamarckiana* cytoplasm, whereas the few *hookeri* plastids transmitted from the father turn green in the alien *lamarckiana* cytoplasm and by sorting-out can give rise to green cell lineages. The constitution of the variegated ^{h}*hookeri.velans* plants can be further analysed by crossing them with *O. syrticola (rigens* ♀*.curvans* ♂*)* as pollen source. The F_1 seedlings will have the genotype ^{h}*hookeri.curvans* but their phenotypic colour will depend on the plastome constitution of the egg cells of their chimeric mother (which can develop green or yellow shoots): green ^{h}*hookeri.velans (hookeri* plastome*)* × *curvans* yields yellow offspring; yellow ^{h}*hookeri.velans (lamarckiana* plastome*)* × *curvans* yields green offspring. The *lamarckiana* plastids which are yellow when combined with the ^{h}*hookeri.velans* genotype thus develop normally and become green when transmitted to ^{h}*hookeri.curvans* cells. RENNER's results show that the changes in phenotype of the plastids in hybrids are not ascribable to the cytoplasm: it is the nature of the plastids themselves that determines their fate when they are combined with alien genomes.

Crosses between *Euoenothera* species have unequivocally demonstrated that this subgenus comprises a number of different plastome types. Each one of these plastome types will permit the development of normal chloroplasts when combined with a restricted number of the possible genotypes; in all other cases bleaching occurs. From the 14 species of RENNER's collection about 400 genome-plastome combinations have been synthesized by STUBBE (1959, 1960). He concluded that these species comprise coalitions of only three basic haploid complexes, which were designated A, B and C. The number of different plastomes was estimated to be five: *hookeri*-plastome (I), *suaveolens*-plastome (II), *lamarckiana*-plastome (III), *parviflora*-plastome (IV) and *argillicola*-plastome (V). According to this classification some of the species mentioned above can be written in the following way: *O. lamarckiana* = AB-III, *O. hookeri* = AA-I, *O. syrticola* = AC-IV.

The disturbances in plastid development that occur when incompatible plastomes and genomes are combined vary in gravity and the hybrid seedlings show every gradation in colour from almost green to white (*cf.* Stubbe, 1959). The extreme non-harmonious combinations result in inviable hydrid embryos. However, most of these lethal combinations can be synthesized as bleached tissues in variegated (chimeric) plants, in which the green parts have the plastome type of their father [Stubbe, 1963(2)]. The physiological disturbances are apparently not as serious in the seedlings as in the embryo, although the non-harmonious cells need nutrients produced by normal photosynthesizing tissues. Schötz (1958) has analyzed the physiology of a great many non-harmonious genome-plastome combinations and found that the bleaching is often periodical, so that during their ontogeny the seedlings pass through a first green period, a phase of disturbance (bleaching) and a second green period. During the second green period further disturbances can appear. Different hybrids enter the bleaching phase at various times during the development of their cotyledons and also the duration of the disturbance varies. The extent of bleaching and the viability of the different hybrids is to some degree dependent on environmental factors. Thus, some hybrids which under normal conditions are lethal can survive if grown in weak light (Schötz and Reiche, 1957). Schötz and Diers have made extensive electron microscope studies of the various aberrant chloroplast structures that occur in the viable bleaching hybrid *velans*. ^{h}*hookeri* with *lamarckiana* plastome, *i.e.* AA-III [Diers and Schötz, 1968, 1969; Diers, Schötz and Bathelt, 1968; Schötz, 1964; Schötz and Diers, 1966, 1968 (1, 2), Schötz, Diers and Bathelt, 1968)].

Plastome mutations occur spontaneously in *Oenothera* with a frequency of about 0.05% (Maly, 1958). Such a "*defect mutation*" (Stubbe, 1958) in the plastome will always prevent the development of normal chloroplasts regardless of the genome constitution of the seedling. The term "*differential mutation*" *(Differenzierungsmutation)* is employed for the kind of mutations which must have occurred in the past during the evolution within the genus and which produced the differences now established between the wild type plastomes in *Oenothera* (Stubbe, 1958).

In *Oenothera* the rate of multiplication of the plastids is controlled by the plastome which becomes apparent when different kinds of plastid are combined in a single cell. For an investigation of this phenomenon Schötz (1954, 1968) employed two lines (with *biennis* and *lamarckiana* plastomes respectively) carrying mutated bleached plastids. Variegated plants carrying one of these mutations produce white branches. Flowers on such shoots were used for crosses with green plants of different species. Hybrid zygotes and embryos having green and mutated plastids gave rise to variegated seedlings. In the progeny from crosses of various green females with the same "white" male the percentage of variegated plants and the proportion between white and green leaf tissues were different. This different ability of the green maternal plastids to compete with the mutant plastids from the pollen was explained as being due to different "strength" *i.e.* relative speed of multiplication of plastids with different plastomes. Reciprocal crosses, *i.e.* "white" female × green male gave similar results. The species tested could be arranged according to the "strength" of their plastomes. This scheme remained unchanged when the plastids were combined with various genotypes. Therefore the speed of multiplication of the plastids must be determined chiefly by the plastome, while the genome only can have a modifying effect. According to Schötz (1968) plastome types I *(hookeri)* and III *(lamarckiana)* are "strong", type II *(suaveolens)* is "average" and types IV *(parviflora)* and V *(argillicola)* are "weak". Stubbe [1960, 1963 (1)] referring to unpublished results by I. Kemper claims a slightly different order of "strength": I > II ≈ III > V > IV.

The shape of starch grains in amyloplasts of pollen is another character controlled by the plastome (Stubbe, 1959, 1960).

f) Euglena

The green flagellate *Euglena gracilis* is presently much used for chloroplast research. Depending on the environment it can grow as an autotroph or as a heterotroph. In the light a *Euglena* cell contains about ten chloroplasts. In darkness the cells do not synthesize chlorophyll. The amount of this pigment originally present is reduced by dilution during the cell divisions and the chloroplasts are replaced by proplastids (Ben-Shaul, Epstein and Schiff, 1965). When illuminated these colourless cells turn green as the proplastids synthesize chlorophyll, enlarge and differentiate into chloroplasts (Epstein and Schiff, 1961; Ben-Shaul, Schiff and Epstein, 1964).

A number of physical and chemical agents can induce irreversible loss of chlorophyll synthesis, *i.e.* bleaching of *Euglena*. Treated cells give rise to colourless strains which never recover the ability to be green. Bleaching agents are *e.g.* high temperatures (Pringsheim and Pringsheim, 1952), ultraviolet light (Pringsheim, 1958; Lyman, Epstein and Schiff, 1961), streptomycin (Provasoli, Hutner and Schatz, 1948) as well as several other antibiotics (see Mego, 1968). Since *Euglena* is an asexual organism, the nature of its mutations cannot be determined by genetical analyses. By proper treatment it is often possible to bleach 100% of the cells in a population. Gibor and Granick [1962 (1)] used a microbeam of UV light to irradiate the cytoplasm while the nucleus was shielded and in this case bleaching readily occurred. On the other hand, irradiation of the nucleus alone could kill the cells but not induce bleaching. These and some other observations demonstrate that the mutable factors responsible for the loss of chlorophyll are localized in the cytoplasm. Available facts are consistent with a localization of these factors (genes) in the plastids (*cf.* Gibor and Granick, 1964). The plastids are known to contain DNA (see p. 62). The action spectrum of bleaching with UV light has a maximum at 260 nm which shows that the targets are molecules of nucleic acid (Lyman, Epstein and Schiff, 1961). Also the bleaching action of streptomycin and other antibiotics may be due to an effect on plastid DNA (Ebringer and Kupkova, 1967; Mego, 1968).

The break-down of chloroplast structures in dividing light-grown *Euglena* cells treated with streptomycin has been studied by electron microscopy (Ophir and Ben-Shaul, 1968). After 11 cell generations the plastids had obtained the same structure as proplastids. After three generations the rate of chlorophyll loss was about 0.5/generation indicating a dilution effect. In non-dividing dark-grown cells exposed to streptomycin and transferred to light the formation of chlorophyll and chloroplasts was inhibited. These cells developed abnormal giant proplastids. Elimination of plastids was not reported in this study. It is often assumed that bleached strains eventually lose all their plastids by dilution and there are electron microscope pictures that might indicate such a loss (SCHIFF and EPSTEIN, 1965). However, other authors have observed proplastids in bleached cells (GIBOR and GRANICK, 1964; KIRK and TILNEY-BASSETT, 1967; MORIBER *et al.*, 1963; SIEGESMUND, ROSEN and GAWLIK, 1962).

C. Nature of the Plastome

1. The Genetic Autonomy of Plastids

The nature of the plastome is generally interpreted according to a theory emanating from BAUR (1909) and further developed by RENNER (1929). This theory states that the plastids are self-duplicating bodies possessing their own genetic material (= plastome). In variegated plants showing extranuclear inheritance there are two kinds of plastid in the leaves: normal chloroplasts and pale (chlorophyll deficient) plastids. It is presumed that a zygote which will ultimately develop into such a variegated plant contains two kinds of proplastid differing in their genetic constitution: one sort will develop into normal chloroplasts, while the other (mutated) sort will eventually be bleached. At the cell divisions the two types are distributed in varying ratios between the daughter cells. This causes sorting-out (segregation) of the plastid types with the production of constant cell lineages with either normal chloroplasts or pale plastids. Before this final result is reached the tissues would contain "mixed cells", having both normal and mutated plastids. The fraction of cells with only one type of plastid will obviously increase with the number of cell divisions taking place, and as a result the pattern of variegation changes and progressively larger pure green and pure bleached tissue sectors are formed (see p. 53). In plant species with the status *albomaculatus* type of inheritance, the zygote obtains all its proplastids from the egg cell, but in the status *paralbomaculatus* type *(e.g. Oenothera, Pelargonium)* some plastids are also contributed by the pollen.

In the electron microscope it has not been possible to distinguish between "mutated" and "wild type" proplastids in gametes or in meristematic or very young parts of variegated plants. However, the theory of plastid inheritance does not require that the two types of plastids should be structurally distinguishable at all stages of ontogenesis, only that they are different in their genetic material and consequently in their reaction norm. The existence of two plastid types (normal chloroplasts and defective plastids) in mature leaves of variegated plastome mutants has been verified by light and electron microscopy. The two types can also occur together in the same cell (Fig. 2). Such mixed cells with bleached, structurally defective plastids lying adjacent to wild type chloroplasts have been observed in the electron microscope in plastome mutants of *Antirrhinum* [DÖBEL and HAGEMANN, 1963; DÖBEL, 1964 (2)], *Nicotiana* (v. WETTSTEIN and ERIKSSON, 1965), *Arabidopsis* [RÖBBELEN, 1966 (2)], *Epilobium* (ANTON-LAMPRECHT, 1966), and *Zea mays* (SHUMWAY and WEIER, 1967).

The existence of mixed cells supports the theory of plastid inheritance. In the *Epilobium* strain the occasional mixed cells were found in a completely bleached shoot, maybe as the result of backmutations. A few normal chloroplasts were noticed exceptionally in cells which otherwise contained degenerated plastids identical in structure with those of the neighbouring cells. The observation in *Epilobium* is important as an indication that individual plastids may undergo mutation (in this case

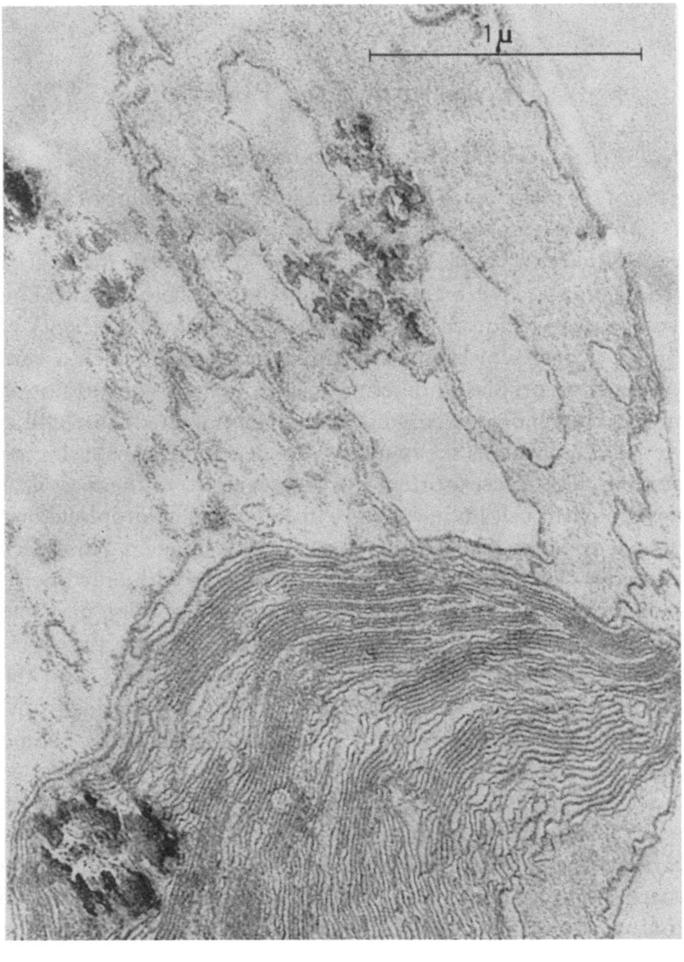

Fig. 2. Chloroplast and bleached mutant plastid with degenerated lamellar system lying side by side in a mixed cell of *Nicotiana tabacum* status *albomaculatus* ($KMnO_4$ fixation) (VON WETTSTEIN and ERIKSSON, 1965)

to produce a structurally normal chloroplast phenotype). If the plastids are genetically autonomous organelles, several plastid "genotypes" should be able to occur together in the same mixed cell. This situation seems, in fact, to have been found in a few instances. STUBBE (1957) synthesized an *Oenothera*-hybrid, in which three plastid types (of which two were mutated) occurred together: green *lamarckiana* plastids, yellow green *suaveolens* plastids and white *suaveolens* plastids. WOODS and DUBUY (1951) observed in *Nepeta*, as the result of gene-induced mutations, up to three different plastid types in addition to wild type chloroplasts in single cells (see p. 55).

2. The Significance of Plastid DNA

Any theory regarding the nature of the plastome must be consistent with the biochemistry of plant cells. According to our present knowledge the native genetic infor-

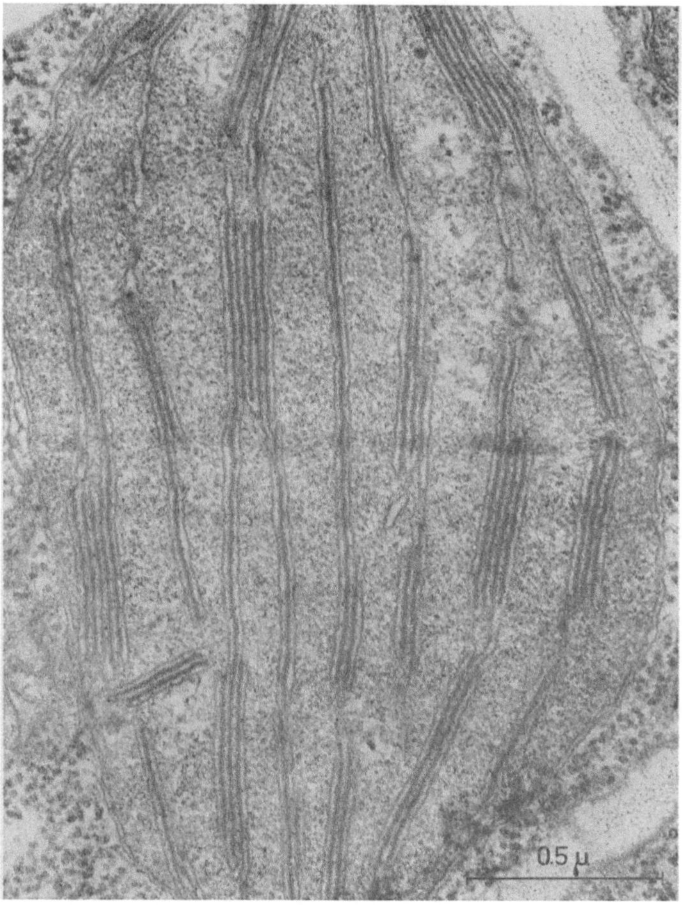

Fig. 3. Section through a tobacco chloroplast with electron transparent regions containing DNA fibrils (Glutaraldehyde-OsO_4 fixation) (VON WETTSTEIN, 1967)

mation of a cell is stored in DNA molecules. Since some kinds of viruses contain information in the form of RNA it has sometimes been suggested that the vehicles for extranuclear inheritance might be some species of RNA (*e.g.* ribosomes). This idea, which implies the existence of autonomous RNA molecules in the cell is, however, not supported by experimental facts. The cellular RNA — ribosomal RNA, soluble (transfer) RNA and messenger RNA — is produced on DNA templates (Hurwitz *et al.*, 1962; Spiegelman and Hayashi, 1963; Woods and Zubay, 1965; Yankofsky and Spiegelman, 1962, 1963). The genetic material of the plastome most probably consists, therefore, of DNA.

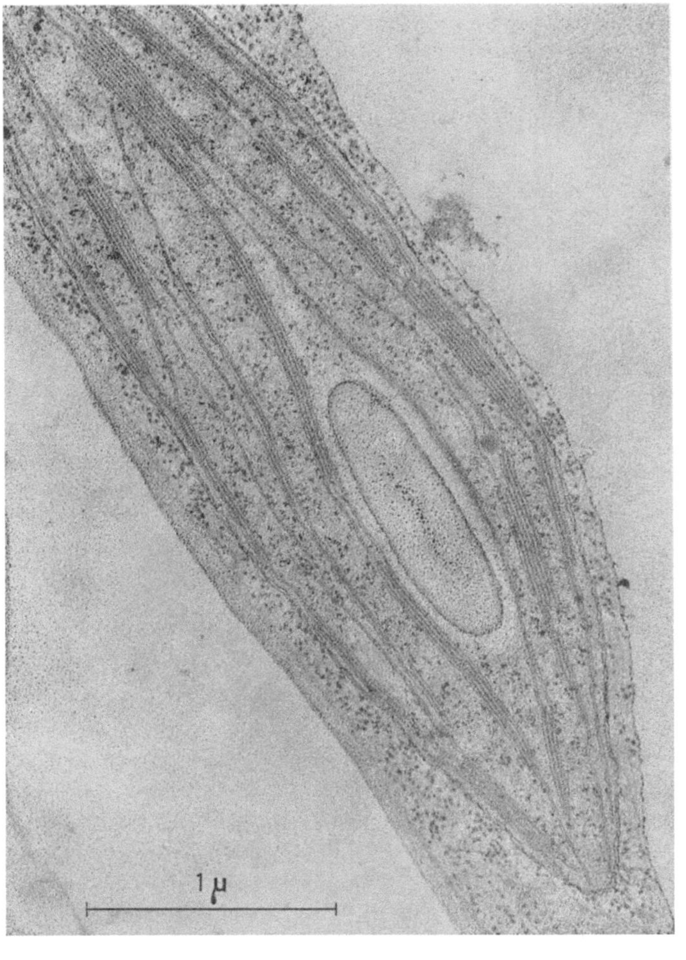

Fig. 4. Ribosomes in chloroplast and in cytoplasm in a mesophyll cell of tobacco (Glutaraldehyde-OsO_4 fixation) (von Wettstein, 1967)

Plastids have been shown to contain double-stranded DNA in the form of histone-free fibrils (Fig. 3), similar in amount and molecular organization to the nucleoplasm DNA of bacteria (Gibor and Granick, 1964; Kislev, Swift and Bo-

GORAD, 1965; KIRK, 1966; PARTHIER and WOLLGIEHN, 1966). The plastid DNA differs from the nuclear and the mitochondrial DNA in base composition and buoyant density (EDELMAN, EPSTEIN and SCHIFF, 1966; PARTHIER and WOLLGIEHN, 1966). The plastids posses the machinery for synthesis of nucleic acids (cf. Fig. 4) and protein [KIRK, 1964 (1, 2); SEMAL *et al.*, 1964; BRAWERMAN, 1962, 1963; EISENSTADT and BRAWERMAN, 1963, 1964; GOFFEAU and BRACHET, 1965; PARTHIER and WOLLGIEHN, 1963, 1966; SPENCER and WILDMAN, 1964; STEPHENSON, THIMANN and ZAMECNIK, 1956]. This does not mean that all the plastid RNA is synthesized in the organelle itself. The origin of the messenger RNA of plastids is of a special interest, since genetic data prove that the development and function of chloroplasts is controlled both by nuclear and extranuclear factors (see p. 65ff.). It has also been calculated for the alga *Acetabularia* that the amount of chloroplast DNA is insufficient to code for all chloroplast proteins (BERGER, 1967). The plastids are thus semi-autonomous organelles.

Some experimental results indicate that bleached *Euglena* cells may have lost their plastid DNA. By density-gradient sedimentation three different DNA fractions (nuclear, plastid and mitochondrial) were established in light-grown and dark-grown wild type cells as well as in cells of some pale green or yellow mutant strains, whereas plastid DNA was not found in colourless mutant strains (EDELMAN, SCHIFF and EPSTEIN, 1965; RAY and HANAWALT, 1965). However, we cannot exclude the possibility that the bleached cells still possess some plastid DNA but that the amount is unsufficient to be detected by the method employed. The plastids of white leaves of a status *albomaculatus* mutant in tobacco contain DNA of the same buoyant density as the DNA of wild type chloroplasts (v. WETTSTEIN, 1967). In the pale plastids of a plastome mutant in maize, DNA fibrils were observed with the electron microscope (SHUMWAY and WEIER, 1967).

3. The Origin of Plastids

The possibility that the plastids have their own genetic material was discussed in a preceding paragraph. If the plastids should be able to carry genetic information they must further be self-duplicating organelles.

A. F. W. SCHIMPER and A. MEYER suggested in the 1880s that plastids arise only by division of pre-existing plastids and are not formed *de novo*. Many algae have at all stages of their life cycle mature chloroplasts, which can be seen to multiply by fission. In the case of phanerogams the situation is more complicated. Here the chloroplasts are developed from proplastids and since morphological criteria seem to be insufficient for a safe identification of very young proplastids, it is doubtful whether the origin of proplastids can be determined by cytological methods. Although plastids in division can be seen in the electron microscope, such observations do not exclude other ways of formation of plastids, as emphasized by v. WETTSTEIN [1959 (1), 1961] and MÜHLETHALER and FREY-WYSSLING (1959).

The existence of plastid DNA is of importance for elucidating the origin of plastids. For biosynthesis of new DNA a DNA primer is required. The DNA-containing plastids must therefore apparently be formed from pre-existing plastids. It is true that the cells have other DNA-containing organelles, namely nuclei and mitochondria, but plastid DNA is chemically different from other kinds of DNA as mentioned earlier. There are indications that in somatic cells DNA synthesis in chloroplasts is independent of nuclear DNA synthesis. In enucleated cells of the alga *Acetabularia*, the number of chloroplasts increases by 15 to 100% in 2 weeks (SHEPHARD, 1965) and in such cells DNA synthesis from $^{14}CO_2$ occurs in the light (GIBOR,

1967). In higher plants it has been found that some cells of young leaves can incorporate ^{3}H-thymidine in chloroplast DNA when there is no incorporation in the nuclei (Wollgiehn and Mothes, 1964; Kislev, Swift and Bogorad, 1965).

It ought to be possible to employ isotope labelling of specific plastid constituents to study the origin of plastids in accordance with the approaches used to trace the origin of mitochondria (*cf.* Luck, 1966). The results obtained with mitochondria indicate that these arise by division of pre-existing mitochondria. That mitochondria actually are formed in this way was proven by Parsons and Rustad (1968) in studies with the protozoan *Tetrahymena pyriformis*. Cells of this organism were labelled with ^{3}H-thymidine after which they were transferred to an unlabelled growth medium. All of the mitochondria appeared to incorporate DNA precursors and they were the only cytoplasmic elements that synthesized DNA. The amount of mitochondrial label was retained for four generations (during this time the number of cells increased by a factor of 16) and this radioactivity remained randomly distributed among all mitochondria. The only way in which a mitochondrion could distribute its label to 16 other mitochondria would be to become these 16 organelles.

Even if the plastids in the cells of an individual plant arise by the division of pre-existing plastids, we have to consider whether they are transmitted to the offspring at sexual reproduction. This point has been raised by Mühlethaler and Bell (1962), who claim that the plastids (and mitochondria) are destroyed during oogenesis and are replaced by new organelles derived from the egg nucleus.

Bell and Mühlethaler (1964) studied the incorporation of ^{3}H-thymidine (fed for 2 days) into archegoniate somatic cells and egg cells in the fern *Pteridium aquilinum*. The distribution of the labelled thymidine between the cell structures was determined by electron microscopic autoradiography. In the somatic cells almost all radioactivity was confined to the nuclei, with some traces associated with plastids and mitochondria, indicating that there was only a small amount of DNA synthesis in the cytoplasmic organelles of these cells. In the mature eggs the situation was quite different. Here, radioactivity was found both in the cytoplasm and in the nucleus, the cytoplasm being the most labelled. The extranuclear radioactivity was localized in plastids and mitochondria. Some radioactivity was also associated with the nuclear evaginations seen in this plant, which according to the authors, detach and give rise to plastids and mitochondria. The results indicate that a rapid plastid and mitochondrial formation (far exceeding that in somatic cells) takes place during oogenesis. However, the origin of these organelles is not revealed by this experiment, since the data seem compatible both with a nuclear origin and with an origin through self-duplication. Mühlethaler and Bell (1962; *cf.* Bell and Mühlethaler, 1962; Bell, Frey-Wyssling and Mühlethaler, 1966) have described in detail how in *Pteridium*, according to their electron microscopical observations, all plastids and mitochondria are destroyed during oogenesis and are replaced by new organelle populations derived from evaginations of the egg nucleus. These authors think that such a reconstitution of the cytoplasm takes place in all plants and animals. In *Pinus laricio* a similar organelle elimination as seen in *Pteridium* and a *de novo* formation of mitochondria and plastids has been suggested (Camefort, 1962). On the other hand, electron microscopical investigations in *Sphaerocarpus* [Diers, 1964, 1965, 1966 (1, 2)], the fern *Dryopteris* (Menke and Fricke, 1964) and the cotton plant, *Gossypium* (Jensen, 1965) did not show any disappearance of the plastids and mitochondria during oogenesis or any *de novo* formation of such organelles from the nucleus. Diers [1966 (1, 2)] measured the number of plastids and mitochondria during development from primary axial cells to mature egg cells in the liverwort *Sphaerocarpus donnellii*. A continuous increase in the number of these organelles was noted during all developmental stages, this increase being more than eight-fold over the whole period. This result seems to exclude the possibility of organelle elimination at any stage of oogenesis in this organism. On the contrary, several of the plastids and mitochondria seemed to be in various stages of fission rather than degenerating. In *Myosurus minimus* no evidence for organelle elimination during oogenesis was found by WOODCOCK and BELL (1968). However, the authors claim to have observed such an elimination in the maturing megaspore and a *de novo* formation of plastids and mitochondria from the nucleus of the same cell.

The views of MÜHLETHALER and BELL are quite inconsistent with the theory of plastid inheritance. The situation can be illustrated with a few examples from the *Oenothera* research (SCHÖTZ, 1962; STUBBE, 1962). We may recall that in *Oenothera* the pollen transmits not only chromosomes but also plastome factors (considered to reside in the plastids). A hybrid is therefore often a mosaic, containing two types of homoplastomatic sectors, one of which has maternal and the other paternal plastids. This property can easily be seen in those cases where one category of plastids is bleached because of incompatibility with the genome (see p. 56). The paternal inheritance of plastids provides a method of transferring plastids from one species to others. Since there are five plastome types in the section *Euoenothera* (see p. 57), combinations between a certain genome and each of these five types can usually be synthesized by appropriate crosses. If the plastids originated from the egg nucleus, it must be explained how individuals with identical nuclei could produce any of five different types of plastid. There is a strain of *O. hookeri* which for 20 generations has carried an *O. biennis* plastome. This type was synthesized in two steps. First, *O. biennis* ♀ was crossed with *O. hookeri* ♂. Hybrids were obtained which had the constitution *albicans*. ^{h}*hookeri* with *biennis* plastids. Such a hybrid was back-crossed with *O. hookeri* as father and yielded ^{h}*hookeri*.^{h}*hookeri* (= *O. hookeri*) plants with *biennis* plastids. The mechanism proposed by MÜHLETHALER and BELL would mean that the ^{h}*hookeri* egg cells of *albicans*. ^{h}*hookeri* plants should have budded off *biennis* plastids and that in all following generations the plants of the new-synthesized *O. hookeri* strain should have continued to produce the alien *biennis* plastids, instead of *hookeri* plastids. The conception of MÜHLETHALER and BELL requires that the plastome factors are localized elsewhere than in the plastids and if that is assumed, a new interpretation of all the experimental data concerning extra-nuclear plastid genetics has to be given.

BAUR's and RENNER's theory of plastid inheritance provides an intelligible explanation of the genetic data and is consistent with the cytology and biochemistry of plants.

D. The Genetic Control of Chloroplast Development and Function

1. Biochemistry of Chloroplast Mutants

Chloroplast mutations interfere with the normal development and function of chloroplasts by affecting the synthesis of individual plastid constituents (WALLES, 1968). These constituents can be divided into two classes. First, there are compounds which are synthesized in plastids. To this group belong, *e.g.* chlorophylls, carotenoids, plastid proteins and plastid lipids. Mutant plants affected in the synthesis of such a compound should be able to grow as heterotrophs on a carbon source such as sugar or acetate to compensate for the loss of photosynthesis. Secondly, there are substances required for chloroplast development which are found also outside plastids, and are of general importance for cell differentiation and metabolism. To this group belong, *e.g.* amino acids and thiamine. Mutant plants that require a compound of this group should be able to grow as green auxotrophs with normal chloroplasts if they are supplied with a substance that can substitute for the missing metabolite.

2. Mutants Affected in Photosynthetic Enzymes

Levine (1960) developed a screening technique to find mutants of the green alga *Chlamydomonas reinhardi* which are unable to assimilate CO_2 but can grow as heterotrophs in a medium containing acetate. All of the mutations found to affect photosynthesis in *Chlamydomonas* are gene mutations (Levine, 1968).

Most of the mutants studied are deficient in enzymes of the photosynthetic electron transport chain (Gorman and Levine, 1965; Levine and Gorman, 1966; Goodenough and Levine, 1969). The two non-allelic mutants *ac*-115 and *ac*-141 lack cytochrome 559 and *ac*-206 lacks cytochrome *f*. In another mutant, *ac*-21, a component may be missing that is required to couple the oxidation and reduction of these two cytochromes. The mutant *ac*-208 has no plastocyanin and the non-allelic types *ac*-80a and F-1 lack an active P 700, a specialized form of chlorophyll *a*.

In some photosynthetic mutants of *Chlamydomonas* the deficiency is in the electron transport system (see p. 150) while in others, an enzyme of the reductive pentose phosphate cycle (see p. 174) may be lacking. In *ac*-20 ribulose 1,5-diphosphate carboxylase is missing (Levine and Togasaki, 1965) and in F-60 phosphoribulokinase is affected (Goodenough and Levine, 1969). The mutants *ac*-40, *ac*-46 and *ac*-59 may be affected in their ability to couple phosphorylation to electron flow or in one of the enzymes catalyzing the conversion of CO_2 to carbohydrate (Levine and Volkmann, 1961).

Normal cells of *Chlamydomonas* possess a single, large cup-shaped chloroplast with a pyrenoid and an eye-spot. This chloroplast contains many stacks of fused discs which resemble the grana of higher plants (Sager and Palade, 1957; Goodenough and Levine, 1969). The chloroplast ultrastructure has been analyzed in nine photosynthetic mutants (Goodenough and Levine, 1969). These mutants have about the same chlorophyll content as the normal type. The total amount of membrane in the mutant chloroplasts appears comparable to that in the wild type and the only obvious difference is an altered stacking pattern of the chloroplast discs seen in some, but not all, of the mutant strains. A normal chloroplast structure, as studied by electron microscopy, was found in a photosynthetic mutant in *Scenedesmus* incapable of evolving oxygen (Bishop, 1962). These results show that the lack of photosynthetic enzymes does not cause drastic alterations in chloroplast ultrastructure.

We are still largely ignorant about the occurrence of photosynthetic mutants in higher plants. One case in *Vicia faba* has been studied by Heber and Gottschalk (1963). This gene mutant has a normal chlorophyll content when grown in moderate light (*e.g.* in a greenhouse), but some bleaching of the chlorophyll takes place in sunlight. The mutant seedlings die when they have exhausted the nutrient of the seed. They consume oxygen both in the light and in the dark and do not assimilate CO_2 photosynthetically. The mutant plants are presumably blocked in the photosynthetic electron transport system between chlorophyll *a* and nicotinamide adenine dinucleotide phosphate.

3. Chlorophyll-Deficient Mutants

Many plant groups synthesize chlorophyll and develop chloroplasts both in dark and in light. Light is, however, a prerequisite for chlorophyll and chloroplast formation in *e.g.* angiosperms (Fig. 5) and *Euglena*. Most "chlorophyll-less" mutants are able to synthesize various amounts of chlorophyll, but this is soon lost by photo-oxidation (see p. 51). There are, however, a few mutations that primarily interfere with chlorophyll synthesis.

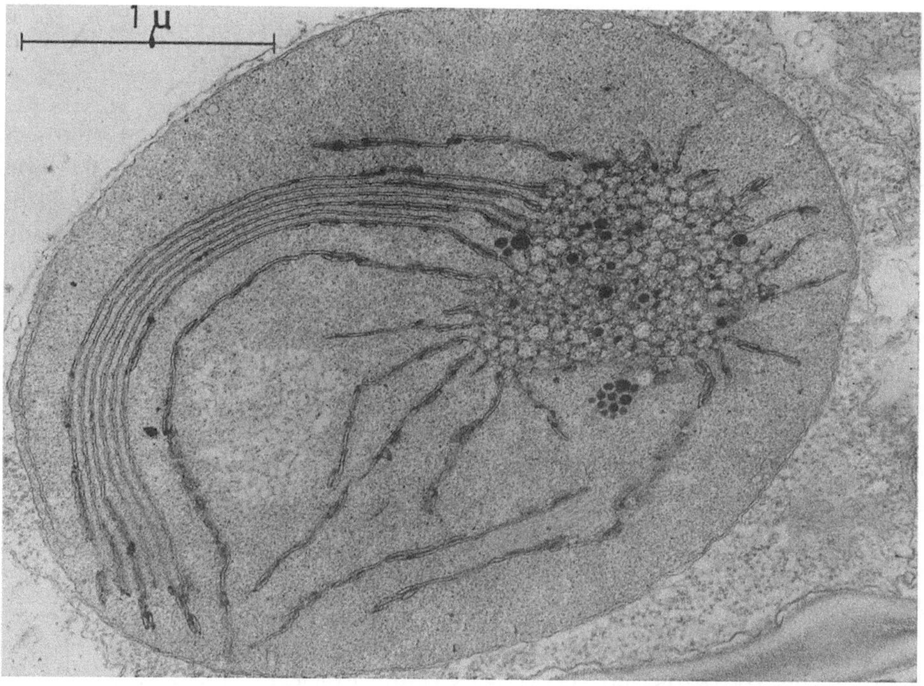

Fig. 5. Etioplast from a dark-grown leaf of wild type barley containing protochlorophyll. A crystalline prolamellar body is interconnected with extensive primary lamellar layers (Glutaraldehyde-OsO_4 fixation). Micrograph: J. E. Boynton and K. W. Henningsen

A number of *Chlorella* mutants with various blocks in chlorophyll synthesis has been analyzed by Granick (see reviews by Granick, 1951, 1954). These mutants accumulate a specific porphyrin intermediate and therefore it is possible to infer which biosynthetic step is affected in each case (see p. 232). The genetics of these mutations is unknown since *Chlorella* is an asexual alga.

Different mutant strains of *Euglena gracilis* — also an asexual alga — vary in their ability to synthesize carotenoids and porphyrins [Gibor and Granick, 1962 (2)]. This property was analyzed by feeding some bleached strains with the porphyrin precursor delta-aminolevulinic acid (ALA). Some strains lack porphyrins whether they are fed ALA or not. These cells have also the lowest concentration of carotenoid pigments. In other strains the cells normally do not synthesize porphyrins, but will do so if they are supplied with ALA. These cells may be unable to produce ALA. A third type of cell is capable of porphyrin synthesis without the addition of ALA. Here the porphyrin biosynthetic chain appears to be intact to the protoporphyrin stage (see p. 237). These cells have the highest carotenoid content found in bleached cells.

In *Chlamydomonas reinhardi* so called "*y*" (yellow) mutants have been found. In contrast to the wild type, these mutants are incapable of chlorophyll synthesis in the dark. Under these conditions the mutant cells form a yellow (carotenoid-containing) plastid which is of the same shape and size as the normal chloroplast but has only scattered vesicles instead of lamellae [Sager and Palade, 1954; Sager, 1958; Hudock and Levine, 1964; Hudock *et al.*, 1964; Ohad, Siekevitz and Palade, 1967 (1, 2)]. Ohad, Siekevitz and Palade [1967 (1)] studied the occurrence of chloroplast enzymes in dark-grown mutant cells and found that they possess the enzymic apparatus required for the synthesis of chloroplast lamellae. The blocked chloroplast differentiation is thus entirely caused by the lack of chlorophyll. Upon illumination the plastids of the dark-grown mutant begin to synthesize chlorophyll and to develop a structurally normal lamellar system. The chemical composition and enzyme activity of the lamellar system remains unchanged during the reconstitution of the chloroplast and chlorophyll is inserted into this system prior to the fusion of discs into grana [Ohad, Siekevitz and Palade, 1967 (2)].

When *y* strains were crossed with the wild type, regular 1:1 segregations occurred at the first meiotic division (Sager and Tsubo, 1962). Linkage with any of the 11 mapped chromosomes could not be detected. These results indicate that the *y* character may be controlled by an extrachromosomal body with regular segregation at meiosis, a behaviour exhibited by the chloroplast!

Y mutants occur spontaneously in *Chlamydomonas* with a frequency of 10^{-4} to 10^{-3}, but the mutation frequency can be considerably increased (up to about 20%) by prolonged cultivation of green cells on an agar medium containing streptomycin (Sager and Tsubo, 1962).

Some recessive mutations affecting chlorophyll synthesis have been identified in higher plants [Boynton and Henningsen, 1967; Röbbelen, 1956; v. Wettstein, 1959 (2)]. The protochlorophyll-less mutant seedlings of *xantha*-10 in barley are unable to synthesize, either in light or in darkness, tubes and prolamellar bodies in their plastids and although large concentric thylakoids are formed, they cannot be aggregated into grana (Fig. 6, v. Wettstein, 1958; v. Wettstein and Eriksson, 1965).

Semi-dominant *aurea* forms constitute a group of yellow green, viable mutants. After self-fertilization they give an F_1 segregation of 1 wild type: 2 *aurea*: 1 *xantha* or *albina* (lethal).

An *aurea* factor in *Picea abies* affects chlorophyll synthesis: the yellow homozygous seedlings lack chlorophyll both when dark-grown and light-grown, while the yellow green heterozygotes have a chlorophyll content which is about 50% of that in corresponding wild type seedlings [Walles, 1967 (2)]. In the homozygous seedlings the plastids are blocked in development prior to the formation of prolamellar bodies. The heterozygous seedlings develop in their plastids small grana consisting of a few thylakoids [v. Wettstein, 1958; Walles, 1967 (2)]. Later the lamellae become bent in an odd way and many of the grana profiles appear semi-circular. These mutant plants possess the ability of spontaneous greening and normalization of the chloroplasts.

An *aurea* mutant in tobacco has a chlorophyll content which is only about one-eighth of that of the wild type (HOMANN and SCHMID, 1967; SCHMID and GAFFRON, 1966, 1967; SCHMID, PRICE and GAFFRON, 1966). At higher light intensities the photosynthetic capacity of these *aurea* plants is superior to that of the wild type, and

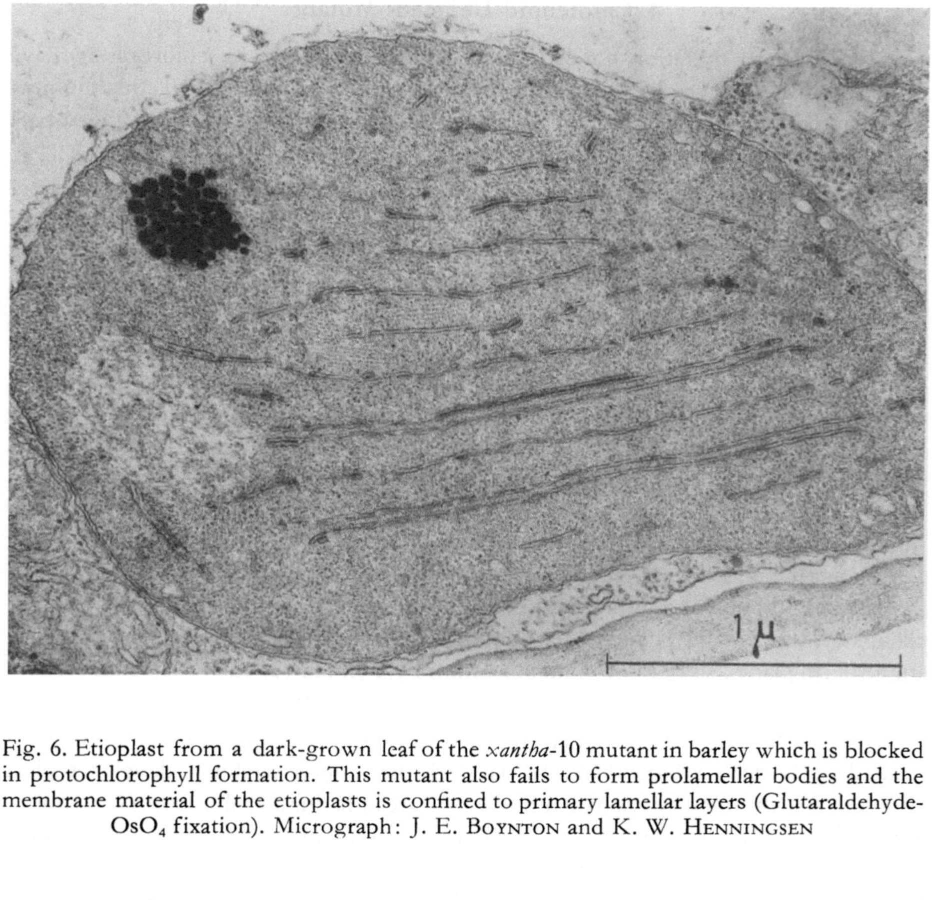

Fig. 6. Etioplast from a dark-grown leaf of the *xantha*-10 mutant in barley which is blocked in protochlorophyll formation. This mutant also fails to form prolamellar bodies and the membrane material of the etioplasts is confined to primary lamellar layers (Glutaraldehyde-OsO_4 fixation). Micrograph: J. E. BOYNTON and K. W. HENNINGSEN

their photosynthesis becomes saturated at a considerably higher light intensity than in green plants. The lamellar system in the *aurea* chloroplasts is much reduced and the grana usually consist of only two to three layers of thylakoids. Homozygous white seedlings have single thylakoids but no grana (SCHMID, 1967). Seedlings homozygous for an *aurea* factor in soybean have plastids of similar organization as those of corresponding mutant homozygotes in spruce and tobacco (SUN, 1963).

Other gene mutations affect the synthesis of chlorophyll *b* (HIGHKIN, 1950; HIGHKIN and FRENKEL, 1962; RÖBBELEN, 1957; HIRONO and RÉDEI, 1963; GOTTSCHALK and MÜLLER, 1964; MÜLLER, 1964). Chlorophyll *b*-less mutants are generally viable; in those which are reported to be lethal the lack of chlorophyll *b* might be a secondary effect of some unknown biochemical lesion. It is uncertain whether chlorophyll *b* is essential for the formation of the normal chloroplast structures. A chlorophyll-*b*-deficient *Arabidopsis* mutant from Rédei's collection was shown to have a normal chloroplast structure (Veleminsky and Röbbelen, 1966), whereas a corresponding mutant in barley had structurally immature chloroplasts with fewer and smaller grana than those in the wild type (Goodchild, Highkin and Boardman, 1966).

4. Carotenoid-Deficient Mutants

Several gene mutations affecting carotenoid synthesis in chloroplasts have been found. Although the carotenoid mutants synthesize chlorophyll, this pigment is photo-oxidized and the mutant seedlings generally develop an *albina* phenotype when growing under normal light conditions (see p. 246).

Seven mutants in maize have been shown to accumulate one or more of the later precursors of carotene (Bachmann *et al.*, 1967). One of these, the *albina* mutant w_3 is presumably blocked in carotenoid biosynthesis between zeta-carotene and neurosporene (Anderson and Robertson, 1961; Robertson, Bachmann and Anderson, 1966). Of two other non-allelic β-carotene-less maize mutants, one is characterized by the accumulation of zeta-carotene and the other by accumulation of lycopene (Faludi-Dániel and Láng, 1964; Faludi-Dániel, Láng and Fradkin, 1966). The *albina* mutant *ghost* in tomato accumulates phytoene (MacKinney, Rick and Jenkins, 1956) but the mutation has apparently an incomplete influence, since the plants often develop both white, yellow and normal green tissues and become variegated (Rick, Thompson and Brauer, 1959). Two non-allelic mutants of *Helianthus annuus* (one *albina* and one *xantha*) are characterized by the lack of β-carotene (Wallace and Habermann, 1959; Habermann, 1960). The yellow mutant contains about half the normal amount of xanthophyll, probably lutein. In regard to its carotenoid biochemistry, this mutant is somewhat reminiscent of dark-grown seedlings of normal angiosperm plants, since these contain xanthophylls (lutein) as their major carotenoid (Kay and Phinney, 1956; Goodwin and Phagpolngarm, 1960; Wolf, 1963).

The ultrastructure of plastids in the two carotene-less mutants of sunflower was studied by Walles (1965, 1966). In the dark the mutant seedlings develop typical etioplasts which contain prolamellar bodies and are indistinguishable from corresponding wild type plastids. During illumination of the mutant seedlings their plastids are capable of undergoing typical structural transformations but morphogenesis is halted after the formation of a few grana (Fig. 7). While the *albina* mutant produces only minute grana, those of the *xantha* mutant can be quite large [*cf.* Walles, 1967 (1)]. During bleaching of the leaves in strong light, their plastids usually attain a more or less amoeboid shape, the thylakoids swell and the grana distintegrate (Fig. 8).

The *albina* mutant w_3 in maize and its "leaky" pale green allele $pastel_{8686}$ are, in regard to plastid development, rather similar to the *Helianthus* mutants and synthesize in darkness normal etioplasts (Bachmann *et al.*, 1967). In the light w_3 seedlings disperse their prolamellar bodies and produce some thylakoids but no grana. On exposure to prolonged light a progressive degradation of the plastid structures occurs. Seedlings of $pastel_{8686}$ form a few large, atypical grana.

The studies of carotenoid-deficient mutants indicate that β-carotene is an indispensible component of normal chloroplasts, necessary for the development of a normal lamellar system, and is also of importance for the protection of chlorophyll and the grana structure against photo-destruction.

The biochemistry and the plastid ultrastructures of some carotenoid-mutants have also been investigated in the algae *Chlorella* and *Chlamydomonas* (CLAES, 1954, 1957; LEFORT, 1962; SAGER, 1958).

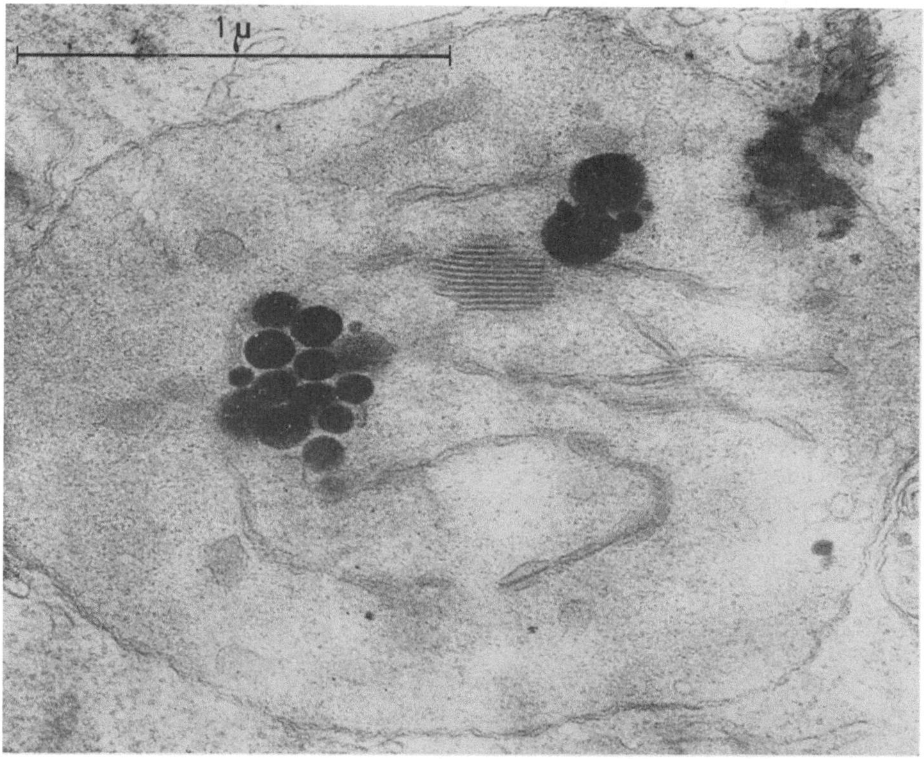

Fig. 7. Chloroplast with grana and groups of globuli in a green leaf of a carotenoid-less *albina* mutant in sunflower (Glutaraldehyde-OsO_4 fixation) (WALLES, 1965)

5. Chloroplast Mutants Defective in General Metabolism

A few lethal chloroplast mutants in barley have been found to respond by becoming green when grown on media containing specific amino acids [ERIKSSON *et al.*, 1961; WALLES, 1963, 1967 (1)].

The leaves of *albina*-7 seedlings vary considerably in phenotype but in typical cases are white with some pale green tissues in their tips. The seedlings respond by producing green leaves when supplied with aspartic acid. After a sufficient amount of green tissue has been induced, the amino acid feeding becomes unnecessary for further growth of the plants. The green tissues can apparently carry out photosynthesis and as a result of this, produce all the substances required for the further growth and development of more chloroplasts. The white leaf tissues of mutant seedlings have typical *albina* plastids, whereas the green tissues possess structurally normal chloroplasts.

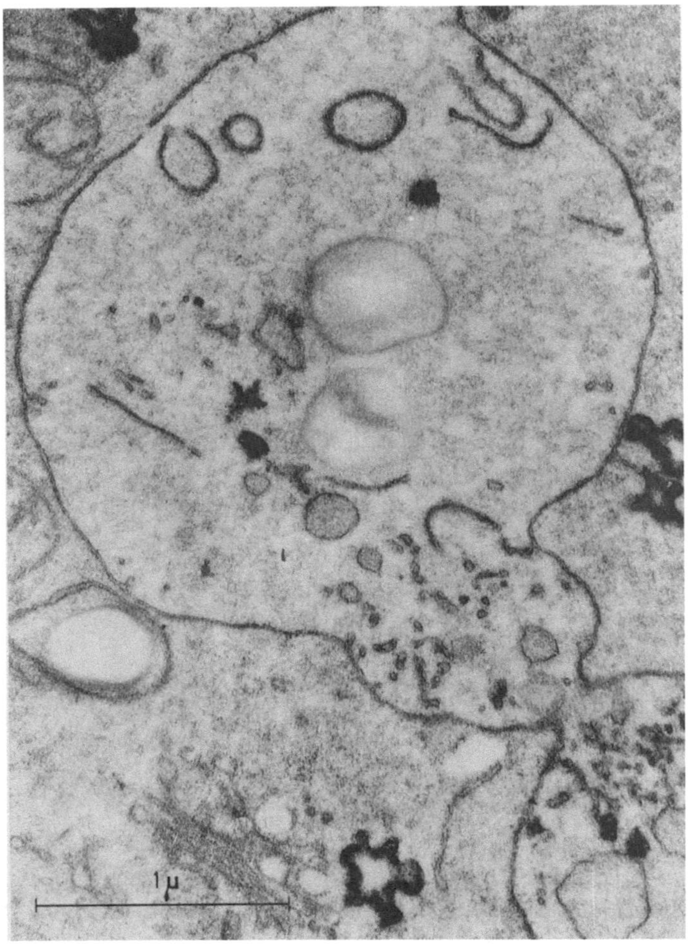

Fig. 8. Degenerated plastid in a bleached leaf of the carotenoid-less *albina* mutant illustrated in Fig. 7 ($KMnO_4$ fixation) (WALLES, 1965)

Seedlings with mutations in the *xantha*-23 (WALLES, 1963) and *xantha*-12 (v. WETTSTEIN, 1960) loci are yellow green. Their chloroplasts contain scattered grana, many of them small and some giant (Fig. 9). The giant grana have both a larger diameter (up to 2 μ) and a higher number of aggregated thylakoids than wild type grana. The mutant seedlings increase their content of chlorophyll considerably upon feeding with leucine, as studied in detail in *xantha*-23. In that experiment the feeding was started when the primary leaf was 2 cm long and had developed abnormal chloroplasts. These were not normalized by the treatment, but in the next leaf chloroplast

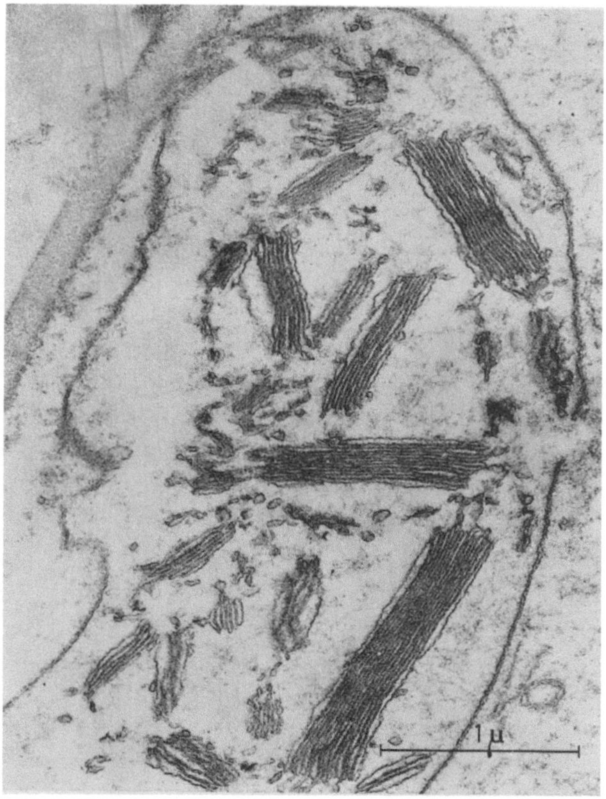

Fig. 9. Chloroplast of the leucine-requiring mutant *xantha*-23 ($KMnO_4$ fixation) (Walles, 1963)

formation from proplastids took place under the influence of the supplied leucine and in this case normal mature chloroplasts were produced (Fig. 10).

In green leaf cells the bulk of the protein — 60 to 75% — (Heber, 1960; Zucker and Stinson, 1962) is found in the chloroplasts. A disturbed protein formation such as would result from impairments in amino acid synthesis must therefore seriously interfere with the chloroplast development in growing leaves.

A thiamine-requiring mutant in *Arabidopsis thaliana* (Langridge, 1955, 1965) was the first case of an auxotrophic chloroplast mutant to be reported in the literature.

Several other thiamine-requiring mutants have since then been investigated in *Arabidopsis* (RÉDEI, 1965) and tomato (LANGRIDGE and BROCK, 1961; BOYNTON, 1966). In regard to the biochemical block, these mutants fall into three categories [BOYNTON, 1966 (1)]: (1) the block is in the biosynthesis of the thiazole moiety of thiamine, (2) the block affects the pyrimidine moiety or (3) the block causes a double require-

Fig. 10. Chloroplast in a secondary leaf of *xantha*-23 grown on leucine medium ($KMnO_4$ fixation) (WALLES, 1967)

ment for both the thiazole and the pyrimidine precursors. Chloroplast ultrastructure was studied in three thiamine-mutants of tomato [BOYNTON, 1966 (2)]. The mutant seedlings are characterized by a progressive chlorosis, as reflected by an initial development of normal chloroplasts which degenerate during the bleaching of the leaves. In the course of this process some of the grana change into a diffuse mass in which individual thylakoids are no longer discernible. In bleached leaf tissue developed subsequently the plastids have a reduced number of grana, which are often diffuse and sometimes abnormally large. Also undifferentiated plastids without grana are found. The plastid degeneration is reversible; if the plants are given thiamine their chloroplasts are normalized. Thiamine, which in the form of thiamine pyrophosphate functions as an essential coenzyme, seems thus to be required for the stabilization of the lamellar structure of chloroplasts.

A few recessive chloroplast mutations influence iron absorption or metabolism and in these cases normalization of the mutant plants can be induced by proper methods of feeding (BELL, BOGORAD and MCILRATH, 1958, 1962; BROWN, HOLMES and TIFFIN, 1958; MACHOLD, 1966; WEISS, 1943). In some *xantha* mutants of tomato, greening can be induced by decreasing the redox potential of the culture solution *e.g.* with hydroxylamine (MACHOLD, 1966, 1967, 1968). This treatment seems to favour formation of Fe^{2+} at the expense of Fe^{3+}. By examining the incorporation of ^{14}C-succinate and ^{14}C-glycine into the intermediates of the biosynthetic pathway of chlorophyll it was found that in one of these mutants a block occurs between coproporphyrinogen and protoporphyrinogen (STEPHAN and MACHOLD, 1969). This block is overcome after activation of the iron metabolism.

6. Chloroplast Mutants with Unidentified Biochemical Defects

a) Gene Mutations

In many cases gene mutations are known to interfere with chloroplast morphogenesis, although the biosynthetic reactions blocked have not been identified [APPELQVIST *et al.*, 1968; DÖBEL, 1964 (1); GOODENOUGH, ARMSTRONG and LEVINE, 1969; HODGE, MCLEAN and MERCER, 1956; LAUDI, 1965; LAUDI and BONATTI, 1965; LEFORT, 1959; MACLACHLAN and ZALIK, 1963; MARUYAMA, 1961; MILLERD, GOODCHILD and SPENCER, 1959; ORSENIGO and MARZIANI, 1965; PAOLILLO and REIGHARD, 1968; RÖBBELEN, 1959, 1966 (1); RÖBBELEN and WEHRMEYER, 1965; v. WETTSTEIN, 1957, 1961, 1967].

Albina mutants have plastids which structurally remain at proplastid stages prior to the formation of thylakoids. The barley mutant *albina*-12 is unable to synthesize prolamellar bodies (v. WETTSTEIN, 1961). *Albina*-17 in the same species can transform the prolamellar bodies in its plastids but cannot disperse them, and it has been suggested that this mutation is in a gene controlling the second step in the formation of the lamellar system in etioplasts of dark-grown seedlings (Fig. 11—12; v. WETTSTEIN, 1967).

Xantha mutants vary with regard to pigment composition and plastid development. They are often able to develop immature or aberrant grana (*cf.* APPELQVIST *et al.*, 1968).

Viridis mutants, some of which are viable, are blocked in later stages of chloroplast differentiation or are, in some instances, able to produce structurally normal chloroplasts (HODGE, MCLEAN and MERCER, 1955; MILLERD, GOODCHILD and SPENCER, 1969).

A *virescens* mutant in barley has in the young stages plastids that contain large vesicles but lack normal discs or grana (MACLACHLAN and ZALIK, 1963). With increasing age the seedlings produce normal chloroplasts and about normal levels of chlorophyll and carotenoids. It has been suggested that the mutation causes a disturbance in a structural plastid component (lipoprotein) during the early stages of seedling growth. Plastid development in *virescens* mutants in maize has been described by PAOLILLO and REIGHARD (1968).

b) Plastome Mutations

The biosynthetic pathway which is affected by the mutation has so far not been identified in any plastome mutant of a higher plant. The present knowledge about the

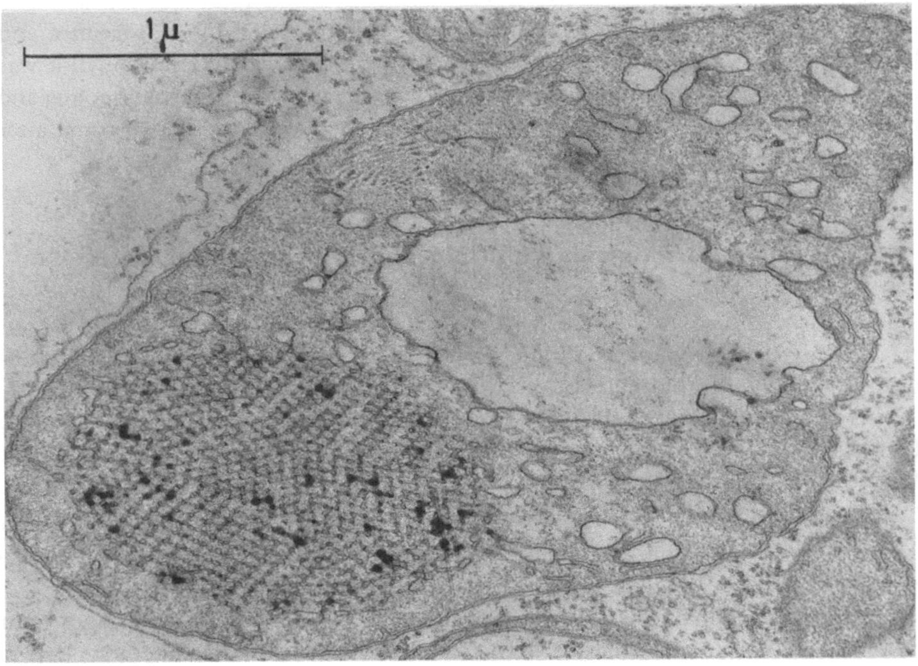

Fig. 11. Etioplast from a dark-grown leaf of the *albina*-17 mutant containing normal amounts of protochlorophyll and a crystalline prolamellar body (Glutaraldehyde-OsO_4 fixation). Micrograph: J. E. BOYNTON and K. W. HENNINGSFN

function of the plastome is therefore restricted to what can be inferred from studies of plastid ultrastructure and physiological behaviour of plastome mutants.

In plastome mutants of *Oenothera* (two different kinds), of *Nicotiana* and of *Humulus japonicus* (see p. 55) (STUBBE and v. WETTSTEIN, 1955; v. WETTSTEIN, 1961; ERIKSSON *et al.*, 1961; v. WETTSTEIN and ERIKSSON, 1965) as well as of *Epilobium* (ANTON-LAMBRECHT, 1966) the mutant leaf tissues are initially green but bleach during their subsequent growth so that mutant sectors of fully expanded leaves are white (or in one of the *Oenothera* mutants studied, yellow green). The bleaching process can to some degree be dependent on environmental conditions (*cf.* v. WETTSTEIN, 1961). During the green period chloroplasts with a normal ultrastructure are developed but during the bleaching of the leaves these chloroplasts degenerate with swelling of the thylakoids, disintegration of the grana and sometimes—as in *Epilobium*—formation of abnormal structures. In three other pale plastome mutants of *Oenothera* chloroplasts in various stages of degeneration are found: some plastids look rather normal although with some swollen grana thylakoids while other ones contain almost only vesicles derived from swollen thylakoids (DOLZMANN, 1968).

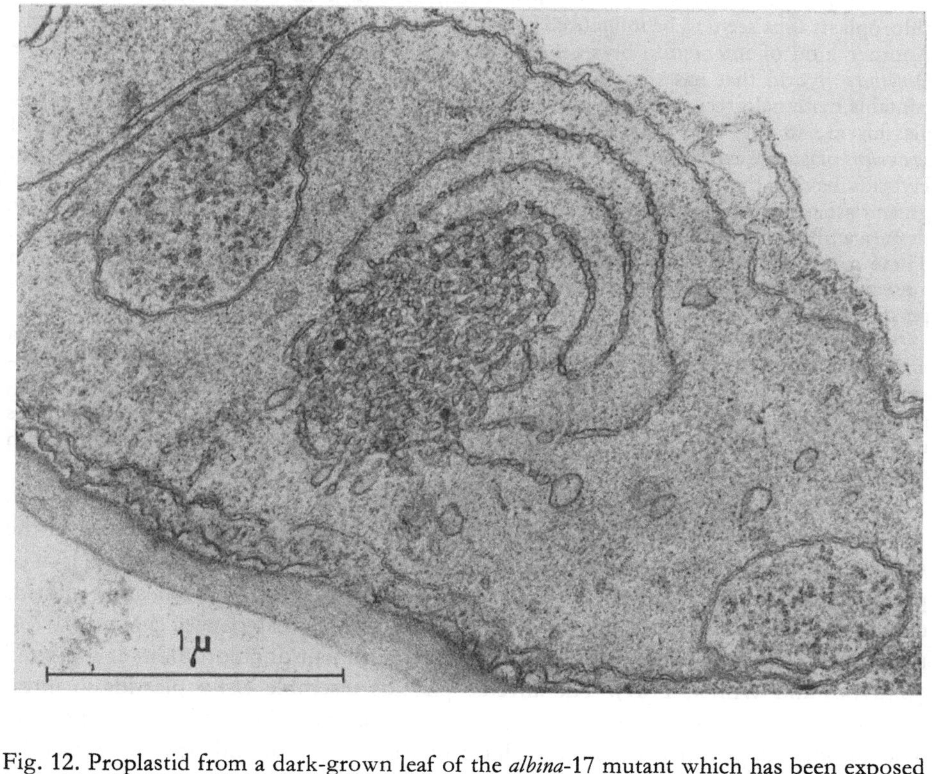

Fig. 12. Proplastid from a dark-grown leaf of the *albina*-17 mutant which has been exposed to light for 24 h. The transformed prolamellar body has failed to disperse and the chlorophyll is completely photo-oxidized (Glutaraldehyde-OsO_4 fixation). Micrograph: J. E. BOYNTON and K. W. HENNINGSEN

The plastids are often of an irregular shape with long projections and invaginations. A fourth mutant investigated contains almost normal chloroplasts and is capable of limited photosynthesis. In an *albomaculatus* form of *Antirrhinum* the plastids are blocked in development shortly after the beginning of grana formation or sometimes at a still earlier stage [DÖBEL, 1964 (2)]. The structurally most advanced plastids possess some minute grana together with a few abnormal giant grana. The formation of giant grana is stimulated if the plants are grown in the greenhouse under reduced illumination. In expanded leaves swelling of the thylakoids may occur. In gene induced *albomaculata* forms in barley (one yellow-striped and one white-striped type;

v. Wettstein, 1961; v. Wettstein and Eriksson, 1965), *Arabidopsis* [Röbbelen, 1966 (2)] and maize (Shumway and Weier, 1967), the mutant plastids are blocked in development at an early stage prior to grana formation and resemble the plastids in some gene mutants of *albina* or *xantha* phenotypes.

In mixed cells (see p. 59) interactions between genetically different plastids have been noticed in some cases. In such cells with normal and mutant plastids the wild type chloroplasts may be somewhat affected. The symptoms vary; there may be a reduction in size of the chloroplasts or some deformations or reductions of the lamellar system [Döbel, 1964 (2); v. Wettstein and Eriksson, 1965; Röbbelen, 1966 (2)]. In these mutants the normal chloroplasts thus seem to be influenced by some substance produced by the mutated plastids. Another kind of interaction between plastids has been described by Stubbe (1958). The *Oenothera* hybrid that has the genome *albicans.velans* is variegated when it simultaneously contains maternal green *suaveolens* plastids and paternal *lamarckiana* plastids. The *lamarckiana* plastids are in this case incompatible with the nucleus and therefore pale. The normal *suaveolens* plastome can be replaced by a mutated plastome giving rise to yellow green plastids. Hybrids having the mutated *suaveolens* plastome have, however, not only pale and yellow green tissues but also normal green areas. The green tissues are found at those places where tissues with *lamarckiana* plastids are lying above or under tissues with *suaveolens* plastids. These green tissues contain *lamarckiana* plastids which consequently seem able to develop a normal phenotype under the influence of some substance which diffuses from the *suaveolens* plastids.

In the bleached cells of *albomaculatus* forms of tobacco and barley, abnormal mitochondria with structurally defective and reduced crista have been noted (v. Wettstein, 1961; Eriksson *et al.*, 1961; v. Wettstein and Eriksson, 1965). The cause of this mitochondrial degeneration is not known. Similar crista abnormalities are found in mitochondria of barley seedlings treated with streptomycin (Kirk and Juniper, 1963) and of tomato plants treated with streptomycin, chloramphenicol or thiouracil (Döbel, 1963).

According to Shumway and Weier (1967) ribosomes cannot be detected in the mutant plastids of *iojap* maize (p. 54) studied with the electron microscope. A lack of plastid ribosomes might provide a mechanism by which thylakoid synthesis in plastids could be blocked. If light-grown wheat seedlings are treated with 3-amino-1,2,4-triazole the leaves become chlorotic and the development of chloroplasts is inhibited (Bartels and Weier, 1969). The affected plastids resemble *albina* plastids in ultrastructure and lack ribosomes, although the cytoplasmic ribosomes appear normal. The authors suggest that the lack of differentiation of the plastids may be caused by the absence of plastid ribosomes.

As regards reaction norm and control of plastid phenotype, the plastome mutations cannot on the basis of the available information be considered different from the gene mutations. It has been found, as already mentioned, that gene-induced *albomaculatus* types in barley, maize and *Arabidopsis*, like many *albina* and *xantha* gene mutants, have plastids which are arrested in development at early stages. The *albomaculatus* type of *Antirrhinum* is blocked at a later stage in chloroplast morphogenesis and seems in this respect somewhat reminiscent of the leucine-requiring *xantha* mutants in barley. Some genetic factors control the maintenance rather than the formation of chloroplast structures. Development of chloroplasts which subsequently degenerate occurs in the plastome mutants of *Oenothera*, *Nicotiana* and *Epilobium* already referred to, as well as in some gene mutants studied by Boynton [1966 (2)], Döbel [1964 (2)], Röbbelen (1959) and Walles (1965, 1966). In weak light carotenoid-deficient mutants of sunflower form chloroplasts with a few grana (Walles, 1965, 1966). In normal light their chlorophyll is bleached and the lamellar structure of the chloroplasts disintegrates. Thiamine-requiring mutant seedlings of tomato develop normal chloroplasts under the influence of the thiamine present in the seed [Boynton, 1966 (2)]. When this supply is exhausted, the plants become chlorotic and their chloroplasts degenerate. Degeneration of chloroplasts is evidently an unspecific symptom which can be due to different metabolic disorders.

E. Concluding Remarks

A survey of the inheritance of plastid phenotypes reveals that this kind of organelle in some respects is controlled by the genome and to some extent by its own genetic apparatus (the plastome).

It is not known with certainty what biosynthetic pathways are controlled by the plastid DNA. It is probable that this DNA codes for the plastid ribosomes and that consequently the synthesis of these can be blocked by mutations in this DNA. Inhibition of plastid ribosomes can be expected to block effectively plastid development at an early stage. However, in some of the plastome mutants chloroplasts are initially produced and only later degenerate.

The chlorophylls (*e.g. a* and *b*, protochlorophyll) and carotenoids are restricted to plastids but mutations that interfere with the biosynthetic pathways of these pigments show Mendelian segregation (at least in higher plants). Studies with *Chlamydomonas* have also revealed that deficiences for different photosynthetic enzymes can be caused by gene mutations. Messenger RNA required for the synthesis of plastid pigments and photosynthetic enzymes might thus migrate from the nucleus to the plastids, where the synthesis of these compounds apparently takes place. The plastids are also dependent on simple soluble metabolites like some amino acids and thiamine, which they presumably obtain from the cytoplasm.

The plastids possess the essential attributes of a prokaryotic cell. Their way of propagation is by division. They have their own kinds of DNA (histone free) and ribosomes and they perform DNA, RNA and protein synthesis. Mutations can occur in their hereditary material causing structural and functional alterations. It has been repeatedly suggested that plastids have evolved from symbiotic organisms organized like blue green algae. If we want to adopt this hypothesis we have to include the additional assumption that these presumed symbiotic cells in the process of becoming plastids underwent a considerable evolution during which their autonomy was restricted and part of the genetic control of their development was assumed by the nucleus of the host cell.

F. Literature

Anderson, I. C., Robertson, D. S.: Role of carotenoids in protecting chlorophyll from photodestruction. Plant Physiol. **35**, 531—534 (1960).

— Carotenoid protection of porphyrins from photodestruction. In: Progress in photobiology (B. C. Christensen, B. Buchmann, Eds.), pp. 477—479. Amsterdam: Elsevier 1961.

Anton-Lamprecht, I.: Beiträge zum Problem der Plastidenabänderung. III. Über das Vorkommen von "Rückmutationen" in einer spontan entstandenen Plastidenschecke von *Epilobium hirsutum*. Z. Pflanzenphysiol. **54**, 417—445 (1966).

Appelqvist, L. A., Boynton, J. E., Henningsen, K. W., Stumpf, P. K., v. Wettstein, D.: Lipid biosynthesis in chloroplast mutants of barley. J. Lipid Res. **9**, 513—524 (1968).

Bachmann, M. D., Robertson, D. S., Bowen, C. C., Anderson, I. C.: Choroplast development in pigment deficient mutants of maize. I. Structural anomalies in plastids of allelic mutants at the w_3 locus. J. Ultrastruct. Res. **21**, 41—60 (1967).

Bartels, P. G., Weier, T. E.: The effect of 3 amino-1, 2, 4-triazole on the ultrastructure of plastids of *Triticum vulgare* seedlings. Amer. J. Bot. **56**, 1—7.

Baur, E.: Das Wesen und die Erblichkeitsverhältnisse der "varietates albomarginatae hort." von *Pelargonium zonale*. Z. Vererbungsl. **1**, 330—351 (1909).

Bell, P. R., Frey-Wyssling, A., Mühlethaler, K.: Evidence for the discontinuity of plastids in the sexual reproduction of a plant. J. Ultrastruct. Res. **15**, 108—121 (1966).

— Mühlethaler, K.: The fine structure of cells taking part in oogenesis in *Pteridium aquilinum*. J. Ultrastruct. Res. **7**, 452—466 (1962).

— Evidence for the presence of deoxyribonucleic acid in the organelles of the egg cells of *Pteridium aquilinum*. J. molec. Biol. **8**, 853—862 (1964).

Bell, W. D., Bogorad, L., McIlrath, W. J.: Response of the yellow-stripe maize mutants (ys_1) to ferrous and ferric iron. Botan. Gaz. **120**, 36—39 (1958).

— Yellow-stripe phenotype in maize. I. Effects of ys_1 locus on uptake and utilization of iron. Botan. Gaz. **124**, 1—8 (1962).

Ben-Shaul, Y., Epstein, H. T., Schiff, J. A.: Studies of chloroplast development in *Euglena*. X. The return of the chloroplast to the proplastid condition during dark adaptation. Canad. J. Botany **43**, 129—136 (1965).

— Schiff, J. A., Epstein, H. T.: Studies of chloroplast development in *Euglena*. VII. Fine structure of the developing plastid. Plant Physiol. **39**, 231—240 (1964).

Berger, S.: RNA synthesis in *Acetabularia*. II. RNA-synthesis in isolated chloroplasts. Protoplasma (Wien) **64**, 13—25 (1967).

Bishop, N. I.: Separation of the oxygen evolving system of photosynthesis from the photochemistry in a mutant of *Scenedesmus*. Nature (Lond.) **195**, 55—57 (1962).

Boynton, J. E.: (1) Chlorophyll-deficient mutants in tomato requiring vitamin B_1. I. Genetics and physiology. Hereditas (Lund) **56**, 171—199 (1966).

— (2) Chlorophyll-deficient mutants in tomato requiring vitamin B_1. II. Abnormalities in chloroplast ultrastructure. Hereditas (Lund) **56**, 238—254 (1966).

— Henningsen, K. W.: The physiology and chloroplast structure of mutants at loci controlling chlorophyll synthesis in barley. Stud. Biophys. **5**, 85—88 (1967).

Brawerman, G.: A specific species of ribosomes associated with the chloroplasts of *Euglena gracilis*. Biochim. biophys. Acta (Amst.) **61**, 313—315 (1962).

— The isolation of a specific species of ribosomes associated with chloroplast development in *Euglena gracilis*. Biochim. biophys. Acta (Amst.) **72**, 317—331 (1963).

Brown, J. C., Holmes, R. S., Tiffin, L. O.: Iron chlorosis in soybeans as related to the genotype of rootstalk. Soil Sci. **86**, 75—82 (1958).

Camefort, H.: L'organisation du cytoplasme dans l'oosphère et la cellule centrale du "*Pinus laricio*" Poir. (var. *austriaca*). Ann. sci. nat. Botan. et biol. végétale, 12e Sér. **3**, 265—291 (1962).

Claes, H.: Analyse der biochemischen Synthesekette für Carotinoide mit Hilfe von *Chlorella*-Mutanten. Z. Naturforsch. **9 b**, 461—469 (1954).

— Biosynthese von Carotinoiden bei *Chlorella*. III. Untersuchungen über die lichtabhängige Synthese von α- und β-Carotin und Xanthophyllen bei der *Chlorella*-Mutante 5/520. Z. Naturforsch. **12 b**, 401—407 (1957).

Cleland, R. E.: The cytogenetics of *Oenothera*. Advanc. Genet. **11**, 147—237 (1962).

Correns, C.: (1) Vererbungsversuche mit blaß(gelb)grünen und buntblättrigen Sippen bei *Mirabilis jalapa*, *Urtica pilulifera* und *Lunaria annua*. Z. Vererbungsl. **1**, 291—329 (1909).

— (2) Zur Kenntnis der Rolle von Kern und Plasma bei der Vererbung. Z. Vererbungsl. **2**, 331—340 (1909).

— (1) Vererbungsversuche mit buntblättrigen Sippen. VIII—XI. S.-B. preuss. Akad. Wiss., Physik.-math. Kl. **11**, 1—31 (1931).

— (2) Über einige Fälle von Buntblättrigkeit. Z. Vererbungsl. **59**, 274—280 (1931).

— Nicht mendelnde Vererbung. Handb. d. Vererbungswissenschaft, Bd. II, H. (F. von Wettstein, Ed.). Berlin: Gebrüder Borntrager Verlag 1937.

Demerec, M.: Behaviour of chlorophyll in inheritance. Cold Spr. Harb. Symp. quant. Biol. **3**, 80—84 (1935).

Diers, L.: Bilden sich während der Oogenese bei Moosen und Farnen die Mitochondrien und Plastiden aus dem Kern? Ber. dtsch. bot. Ges. **77**, 369—371 (1964).

— Elektronenmikroskopische Untersuchungen über die Eizellbildung und Eizellreifung des Lebermooses *Sphaerocarpus donnellii*. Aust. Z. Naturforsch. **20 b**, 795—801 (1965).

— (1) Über die Vermehrung von Plastiden und Mitochondrien während der Oogenese von *Sphaerocarpus*. In: Probleme der biologischen Reduplikation (P. SITTE, Ed.), p. 227—243. Berlin-Heidelberg-New York: Springer 1966.
— (2) On the plastids, mitochondria, and other cell constituents during oögenesis of a plant. J. Cell Biol. **28**, 527—543 (1966).
— SCHÖTZ, F.: Räumliche Beziehungen zwischen osmiophilen Granula und Thylakoiden. Z. Pflanzenphysiol. **58**, 252—265 (1968).
— Über ring- und schalenförmige Thylakoidbildungen in den Plastiden. Z. Pflanzenphysiol. **60**, 187—210 (1969).
— — BATHELT, H.: Zur Frage des Aufbaus von Thylakoiden aus Vesikeln. Planta **80**, 211—226 (1968).
DOLZMANN, P.: Photosynthese-Reaktionen einiger Plastom-Mutanten von *Oenothera*. III. Strukturelle Aspekte. Z. Pflanzenphysiol. **58**, 300—309 (1968).
DULIEU, H.: Sur les différents types de mutations extranucléaires induites par le méthane sulfonate d'éthyle chez *Nicotiana tabacum* L. Mutation Res. **4**, 177—189 (1967).
DÖBEL, P.: Untersuchung der Wirkung von Streptomycin-, Chloramphenicol- und 2-Thio-uracil-Behandlung auf die Plastidenentwicklung von *Lycopersicon esculentum* MILLER. Biol. Zbl. **82**, 275—295 (1963).
— (1) Die Plastiden einer nach Pfropfung schwach ergrünungsfähigen *albina*-Mutante der Tomate. Kulturpflanze **12**, 153—161 (1964).
— (2) Über die Plastiden einer Herkunft des status *albomaculatus* von *Antirrhinum majus* L. Z. Vererbungsl. **95**, 226—235 (1964).
— HAGEMANN, R.: Elektronenmikroskopischer Nachweis echter Mischzellen bei *Antirrhinum majus* status *albomaculatus*. Biol. Zbl. **82**, 749—751 (1963).
EBRINGER, L., KUPKOVÁ, H.: Antibiotics and apochlorosis. II. Macromolecular synthesis and the bleaching effect of streptomycin. Folia microbiol. (Praha) **12**, 36—41 (1967).
EDELMAN, M., EPSTEIN, H. T., SCHIFF, J. A.: Isolation and characterization of DNA from the mitochondrial fraction of *Euglena*. J. molec. Biol. **17**, 463—469 (1966).
— SCHIFF, J. A., EPSTEIN, H. T.: Studies of chloroplast development in *Euglena*. XII. Two types of satellite DNA. J. molec. Biol. **11**, 769—774 (1965).
EISENSTADT, J. M., BRAWERMAN, G.: The incorporation of amino acids into the protein of chloroplasts and chloroplast ribosomes of *Euglena gracilis*. Biochim. biophys. Acta (Amst.) **76**, 319—321 (1963).
— The protein-synthesizing systems from the cytoplasm and the chloroplasts of *Euglena gracilis*. J. molec. Biol. **10**, 392—402 (1964).
EPSTEIN, H. T., SCHIFF, J. A.: Studies of chloroplast development in *Euglena*. IV. Electron and fluorescence microscopy of the plastid and its development into a mature chloroplast. J. Protozool. **8**, 427—432 (1961).
ERIKSSON, G., KAHN, A., WALLES, B., VON WETTSTEIN, D.: Zur makromolekularen Physiologie der Chloroplasten III. Ber. dtsch. bot. Ges. **74**, 221—232 (1961).
FALUDI-DÁNIEL, Á., LÁNG, F.: Characteristics of chloroplast mutants with abnormal carotenoid synthesis. Ann. Univ. Sci. Budapest **7**, 76—80 (1964).
— LÁNG, F., FRADKIN, L. I.: The state of chlorophyll *a* in leaves of carotenoid mutant maize. In: Biochemistry of chloroplasts, Vol. I, 269—274 (T. W. GOODWIN, Ed.). London-New York: Academic Press 1966.
GIBOR, A.: DNA synthesis in chloroplasts. In: Biochemistry of chloroplasts, Vol. II, 321—328 (T. W. GOODWIN, Ed.). London-New York: Academic Press 1967.
— GRANICK, S.: (1) Ultraviolet sensitive factors in the cytoplasm that affect the differentiation of *Euglena* plastids. J. Cell Biol. **15**, 599—603 (1962).
— (2) The plastid system of normal and bleached *Euglena gracilis*. J. Protozool. **9**, 327—334 (1962).
— Plastids and mitochondria: Inheritable systems. Science **145**, 890—897 (1964).
GOFFEAU, A., BRACHET, J.: Deoxyribonucleic acid-dependent incorporation of amino acids into the proteins of chloroplasts isolated from anucleate *Acetabularia* fragments. Biochim. biophys. Acta (Amst.) **95**, 302—313 (1965).
GOODCHILD, D. J., HIGHKIN, H. R., BOARDMAN, N. K.: The fine structure of chloroplasts in a barley mutant lacking chlorophyll *b*. Exp. Cell Res. **43**, 684—688 (1966).

Goodenough, U. W., Armstrong, J. J., Levine, R. P.: Photosynthetic properties of ac-31, a mutant strain of *Chlamydomonas reinhardi* devoid of chloroplast membrane stacking. Plant Physiol. **44**, 1001—1012 (1969).

— Levine, R. P.: Chloroplast ultrastructure in mutant strains of *Chlamydomonas reinhardi* lacking components of the photosynthetic apparatus. Plant Physiol. **44**, 990—1000 (1969).

Goodwin, T. W., Phagpolngarm, S.: Studies in carotenogenesis. 28. The effect of illumination on carotenoid synthesis in French-bean *(Phaseolus vulgaris)* seedlings. Biochem. J. **76**, 197—199 (1960).

Gorman, D. S., Levine, R. P.: Cytochrome *f* and plastocyanin: their sequence in the photosynthetic electron transport chain of *Chlamydomonas reinhardi*. Proc. nat. Acad. Sci. (Wash.) **54**, 1665—1669 (1965).

Gottschalk, W., Müller, F.: Quantitative Pigmentuntersuchungen an strahleninduzierten Chlorophyllmutanten von *Pisum sativum*. II. Die fertilen und sterilen Chlorophyllmutanten. Planta **62**, 1—21 (1964).

Granick, S.: Biosynthesis of chlorophyll and related pigments. Ann. Rev. Plant Physiol. **2**, 115—144 (1951).

— Biosynthesis and function of heme and chlorophyll. Record Chem. Progr. (Kresge-Hooker Sci. Lib.) **15**, 27—35 (1954).

Habermann, H. M.: Spectra of normal and pigment-deficient mutant leaves of *Helianthus annuus* L. Physiol. Plantarum **13**, 718—725 (1960).

Hagemann, R., Scholz, F.: Ein Fall geninduzierter Mutationen des Plasmotypus bei Gerste. Züchter **32**, 50—59 (1962).

Heber, U.: Vergleichende Untersuchungen an Chloroplasten, die durch Isolierungs-Operationen in nicht-wäßrigem und in wäßrigem Milieu erhalten wurden. I. Ermittlung des Verhältnisses von Chloroplastenprotein zum Gesamtprotein der Blattzelle. Z. Naturforsch. **15 b**, 95—99 (1960).

— Gottschalk, W.: Die Bestimmung des genetisch fixierten Stoffwechselblockes einer Photosynthese-Mutante von *Vicia faba*. Z. Naturforsch. **18 b**, 36—44 (1963).

Highkin, H. R.: Chlorophyll studies on barley mutants. Plant Physiol. **25**, 294—366 (1950).

— Frenkel, A. W.: Studies of growth and metabolism of a barley mutant lacking chlorophyll *b*. Plant Physiol. **37**, 814—820 (1962).

Hirono, Y., Rédei, G. P.: Multiple allelic control of chlorophyll *b* level in *Arabidopsis thaliana*. Nature (Lond.) **197**, 1324—1325 (1963).

Hodge, A. J., McLean, J. D., Mercer, F. V.: Ultrastructure of the lamellae and grana in the chloroplasts of *Zea mays* L. J. biophys. biochem. Cytol. **1**, 605—614 (1955).

— A possible mechanism for the morphogenesis of lamellar systems in plant cells. J. biophys. biochem. Cytol. **2**, 597—608 (1956).

Homann, P. H., Schmid, G. H.: Photosynthetic reactions of chloroplasts with unusual structures. Plant Physiol. **42**, 1619—1632 (1967).

Hudock, G. A., Levine, R. P.: Regulation of photosynthesis in *Chlamydomonas reinhardi*. Plant Physiol. **39**, 889—897 (1964).

— McLeod, G. C., Moravkova-Kiely, J., Levine, R. P.: The relation of oxygen evolution to chlorophyll and protein synthesis in a mutant strain of *Chlamydomonas reinhardi*. Plant Physiol. **39**, 898—903 (1964).

Hurwitz, J., Furth, J. J., Malami, M., Alexander, M.: The role of deoxyribonucleic acid in ribonucleic acid synthesis, III. The inhibition of the enzymatic synthesis of ribonucleic acid and deoxyribonucleic acid by actinomycin D and proflavin. Proc. nat. Acad. Sci. (Wash.) **48**, 1222—1230 (1962).

Imai, Y.: (1) Recurrent auto- and exomutation of plastids resulting in tricolored variegation of *Hordeum vulgare*. Genetics **21**, 752—757 (1936).

— (2) Chlorophyll variegations due to mutable genes and plastids. Z. Vererbungsl. **71**, 61—83 (1936).

— Is the variegation of *Humulus japonica* due to defect cytoplasm? Z. Vererbungsl. **73**, 598—600 (1937).

Jensen, W.: The ultrastructure and composition of the egg and central cell of cotton. Amer. J. Bot. **52**, 781—797 (1965).

KANDLER, O., SCHÖTZ, F.: Untersuchungen über die photooxydative Farbstoffzerstörung und Stoffwechselhemmung bei Chlorellamutanten und panaschierten Oenotheren. Z. Naturforsch. **11 b**, 708—718 (1956).

KAPPERT, H.: Über einen Fall plasmonbedingter Weißscheckigkeit bei *Plantago major*. Ber. dtsch. bot. Ges. **66**, 123—133 (1953).

KAY, R. E., PHINNEY, B.: Plastid pigment changes in the early seedling leaves of *Zea mays* L. Plant Physiol. **31**, 226—231 (1956).

KIRK, J. T. O.: (1) DNA dependent RNA synthesis in chloroplast preparations. Biochem. biophys. Res. Commun. **14**, 393—397 (1964).

— (2) Studies on RNA synthesis in chloroplast preparations. Biochem. biophys. Res. Commun. **16**, 233—238 (1964).

— Nature and function of chloroplast DNA. In: Biochemistry of chloroplasts, Vol. I, pp. 319—340 (T. W. GOODWIN, Ed.). London-New York: Academic Press 1966.

— JUNIPER, B. E.: The effect of streptomycin on the mitochondria and plastids of barley. Exp. Cell Res. **30**, 621—623 (1963).

— TILNEY-BASSETT, R. A. E.: The plastids. London-San Francisco: Freeman & Co. 1967.

KISLEV, N., SWIFT, H., BOGORAD, L.: Nucleic acids of chloroplasts and mitochondria in Swiss chard. J. Cell Biol. **25**, 327—344 (1965).

KOSKI, V. M., SMITH, J. H. C.: Chlorophyll formation in a mutant, white seedling-3. Arch. Biochem. **34**, 189—195 (1951).

LANGRIDGE, J.: Biochemical mutations in the crucifer *Arabidopsis thaliana* (L) HEYNH. Nature (Lond.) **176**, 260 (1955).

— Temperature-sensitive, vitamin-requiring mutants of *Arabidopsis thaliana*. Aust. J. biol. Sci. **18**, 311—321 (1965).

— BROCK, R. D.: A thiamine-requiring mutant of the tomato. Aust. J. biol. Sci. **14**, 66—69 (1961).

LAUDI, G.: Studio dell'infrastruttura dei chloroplasti di un mutante clorofilliano di *Lycopersicum esculentum* cv. Sioux. Genet. agrar. **19**, 327-337 (1965).

— BONATTI, P.: Ultrastruttura dei plastidi dei cotiledoni di una plantula albina di *Larix decidua*. Ann. fac. agrar. **5**, 1—11 (1965).

LEFORT, M.: Êtude de quelques mutants chlorophylliens induits chez *Lycopersicum esculentum* MILL. Rev. cytol. et biol. végétales **20**, 1—160 (1959).

— Êtude de l'infrastructure plastidiale d'un mutant chlorophyllien du *Chlorella vulgaris*. Compt. rend. **254**, 2414—2416 (1962).

LEVINE, R. P.: A screening technique for photosynthetic mutants in unicellular algae. Nature (Lond.) **188**, 339—340 (1960).

— Genetic dissection of photosynthesis. Science **162**, 768—771 (1968).

— GORMAN, D. S.: Photosynthetic electron transport chain of *Chlamydomonas reinhardi*. III. Light-induced absorbance changes in chloroplast fragments of the wild type and mutant strains. Plant Physiol. **41**, 1293—1300 (1966).

— TOGASAKI, R. K.: A mutant strain of *Chlamydomonas reinhardi* lacking ribulose diphosphate carboxylase activity. Proc. nat. Acad. Sci. (Wash.) **53**, 987—990 (1965).

— VOLKMANN, D.: Mutants with impaired photosynthesis in *Chlamydomonas reinhardi*. Biochem. biophys. Res. Commun. **6**, 264—269 (1961).

LUCK, D. J. L.: The biogenesis of mitochondria in *Neurospora*—A summary of present findings. In: Probleme der biologischen Reduplikation, p. 314—324 (P. SITTE, Ed.). Berlin-Heidelberg-New York: Springer 1966.

LYMAN, H., EPSTEIN, H. T., SCHIFF, J. A.: Studies of chloroplast development in *Euglena*. I. Inactivation of green colony formation by ultraviolet light. Biochim. biophys. Acta (Amst.) **50**, 301—309 (1961).

MACHOLD, O.: Untersuchungen an stoffwechseldefekten Mutanten der Kulturtomate. II. Einfluß des Eisenstoffwechsels auf die Ausbildung des Chlorophylldefekts. Flora (Jena) **157**, 183—199 (1966).

— Untersuchungen an stoffwechseldefekten Mutanten der Kulturtomate. IV. Einfluß des Redoxpotentials der Nährlösung auf den Chlorophyllgehalt verschiedener Mutanten. Kulturpflanze **15**, 75—83 (1967).

Machold, O.: Einfluß der Ernährungsbedingungen auf den Zustand des Eisens in den Blättern, den Chlorophyllgehalt und die Katalase- sowie Peroxydaseaktivität. Flora (Jena) **159**, 1—25 (1968).
MacKinney, G., Rick, C. M., Jenkins, J. A.: The phytoene content of tomatoes. Proc. nat. Acad. Sci. (Wash.) **42**, 404—408 (1956).
MacLachlan, S., Zalik, S.: Plastid structure, chlorophyll concentration, and free amino acid composition of a chlorophyll mutant of barley. Canad. J. Botany **41**, 1053—1062 (1963).
Maly, R.: Die Mutabilität der Plastiden von *Antirrhinum majus* L. *Sippe 50*. Z. Vererbungsl. **89**, 692—696 (1958).
Maruyama, K.: Electron microscope observation on the development of chloroplasts of *Avena* and chlorophyll deficient mutants. Cytologia (Tokyo) **26**, 105—115 (1961).
Mego, J. L.: Inhibitors of the chloroplast system in *Euglena*. In: The biology of *Euglena*, Vol. 2, pp. 351—381 (D. E. Buetow, Ed.). New York-London: Academic Press 1968.
Menke, W., Fricke, B.: Beobachtungen über die Entwicklung der Archegonien von *Dryopteris filix mas*. Z. Naturforsch. **19 b**, 520—524 (1964).
Meyer, A.: Über Krystalloide der Trophoplasten und über die Chromoplasten der Angiospermen. Botan. Z. **41**, 489—498, 505—514, 525—531 (1883).
Michaelis, P.: Über die Vererbung von Plastidmerkmalen. Protoplasma (Wien) **48**, 403—418 (1957).
— Untersuchungen zur Mutation plasmatischer Erbträger, besonders der Plastiden. I. Flora (Jena) **51**, 600—634 (1958).
Millerd, A., Goodchild, D. J., Spencer, D.: Studies on a maize mutant sensitive to low temperature. II. Chloroplast structure, development, and physiology. Plant Physiol. **44**, 567—583 (1969).
Moriber, L. G., Hershenov, B., Aaronson, S., Bensky, B.: Teratological chloroplast structures in *Euglena gracilis* permanently bleached by exogenous physical and chemical agents. J. Protozool. **10**, 80—86 (1963).
Mühlethaler, K., Bell, P. R.: Untersuchungen über die Kontinuität von Plastiden und Mitochondrien in der Eizelle von *Pteridium aquilinum* (L.) Kuhn. Naturwissenschaften **49**, 63—64 (1962).
— Frey-Wyssling, A.: Entwicklung und Struktur der Proplastiden. J. biophys. biochem. Cytol. **6**, 507—512 (1959).
Müller, F.: Untersuchungen über Chloroplastenfarbstoffe und Assimilationsstärke Chlorophyll *b*-freier durch Röntgenbestrahlung induzierter Mutanten von *Pisum sativum*. Planta **63**, 65—82 (1964).
Ohad, I., Siekevitz, P., Palade, G. E.: (1) Biogenesis of chloroplast membranes. I. Plastid dedifferentiation in a dark-grown algal mutant *(Chlamydomonas reinhardi)*. J. Cell Biol. **35**, 521—552 (1967).
— (2) Biogenesis of chloroplast membranes. II. Plastid differentiation during greening of a dark-grown algal mutant *(Chlamydomonas reinhardi)*. J. Cell Biol. **35**, 553—584 (1967).
Ophir, I., Ben-Shaul, Y.: Kinetics of fine structure changes in *Euglena* cells bleached by streptomycin. Fourth European Regional Conference on Electron Microscopy, Vol. 2, pp. 419—420, Rome 1968.
Orsenigo, M., Marziani, G.: Observations on plastid evolution in a "golden leaf" mutant of maize. A preliminary note. Atti del V Congr. Ital. Microscop. Elettronica, pp. 153—157, Bologna 1965.
Paolillo, D. J., Jr., Reighard, J. A.: The plastids of virescent corn mutants. Trans. Amer. Microscop. Soc. **87**, 54—59 (1968).
Parsons, J. A., Rustad, R. C.: The distribution of DNA among dividing mitochondria of *Tetrahymena pyriformis*. J. Cell Biol. **37**, 683—693 (1968).
Parthier, B., Wollgiehn, R.: Zur Frage der Aminosäure-Inkorporation in RNS-reiche Partikeln aus Chloroplasten. Naturwissenschaften **50**, 598—599 (1963).
— Nucleinsäure und Proteinsynthese in Plastiden. In: Probleme der biologischen Reduplikation, pp. 244—272 (P. Sitte, Ed.). Berlin-Heidelberg-New York: Springer 1966.
Pringsheim, E. G.: Die Apoplastidie bei *Euglena gracilis*. Rev. Algol. **4**, 41—56 (1958).
— Pringsheim, O.: Elimination of chromatophores and eye-spots in *Euglena gracilis*. New Phytologist **51**, 65—76 (1952).

Provasoli, L., Hutner, S. H., Schatz, A.: Streptomycin-induced chlorophyll-less races of *Euglena*. Proc. Soc. exp. Biol. (N. Y.) **69**, 279—282 (1948).

Ray, D. S., Hanawalt, C.: Satellite DNA components in *Euglena gracilis* cells lacking chloroplasts. J. molec. Biol. **11**, 760—768 (1965).

Rédei, G. P.: Genetic blocks in the thiamine synthesis of the angiosperm *Arabidopsis*. Amer. J. Bot. **52**, 834—841 (1965).

Renner, O.: Die Scheckung der Oenotheıenbastarde. Biol. Zbl. **44**, 309—336 (1924).

— Artbastarde bei Pflanzen. Handb. d. Vererbungswissenschaft, Bd. II A. Berlin: Gebrüder Borntraeger Verlag 1929.

— Die pflanzlichen Plastiden als selbständige Elemente der genetischen Konstitution. Ber. Verhandl. sächs. Akad. Wiss. Leipzig, Math.-phys. Kl. **86**, 241—266 (1934).

Rhoades, M. M.: Genic induction of an inherited cytoplasmic difference. Proc. nat. Acad. Sci. (Wash.) **29**, 327—329 (1943).

— Plastid mutations. Cold Spr. Harb. Symp. quant. Biol. **11**, 202—207 (1946).

Rick, C. M., Thompson, A. F., Brauer, O.: Genetics and development of an unstable chlorophyll deficiency in *Lycopersicon esculentum*. Amer J. Bot. **46**, 1—11 (1959).

Röbbelen, G.: Über die Protochlorophyllreduktion in einer Mutante von *Arabidopsis thaliana* (L.) Heynh. Planta **47**, 532—546 (1956).

— Eine Blattfarbmutante ohne Chlorophyll *b* von *Arabidopsis thaliana* (L.) Heynh. Naturwissenschaften **44**, 288 (1957).

— Untersuchungen über die Entwicklung der submikroskopischen Chloroplastenstruktur in Blattfarbmutanten von *Arabidopsis thaliana*. Z. Vererbungsl. **90**, 503—506 (1959).

— Plastommutationen nach Röntgenbestrahlung von *Arabidopsis thaliana* (L.) Heynh. Z. Vererbungsl. **93**, 25—34 (1962).

— Plastom mutations. — Arabidopsis Research, p. 100—105. Rep. Intern. Symp. Univ. Göttingen 1965.

— (1) Gestörte Thylakoidbildung in Chloroplasten einer *Xantha*-Mutante von *Arabidopsis thaliana* (L.) Heynh. Planta **69**, 1—26 (1966).

— (2) Chloroplastendifferenzierung nach geninduzierter Plastommutation bei *Arabidopsis thaliana* (L.) Heynh. Z. Pflanzenphysiol. **55**, 387—403 (1966).

— Wehrmeyer, W.: Gestörte Granabildung in chloroplasten einer *Chlorina*-Mutante von *Arabidopsis thaliana* (L.) Heynh. Planta **65**, 105—128 (1965).

Robertson, D. S., Bachman, M. D., Anderson, I. C.: Biochemical and plastid alterations in allelic mutants of the w_3 locus in maize. Genetics **54**, 357—358 (1966).

Sager, R.: The architecture of the chloroplast in relation to its photosynthetic activities. Brookhaven Symposia in Biol. **11**, 101—117 (1958).

— Palade, G.: Chloroplast structure in green and yellow strains of *Chlamydomonas*. Exp. Cell Res. **7**, 584—588 (1954).

— Structure and development of the chloroplast in *Chlamydomonas*. I. The normal green cell. J. biophys. biochem. Cytol. **3**, 463—488 (1957).

— Tsubo, Y.: Mutagenic effects of streptomycin in *Chlamydomonas*. Arch. Mikrobiol. **42**, 159—175 (1962).

Schiff, J. A., Epstein, H. T.: The continuity of the chloroplast in *Euglena*. In: Reproduction: subcellular, and cellular, pp. 131—189 (M. Locke, Ed.). New York-London: Academic Press 1965.

Schimper, A. F. W.: Über die Entwicklung der Chlorophyllkörner und Farbkörper. Botan. Zeit. **41**, 105—112, 121—131, 137—146, 153—162 (1883).

— Untersuchungen über die Chlorophyllkörner und die ihnen homologen Gebilde. Jb. wiss. Botan. **16**, 1—247 (1885).

Schmid, G. H.: Photosynthetic capacity and lamellar structure in various chlorophyll-deficient plants. J. Microscopie **6**, 485—498 (1967).

— Gaffron, H.: Chloroplast structure and the photosynthetic unit. Brookhaven Symposia in Biol. **19**, 380—392 (1966).

— Light metabolism and chloroplast structure in chlorophyll-deficient tobacco mutants. J. gen. Physiol. **50**, 563—582 (1967).

— Price, M., Gaffron, H.: Lamellar structure in chlorophyll deficient but normally active chloroplasts. J. Microscopie **5**, 205—212 (1966).

Schötz, F.: Über Plastidenkonkurrenz bei *Oenothera*. Planta **43**, 182—240 (1954).
— Über die photosynthetische Leistungsfähigkeit einiger panaschierter Oenotheren. Phot. u. Forsch. **7**, 12—16 (1956).
— Periodische Ausbleichungserscheinungen des Laubes bei *Oenothera*. Planta **52**, 351—392 (1958).
— Zur Kontinuität der Plastiden. Planta **58**, 333—336 (1962).
— Elektronenmikroskopische Untersuchungen an den Plastiden eines Oenotheren-Bastards mit disharmonischer Genom-Plastom-Kombination. Ber. dtsch. bot. Ges. **77**, 372—378 (1964).
— Über Plastidenkonkurrenz bei *Oenithera* II. Biol. Zbl. **87**, 33—61 (1968).
— Diers, L.: Über die Bildung von Stromasubstanz und Thylakoiden im Raum zwischen dem äußeren und inneren Teil der Plastiden-Doppelmembran. Planta **69**, 258—287 (1966).
— (1) Beeinflussung der Thylakoidbildung durch Disharmonie zwischen Genom und Plastom. Protoplasma (Wien) **65**, 335—348 (1968).
— (2) Über das Vorkommen kleiner netzartiger Strukturen in den Plastiden. Planta **79**, 312—318 (1968).
— — Bathelt, H.: Über den Einfluß einer Genom-Plastom-Disharmonie auf die Thylakoidanordnung. Z. Naturforsch. **23**, 1248—1252 (1968).
— Reiche, G. A.: Untersuchungen an panaschierten Oenotheren II. Über die Lebensfähigkeit von Bastarden, die ausschließlich die durch den Bastardkern geschädigte Plastidensorte enthalten. Z. Naturforsch. **12 b**, 757—764 (1957).
Semal, J., Spencer, D., Kim, Y. T., Wildman, S. G.: Properties of a ribonucleic acid synthesizing system in cell-free extracts of tobacco leaves. Biochim. biophys. Acta (Amst.) **91**, 205—216 (1964).
Shephard, D. C.: Chloroplast multiplication and growth in the unicellular alga *Acetabularia mediterranea*. Exp. Cell Res. **37**, 93—110 (1965).
Shumway, L. K., Weier, T. E.: The chloroplast structure of iojap maize. Amer. J. Bot. **54**, 773—780 (1967).
Siegesmund, K. A., Rosen, W. G., Gawlik, S. R.: Effects of darkness and of streptomycin on the fine structure of *Euglena gracilis*. Amer. J. Bot. **49**, 137—145 (1962).
Smith, J. H. C., Durham, L. J., Wurster, C. F.: Formation and bleaching of chlorophyll in albino corn seedlings. Plant Physiol. **34**, 340—345 (1959).
Spencer, D., Wildman, S. G.: The incorporation of amino acids into protein by cell-free extracts from tobacco leaves. Biochemistry **3**, 954—959 (1964).
Spiegelman, S., Hayashi, M.: The present status of the genetic information and its control. Cold Spr. Harb. Symp. quant. Biol. **28**, 161—181 (1963).
Stephan, U. W., Machold, O.: Über die Funktion des Eisens bei der Porphyrin- und Chlorophyll-Biosynthese. Z. Pflanzenphysiol. **61**, 98—113 (1969).
Stephenson, M. L., Thimann, K. V., Zamencnik, P. C.: Incorporation of C^{14}-amino acids into protein of leaf disks and cell-free fractions of tobacco leaves. Arch. Biochem. **65**, 194—209 (1956).
Stubbe, W.: Dreifarbenpanaschierung bei *Oenothera*. I. Entmischung von drei in der Zygote vereinigten Plastidomen. Ber. dtsch. bot. Ges. **70**, 221—226 (1957).
— Dreifarbenpanaschierung bei *Oenothera*. II. Wechselwirkungen zwischen Geweben mit zwei erblich verschiedenen Plastidensorten. Z. Vererbungsl. **89**, 189—203 (1958).
— Genetische Analyse des Zusammenwirkens von Genom und Plastom bei *Oenothera*. Z. Vererbungsl. **90**, 288—298 (1959).
— Untersuchungen zur genetischen Analyse des Plastoms von *Oenothera*. Z. Botan. **48**, 191—218 (1960).
— Sind Zweifel an der genetischen Kontinuität der Plastiden berechtigt? Eine Stellungsnahme zu den Ansichten von Mühlethaler und Bell. Z. Vererbungsl. **93**, 175—176 (1962).
— (1) Die Rolle des Plastoms in der Evolution der Oenotheren. Ber. dtsch. bot. Ges. **76**, 154—167 (1963).
— (2) Extrem disharmonische Genom-Plastom-Kombinationen und väterliche Plastidenvererbung bei *Oenothera*. Z. Vererbungsl. **94**, 392—411 (1963).

— von Wettstein, D.: Zur Struktur erblich verschiedener Chloroplasten von *Oenothera*. Protoplasma (Wien) **45**, 241—250 (1955).

Sun, C. N.: The effect of genetic factors on the submicroscopic structure of soybean chloroplasts. Cytologia (Tokyo) **28**, 257—263 (1963).

Veleminský, J., Röbbelen, G.: Beziehungen zwischen Chlorophyllgehalt und Chloroplastenstruktur in einer *Chlorina*-Mutante von *Arabidopsis thaliana* (L.) Heynh. Planta **68**, 15—35 (1966).

Wallace, R. H., Habermann, H. M.: Genetic history and general comparisons of two albino mutations of *Helianthus annuus*. Amer. J. Bot. **46**, 157—162 (1959).

Walles, B.: Macromolecular physiology of plastids IV. On amino acid requirements of lethal chloroplast mutants in barley. Hereditas (Lund) **50**, 317—344 (1963).

— Plastid structures of carotenoid-deficient mutants of sunflower (*Helianthus annuus* L.). I. The white mutant. Hereditas (Lund) **53**, 247—256 (1965).

— Plastid structures of carotenoid-deficient mutants of sunflower (*Helianthus annuus* L.). II. The yellow mutant. Hereditas (Lund) **56**, 131—136 (1966).

— (1) Use of biochemical mutants in analyses of chloroplast morphogenesis. In: Biochemistry of chloroplasts, Vol. II, 633—653 (T. W. Goodwin, Ed.). London-New York: Academic Press 1967.

— (2) The homozygous and heterozygous effects of an *aurea* mutation on plastid development in spruce [*Picea abies* (L.) Karst.]. Stud. For. Suec. **60**, 1—120 (1967).

— Biochemical mutants affected in chloroplast morphogenesis. Hereditas (Lund) **59**, 346—352 (1968).

Weiss, M. G.: Inheritance and physiology of efficiency in iron utilization in soybeans. Genetics **28**, 253—268 (1943).

von Wettstein, D.: Chlorophyll-Letale und der submikroskopische Formwechsel der Plastiden. Exp. Cell Res. **12**, 427—506 (1957).

— The formation of plastid structures. Brookhaven Symposia in Biol. **11**, 138—159 (1958).

— (1) Developmental changes in chloroplasts and their genetic control. In: Develop. Cytol., p. 123—160 (D. Rudnick, Ed.). New York: Ronald Press Company 1959.

— (2) Spectrophotometric studies of chlorophyll mutants in barley. Carnegie Inst. Wash. Yearbook **58**, 338—339 (1959).

— Multiple allelism in induced chlorophyll mutants. II. Error in the aggregation of the lamellar discs in the chloroplast. Hereditas (Lund) **46**, 700-708 (1960).

— Nuclear and cytoplasmic factors in development of chloroplast structure and function. Canad. J. Botany **39**, 1537—1545 (1961).

— Chloroplast structure and genetics. In: Harvesting the sun. Photosynthesis in plant life, p. 153—190 (A. San Pietro, F. A. Greer, T. I. Army, Eds.). New York-London: Academic Press 1967.

— Eriksson, G.: The genetics of chloroplasts. Proc. 11th Intern. Congr. Genet. The Hague 1963, Vol. 3, pp. 591—612 (1965).

Winge, Ö.: The chromosomes. Their numbers and general importance. C. R. Lab. Carlsberg **13**, 131—275 (1917).

— On the non-mendelian inheritance in variegated plants. C. R. Lab. Carlsber **14**, 1—20 (1919).

Wolf, F. T.: Effects of light and darkness on biosynthesis of carotenoid pigments in wheat seedlings. Plant Physiol. **38**, 649—652 (1963).

Wollgiehn, R., Mothes, K.: Über die Incorporation von ^{3}H-Thymidin in die Chloroplasten-DNS von *Nicotiana rustica*. Exp. Cell Res. **35**, 52—57 (1964).

Woodcock, C. L. F., Bell, P. R.: Features of the ultrastructure of the female gametophyte of *Myosurus minimus*. J. Ultrastruct. Res. **22**, 546—563 (1968).

Woods, M. W., Dubuy, H. G.: Hereditary and pathogenic nature of mutant mitochondria in *Nepeta*. J. nat. Cancer Inst. **11**, 1105—1151 (1951).

Woods, P. S., Zubay, G.: Biochemical and autoradiographic studies of different RNA's: Evidence that transfer RNA is chromosomal in origin. Proc. nat. Acad. Sci. (Wash.) **54**, 1705—1712 (1965).

Yankofsky, C., Spiegelman, S.: (1) The identification of the ribosomal RNA cistron by sequence complementarity. I. Specificity of complex formation. Proc. nat. Acad. Sci. (Wash.) **48**, 1069—1078 (1962).

— (2) The identification of the ribosomal RNA cistron by sequence complementarity. II. Saturation of a competitive interaction at the RNA cistron. Proc. nat. Acad. Sci. (Wash.) **48**, 1466—1472 (1962).
— Distinct cistrons for the two ribosomal RNA components. Proc. nat. Acad. Sci. (Wash.) **49**, 539—544 (1963).
Zucker, M., Stinson, H. T.: Chloroplasts as the major protein-bearing structures in *Oenothera* leaves. Arch. Biochem. **96**, 637—644 (1962).

Nucleic Acids and Information Processing in Chloroplasts*

Christopher L. F. Woodcock and Lawrence Bogorad

A. Introduction

The earliest indications that plastids might contain nucleic acids came in the 1950's and were principally the result of histochemical investigations. By the mid-1960's improvements in technique, and the application of biochemical methods resulted in conclusive demonstrations that plastids contained both DNA and RNA. These findings naturally led to the hypothesis that plastids possess complete information storage and processing systems and in the last few years this hypothesis has been put to test. At present the indications are that plastids not only possess DNA but contain systems capable of forming DNA, of using a DNA template for the synthesis of RNA and of using polyribonucleotides as templates for the incorporation of amino acids into polypeptides. However, a more refractory problem than that of identifying the components and reactions of the information processing system has been the evaluation of its significance in the "life" of the plastid. The techniques necessary to answer such key questions as how much information does the plastid DNA contain, how much of it is put to use and which proteins are manufactured, are now becoming available. Progress can be anticipated in these areas in the next few years. Until such questions are satisfactorily resolved, the degree of autonomy of plastids within the cell will remain one of the more tantalizing problems in the plant kingdom, as will the possibility of achieving the ancient goal of culturing plastids *in vitro*.

Another line of inquiry stimulated by the discovery of plastid nucleic acids has been the comparison of the DNA and RNA species found in plastids with those of other cellular components. Many of the systems operating in plastids have been found to resemble those of procaryotic organisms and this has obvious implications concerning the possible origin and evolution of plastids.

In this chapter, both the functional and comparative aspects of plastid nucleic acids will be discussed using the information storage and processing system: DNA→ RNA→ protein as a framework. For more detailed accounts of many of the following topics, the reader is referred to the excellent reviews of Gibor and Granick (1964), Granick and Gibor (1967), Kirk and Tilney-Bassett (1967) and Smillie and Scott (1969).

B. Information Storage

Early work on the DNA of plastids was concerned solely with demonstrating its presence. Two approaches were used: *in situ* studies in which a DNA stain or other marker was employed, and assays of DNA extracted from isolated plastids. Up

* The preparation of this chapter was supported in part by the National Institute of General Medical Sciences grants No. GM-06637 and GM-14991.

until the early 1960's, conflicting data were obtained using both these approaches, but since then advances in technique have resulted in considerable progress in both areas.

1. In situ Studies

Results of the application of classical DNA stains such as Feulgen and methyl green are shown in Table 1. It is clear that sufficient DNA is present in the plastids of certain plants to be detected by these staining procedures. In two cases, *Chlamydomonas moewusii* (Ris and Plaut, 1962) and *Beta vulgaris* (Kislev *et al.*, 1965), corroborative evidence was obtained using other *in situ* techniques (see Table 1). Failure to detect DNA using the Feulgen technique may be ascribed to insufficient DNA, to its presence in an "unstainable" form or to its absence. The failure of certain plant nuclei to react with the Feulgen stain (Pavulans, 1950) testifies to the unreliability of this stain in certain cases.

Increased sensitivity of DNA detection compared with Feulgen staining is obtained with the acridine orange fluorescence technique as can be seen from the published micrographs of *Chlamydomonas* plastid DNA by Ris and Plaut (1962). Although we have confirmed these results, and applied the techniques successfully to *Acetabularia* plastids (Woodcock and Bogorad, 1970) both *Nitella* and *Spinacia oleracea* gave negative results (Woodcock, 1970). Interpretation of negative data is again complicated by the questionable reliability of the method; Randall and Disbrey (1965) found acridine orange fluorescence of the basal bodies of *Tetrahymena pyriformis* to disappear at a certain stage in the division cycle and considered that a

Table 1. *Detection of DNA in plastids — in situ studies*

Methods	Species	Control	Result	Author
Feulgen stain	*Selaginella savatieri*	Hot TCA[a]	+	Chiba (1951)
	Tradescantia fluminiensis	Hot TCA	+	Chiba (1951)
	Rhoeo discolor	Hot TCA	+	Chiba (1951)
	Agapanthus umbellatus	Hot TCA	+	Metzner (1952)
	All four species above		−	Littau (1958)
	Vicia faba		−	Vosa and Kirk (1963)
	Daucus carota (chromoplasts)		−	Straus (1954)
	Chlorophytum comosum	DNase	+	Spiekerman (1957)
	Helianthus tuberosus	DNase	+	Spiekerman (1957)
	Chlorella vulgaris	DNase	+	Flaumenhaft *et al.* (1960)
	Chlamydomonas moewussii	DNase	+	Ris and Plaut (1962)
	Euglena gracilis		−	Vosa and Kirk (1963)
Feulgen stain modifications	*Beta vulgaris*	DNase	+	Kislev *et al.* (1965)
	Euglena gracilis		−	Givner and Moriber (1964)
Methyl green	*Selaginella savatieri*	Hot TCA	+	Chiba (1951)
	Tradescantia fluminiensis	Hot TCA	+	Chiba (1951)
	Rhoeo discolor	Hot TCA	+	Chiba (1951)
	Agapanthus umbellatus	Hot TCA	+	Chiba (1951)

Table 1 (continued)

Method	Species	Control	Result	Author
Acridine orange fluorescence	*Chlamydomonas moewusii*	DNase	+	RIS and PLAUT (1962)
	Acetabularia mediterranea	DNase	+	WOODCOCK and BOGORAD (1970)
	Polyphysa cliftoni	DNase	+	WOODCOCK and BOGORAD (1970)
Radio-autography	*Spirogyra*	none	+	STOCKING and GIFFORD (1959)
	Pteridium aquilinum (proplastids)	DNase	+	BELL (1961); BELL and MUHLETHALER (1964)
	Nicotiana rustica	DNase	+	WOLLGIEHN and MOTHES (1964)
	Zea mays	Radioactivity incorporated to plastids, but not removed by DNase	–	LIMA DA FARIA and MOSES (1965)
	Beta vulgaris	DNase	+	KISLEV *et al.* (1965)
	Dictyota	DNase, hot TCA	+	STEFFENSEN and SHERIDAN (1965)
	Padina	DNase, hot TCA	+	STEFFENSEN and SHERIDAN (1965)
	Bryopsis	DNase, hot TCA	+ +	STEFFENSEN and SHERIDAN (1965)
	Euglena gracilis	DNase	+	SAGAN *et al.* (1964)
	Acetabularia mediterranea	hot TCA	+	SHEPHARD [1965 (1)]
Electron microscopy	*Chlamydomonas moewusii*	DNase	+	RIS and PLAUT (1962)
	Beta vulgaris	DNase	+	KISLEV *et al.* (1965)
	Acetabularia mediterranea	DNase (but fibrils also removed by DNase-free digestion)	+	PUISEUX-DAO (1967)
	Acetabularia mediterranea	DNase	+	WOODCOCK and BOGORAD (1970)
	Laurencia spectabilis	DNase	+	BISALPUTRA and BISALPUTRA (1967)
	Egrezia	DNase	+	BISALPUTRA and BISALPUTRA (1967)
	Spinacia	DNase	+	WOODCOCK and FERNANDEZ-MORAN (1968)
	Ochromonas danica	Localization of ^{3}H thymidine label	+	GIBBS (1968)
	Sphaceloriap	DNase	+	BISALPUTRA and BISALPUTRA (1969)
	Beta vulgaris	DNase	+	HERRMANN and KOWALLIK (1970)

[a] Trichloroacetic acid.

change in form from a condensed to a more dispersed state or complexing with protein could explain the phenomenon.

Radioautography following administration of tritiated thymidine has been frequently used to localize plastid DNA and has given positive results in most of the published reports (Table 1). The importance of adequate controls was reinforced by the work of LIMA DE FARIA and MOSES (1964) who found incorporation of thymidine into plastids which could not be removed with DNase.

Radioautographic methods are more flexible than staining techniques in that the range of sensitivity can be varied at will by altering the dose of isotope administered, and the exposure time of the photographic emulsion. Thus, although staining techniques failed to reveal DNA in *Euglena gracilis* (GIVNER and MORIBER, 1964), autoradiography after incorporation of tritiated adenine or guanine was successful (SAGAN *et al.*, 1964).

RIS and PLAUT (1962) extended their studies on *Chlamydomonas moewusii* plastid DNA by correlating the Feulgen and acridine orange staining material with characteristic areas seen in thin sections with the electron microscope. These fibrillar areas were similar in appearance to the "nucleoplasm" of bacteria, and the smallest fibril diameter was 20 to 25 Å, that expected of a DNA double-helix. Since then, the presence of similar DNase-sensitive fibrillar aggregates has been demonstrated in a wide range of plants (Table 1). Not mentioned in the table, however, are the many papers in which the presence of these characteristic areas in plastids have been noted, but not tested for DNase sensitivity.

Thus, despite the equivocal results of early work, *in situ* studies have demonstrated the presence of DNA in the plastids of a wide variety of plant species. None of these techniques, however, allow quantitation, and it has been studies on extracted plastid DNA to which most attention has been given in recent years.

2. Studies on Extracted DNA

Isolation of plastids followed by extraction and testing for the presence of DNA has been extensively used as a detection technique (Table 2). Early work again produced conflicting results, and even in cases where DNA was successfully detected in isolated plastids, a major criticism was the possible contamination of the plastids with nuclear DNA. In order to reduce this contamination, techniques for purifying plastids more thoroughly than by differential centrifugation were devised, and included flotation and density gradient centrifugation. Although these methods were successful in removing cytologically detectable nuclei and nuclear fragments, the possibility that the very small amounts of DNA detected were due to contamination was not excluded. More recently, experiments to assess the likelihood of nuclear contamination by deliberately introducing foreign DNA into the chloroplast preparations have shown that little of the contaminant becomes adsorbed (POLLARD, 1964; IWAMURA, 1969; BARD and GORDON, 1969).

In 1963, CHUN, VAUGHAN, and RICH published the first account of the application of cesium chloride isopycnic centrifugation to plant DNA and their demonstration that several components differing in buoyant density were present, stimulated much interest in the purification and characterization of plastid DNA.

Table 2. *Detection of DNA in isolated plastids*

Plant	Plastid purification	DNA assay	Result	Comment	Author
Nicotiana tabacum	Differential centrifugtion	Diphenylamine, cysteine sulphuric acid and tryptophan-perchloric acid reactions	–	DNA detection not primary objective	HOLDEN (1952)
		Diphenylamine reaction, UV absorption	+	Adsorption of broken nuclei to plastids suspected	MCCLENDON (1952)
				Cytochemical tests showed some contamination	JAGENDORF and WILDMAN (1954)
		UV absorption of hydrolysis product	+	Preparation not checked for contamination	COOPER and LORING (1957)
Nicotiana tabacum, *Spinacia oleracea*	Differential centrifugation	Diphenylamine reation	+	Nuclear fragments reduced to 1/600 plastids	CHIBA and SUGAHARA (1957)
Cichorium endivia	Differential centrifugation	Analysis of hydrolysis products	–		KERN (1959)
Zea mays	Differential and density gradient centrifugation	Phosphorus assay on extracted nucleic acids	–	Little difference between centrifugation methods	ORTH and CORNWALL (1963)
Vicia faba	Density gradient centrifugation	Diphenylamine reaction	+	No contamination detected cytologically	KIRK (1963)
Acetabularia mediterranea	Density gradient centrifugation	Fluorimetry	+	Nuclear contamination excluded	BALTUS and BRACHET (1963)
			+		GIBOR and IZAWA (1963)
Spinacia oleracea	Flotation and density gradient centrifugation	Diphenylamine, cysteine sulphuric acid and indole reaction	+	Contamination checked by deliberately adding nuclear material	POLLARD (1964)
	Non-aqueous preparation	Diphenylamine reaction UV absorption	+	No contamination detected cytologically	BIGGINS and PARK (1964)
Antirrhinum majus, *Allium porrum*	Non-aqueous preparation	Thymine assay after formic acid hydrolysis	–	KIRK (1967) has suggested that this negative result was due to adsorption of thymine by charcoal	RUPPEL (1964)
Antirrhinum majus	Non-aqueous preparation	Diphenylamine reaction	+		RUPPEL and VAN WYCK (1965)

3. Preparation and Characterization

Although published accounts of the preparation of plastid DNA have varied greatly in procedural details, the following main steps have been consistently followed:

1. Breaking of cells in a medium hypertonic or isotonic with the plastid stroma.
2. Low-speed differential centrifugation to remove cell walls and whole nuclei.
3. High speed sucrose density gradient centrifugation or flotation to purify plastids.
4. Lysis of purified plastids.
5. Extraction, deproteinization, and purification of DNA.
6. Analysis of DNA components by CsCl centrifugation.
7. Collection of the plastid DNA fraction, and removal of CsCl. In Step 6, the buoyant densities of the different DNA species may be determined, and from these figures the proportions of adenine and thymine to cytosine and guanine can be calculated (SCHILDKRAUT *et al.*, 1962) if no unusual bases are present.

DNAs differing in density by 0.010 or 0.020 gm/cc can be separated from one another by one or more cycles of preparative CsCl density gradient centrifugation. From the purified DNA, the base ratios can also be determined directly by analysis of the hydrolysis products. Other physical properties which may be determined with the purified product are the hyperchromicity on heat melting which provides information on the strandeness of the DNA, the melting temperature (T_m) which varies with base composition, and the rate of reannealing, or renaturation after melting. The latter is influenced by two factors, the length of the molecule, and the number of repeating sequences—long molecules with few repeating sequences anneal only slowly, while short identical molecules reanneal rapidly.

In cases where a satellite DNA peak is clearly separated from the nuclear peak, the absolute amount of the satellite DNA species may be estimated. From the CsCl density gradient profile of a whole cell extract, the ratio of satellite to nuclear DNA can be determined and if the amount of DNA per nucleus is known that in the satellite may be calculated. If it is established that the satellite is of plastid origin, then the amount of DNA per plastid can be found (provided that the number of plastids per cell is also known).

4. Physical Properties of Plant DNAs

The physical properties of DNA species isolated from a number of plants is shown in Table 3. In many cases, the nuclear DNA, comprising by far the largest fraction in whole cell extracts, was accompanied by one or two fractions which banded separately in CsCl, i.e. as satellites of the major band.

Despite the many inconsistencies in the assembled data, three features emerge which are common to all the DNAs attributed to plastids. Firstly, on heating, all "plastid" DNAs were shown to exhibit the hyperchromicity (measured as the increase in absorption at 260 nm) characteristic of double-stranded DNA.

Secondly, after heat melting, "plastid" DNAs renaturated rapidly (measured as the decrease in absorption at 260 nm) compared with nuclear DNAs, indicating a large difference in molecular complexity between the two. Such a difference provides a useful tool in distinguishing plastid from nuclear DNA and in separating the two types. Since denatured DNA has a greater buoyant density than the native form (of the order of 0.015 g/cc), after heat melting and cooling, nuclear DNA retains its higher density, while the density of the plastid DNA returns towards its original native value.

Thirdly, the absence of the base 5-methyl-cytosine (5 Me C) has been reported (with one exception) from all "plastid" DNAs analyzed. In the exceptional case, a species of plastid DNA from *Spinacia oleracea* with the same buoyant density as nuclear DNA was shown by BARD and GORDON (1969) to contain 1.6% 5-methyl cytosine compared with the 3.6% found in nuclear DNA. It is possible in this case that the 5-methyl cytosine of the "plastid" DNA may have come from cross-contaminating nuclear DNA.

However, mitochondrial DNAs are also double-stranded and show rapid renaturation (see e.g. NASS, 1969), and there is no evidence that they possess 5-methyl cytosine (TEWARI *et al.*, 1966). Thus, none of these properties are diagnostic for plastid DNA. The consistent features of satellite DNAs are overshadowed by a large number of differences not only between different species of plants but also between different groups working with the same species. As can be seen from Table 3, a natural division occurs between flowering plants and algae with respect to the buoyant density of their nuclear and satellite DNAs and these two groups will be considered separately.

a) Higher Plants

Between 1963 and 1967 reports of the buoyant densities and other physical properties of higher plant DNAs showed substantial agreement. The density of the nuclear DNAs fell between 1.689 and 1.697 g/cc and one major satellite of greater density (1.702 to 1.709 g/cc) was reported.

Since the ratio of satellite to nuclear DNA increased as the chloroplasts were progressively purified, it seemed reasonable to assume that the heavy band originated in the chloroplast. The differences between the buoyant density measurements of the same species by independent workers were small, and were probably within the limits of experimental error. Only two reports did not fit into this pattern. KIRK (1963) found very similar base ratios for nuclear (40% GC) and chloroplast (37% GC) DNA fractions of *Vicia faba*, but the densities of the DNAs were not determined. The criterion for purity of the chloroplast fraction was that no appreciable amount of DNA was lost in a series of sucrose gradients. The second set of conflicting data came from SUYAMA and BONNER (1966), working with *Brassica rapa* who also found very little difference between nuclear and chloroplast DNAs (0.003 g/cc). They did, however, obtain a "heavy" satellite at 1.706 g/cc which they attributed to mitochondria. DNA with identical density (1.706 g/cc) was also obtained from mitochondrial fractions of *Phaseolus aureus* hypocotyls, *Allium cepa* bulb, and *Ipomoea batatas* root. A possible objection here was the lack of electron microscopic examination of the fractions; proplastids are present in all these tissues, are often indistinguishable from mitochondria in the light microscope, and might have contributed significant amounts of DNA.

The latter part of 1967 marked the end of this relatively consistent pattern and the start of a period during which the accepted values for the buoyant density of higher plant plastid DNA were questioned and reevaluated. As will be seen, this uncertainty has yet to be fully resolved. Characteristic of post-1967 work has been an inability to

Table 3. *Properties of plant DNAs*

Plant	Nuclear DNA					Other DNAs					Authors
	Density g/cc	% GC	5 Me C present	Hyper-chromocity	Rapid re-naturation	Density g/cc	% GC	5 Me C present	Hyper-chromicity	Rapid re-naturation	
Nicotian[a]	1.690					1.703 c					Shipp *et al.* (1965)
	1.697	40	+	+	−	1.702 c	42	−	+	+	Tewari and Wildman (1966)
		38[a]					43[a]				
		36[a]					41[a]				
	1.696					1.706 c			+		Green and Gordon (1967)
						1.711					
	1.697	35	+	+	−	1.697 c	41	−	+	+	Whitfeld and Spencer (1968)
Vicia faba		39.4					37.4				Kirk (1963)
	1.696					1.696 c			+	+	Kung and Williams (1969)
	1.695					1.697 c					Wells and Birnstiel (1967, 1969)
Spinacia oleracea	1.695	36[a]		+		1.705 c	46[a]		+		Chun *et al.* (1963)
						1.719	60[a]		+		
	1.694	34	+	+	−	1.696 c	40	−	+	+	Whitfeld and Spencer (1968)
	1.695	34	+			1.696 c	37	+	+	±	Bard and Gordon (1969)
						1.706 c	47	−	+	+	
	1.694					1.697					Wells and Birnstiel (1969)
						1.706					
Beta vulgaris (Swiss chard)	1.695	36[a]				1.705 c	46[a]				Chun *et al.* (1963)
						1.719	60[a]				
	1.698	30[a]				1.700 c	41				Kislev *et al.* (1965)
Phaseolus vulgaris	1.694					1.694 c					Beridze *et al.* (1967)
	1.703					1.703 c					
	1.693					1.695					Wolstenholme and Gross (1968)
		37.34					37.73				Baxter and Kirk (1969)
Brassica rapa	1.692					1.695 c					Suyama and Bonner (1966)
						1.706 m					

Tagetes patula	1.692					1.702 c					Green and Gordon ((1967)
						1.707					
Antirrhinum majus	1.697					1.709 c					
Dianthus caryophyllus	1.195					1.706 c					
Ranunculus repens	1.697					1.703 c					
Equisetum sp.	1.697					1.713 c					
Lettuce	1.694	43	+	+	−	1.697 c	44	−	+	+	Wells and Birnstiel (1969)
						1.606 m					
Sweet pea	1.695					1.697 c					
Antirrhinum majus	1.689					1.689 c					Ruppel (1967)
Euglena gracilis	1.708					1.688 c					Leff *et al.* (1963)
	1.707	48[a]				1.685 c	26[a]				Edleman *et al.* (1964)
		55[b]					30[a]				
	1.707	51	+			1.685 c	24	−			Ray and Hanawalt (1964)
	1.708		+	+		1.684 c		−	+		Brawerman and Eisenstadt [1964 (1)]
						1.692					
	1.707					1.681 c					Richards (1967)
Chlamydomonas reinhardi	1.628	62				1.702 c	39				Sager and Ishida (1963)
	1.723	64[a]		+		1.695 c	36[a]		+		Chun *et al.* (1963)
	1.723	64[a]		+		1.692 c	33[a]		+	+	Chiang and Sueoka (1967)
	1.721					1.694 c					Leff *et al.* (1963)
Chlorella ellipsoidea	1.716	57[a]		+		1.695 c	36[a]		+		Chun *et al.* (1963)
	1.717					1.692 c					Iwamura and Kuwashima (1969)
						1.717 c					
Acetabularia mediterranea	1.702					1.704 c	45[a]				Green *et al.* (1967)
						1.695 c					Gibor (1967)

c = Attributed to chloroplasts. m = Attributed to mitochondria.
[a] Calculated from the buoyant density. [b] Calculated from the melting temperature.

distinguish between nuclear and chloroplast DNA on the basis of buoyant density. Identification of the chloroplast component has therefore been on the basis of renaturation characteristics and the presence or absence of 5-methyl cytosine.

One particularly puzzling case is that of *Nicotiana* on which four independent studies have been made. In three instances, Shipp *et al.* (1965), Tewari and Wildman (1966), and Green and Gordon (1967), a DNA species satellite to the nuclear DNA was found, but in the fourth (Whitfeld and Spencer, 1968) both nuclei and chloroplasts had DNAs of the same density. In this case, it is difficult to ascribe the differences in result to technique since Whitfeld and Spencer were using the same procedures as Tewari and Wildman, even to the same variety of *Nicotiana*.

Bard and Gordon (1969), using zonal centrifugation to purify *Spinacia oleracea* chloroplasts, attributed two DNA species to the latter. One had the same density as nuclear DNA, 1.695 g/cc, the other (comprising 35% of the total chloroplast DNA) had a density of 1.706 g/cc.

Wells and Birnstiel (1969) found a slight density difference between nuclear and chloroplast DNAs of four higher plants. A second satellite at 1.706 g/cc was thought to be of mitochondrial origin. An innovation in technique here was the use of low concentrations of DNase during the chloroplast isolation procedures to remove nuclear contaminants. In the hands of other groups (Bard and Gordon, 1969) this procedure had resulted in the loss of all DNA.

How can the pre- and post-1967 results be reconciled? Kirk (1970) in a recent review, concluded that the "real" plastid DNA had a buoyant density close to that of nuclear DNA and that the component of density in the region of 1.706 g/cc was most likely contributed by mitochondria. Although the explanation is attractive, it seems too early to dismiss all the pre-1967 "plastid" satellites as artifacts. It is certainly plausible that a typical chloroplast preparation contains both mitochondria and nucleases and that the latter will tend to affect the chloroplast DNA more than the mitochondrial DNA. But in that case, Suyama and Bonner (1966) should have found a large proportion of mitochondrial DNA (at 1.706 g/cc) in their relatively crude chloroplast preparation from *Brassica rapa*, whereas in fact, there was a barely detectable shoulder at this density. Also, it has to be assumed that the single band at 1.702 g per cc in the chloroplast preparations of Tewari and Wildman (1966) was truly plastid DNA (what happened to the mitochondrial DNA here?) and that the density measurements erred on the heavy side. The heavier "plastid" DNA found by Bard and Gordon (1969) must also be accounted for. Kirk (1970) has suggested that the lack of a preliminary low-speed centrifugation, and the large volume of the final chloroplast band (15% of the zonal gradient) would account for the massive mitochondrial contamination necessary, but in an experiment to increase mitochondrial contamination, an extra band at 1.719 g/cc was obtained. That a single species of bacteria could contaminate both these preparations and those of Chun *et al.* (1963) to such an extent as to cause a distinct DNA band seems unlikely.

Thus, we feel that further work should be awaited before accepting that chloroplasts from higher plants contain a single, constant DNA species, with a density of about 1.696 g/cc. As Kirk (1970) suggests, future studies would benefit from measurements of cytochrome oxidase activity in preparations showing bands in the 1.706 g/cc region. Also it should be possible in cases where a single DNA of 1.695 g/cc is obtained (e.g., WHITFELD and SPENCER, 1968; KUNG and WILLIAMS, 1969) deliberatelλ to enrich for mitochondria, and then obtain a band at 1.706 g/cc.

In the meantime, are there alternative explanations of the conflicting density measurements? Differences in preparative technique are unlikely to produce such variable data, unless it is assumed that two or more populations of plastids are present in higher plants and that small differences in plastid isolation methods could enrich for a particular population. Since sucrose density gradients of plastid preparations usually give several bands, this may not be unreasonable to expect. However, we then have to postulate that the different populations of plastids contain DNAs of differing density. Equally unattractive (at least from an aesthetic point of view) is the possibility that the chloroplasts themselves form a single population but that each contains more than one density mode of DNA. Thus, there seems to be no explanation which reduces the conflicting density data to a uniform system as elegantly as that of KIRK (1970). It remains to be seen whether the most attractive explanation is the correct one*.

b) Algal Plastid DNAs

Among the algae, the process of identifying and characterizing chloroplast DNA has progressed steadily; there is little of the confusion which is still surrounding the higher plant data.

In *Euglena gracilis* agreement between groups of authors has been exceptionally good, buoyant density measurements of nuclear DNA ranged from 1.707 g/cc to 1.708 g/cc, and for chloroplast DNA from 1.681 g/cc to 1.688 g/cc [LEFF *et al.*, 1963; RAY and HANAWALT, 1964; EDELMAN *et al.*, 1964; BRAWERMAN and EISENSTADT, 1964 (1); RICHARDS, 1967]. A third DNA species of intermediate density but unknown origin was reported by BRAWERMAN and EISENSTADT [1964 (1)]. There is also substantial agreement on the buoyant densities of *Chlamydomonas reinhardi* DNAs, data for the chloroplast component ranging from 1.692 g/cc to 1.695 g/cc, and for the nuclear DNA from 1.721 g/cc to 1.723 g/cc (CHUN *et al.*, 1963; LEFF *et al.*, 1963; CHIANG and SUEOKA, 1967). SAGER and ISHIDA (1963) obtained values approximately 0.005 g/cc higher than these.

In the other two algae investigated, there is less agreement. Besides the nuclear (1.716 g/cc) and chloroplast (1.695 g/cc) components of *Euglena gracilis* reported by CHUN *et al.* (1963). IWAMURA and KUWASHIMA (1969) considered that a third component was the major chloroplast DNA. This had the same density as nuclear DNA (1.717 g/cc) and as KIRK (1970) suggests, the evidence that it is indeed of chloroplast origin is not compelling.

Conflicting data has also been obtained from *Acetabularia* (GIBOR, 1967; GREEN *et al.*, 1967). The several problems which arise when working with this organism are discussed in a later section covering all aspects of information storage and processing in *Acetabularia* (see p. 112).

* *Note added in proof:* WELLS and INGLE [(Plant Physiol. **46**, 178—179 (1970)] have provided additional evidence that plastid DNAs from different species are of relatively constant buoyant density (1.694—1.698 g/cc) while the corresponding nuclear DNAs are more variable.

5. Coding Capacity of Chloroplast DNA

The information content of a given DNA is proportional to the length of non-repetitive base sequences. Thus, to establish the coding capacity of chloroplast DNA, it is insufficient simply to determine the amount of DNA per chloroplast; the number of identical molecules and/or identical sequences in the same molecule must also be known. Although the amount per plastid has been estimated for a number of plants (Table 4), only two attempts at measuring base sequence redundancy have been made (WELLS and BIRNSTIEL, 1969; STUTZ, 1970).

The estimates of the amount of DNA per plastid fall within the 2 orders of magnitude between bacteriophage and bacterial DNA content and are 3 to 6 orders of

Table 4. *Estimates of plastid DNA content*

Plant	Basis of estimation	DNA g/plastid	Author
Acetabularia mediterranea	Fluorimetry	1×10^{-16}	GIBOR and IZAWA (1963)
	Electron microscopy	Varied-up to 3×10^{-15}	WOODCOCK and BOGORAD (1970)
Vicia faba	Diphenylamine assay	2×10^{-15}	KIRK (1963, 1967)
Euglena gracilis	CsCl gradients	1.1×10^{-14}	BRAWERMAN and EISENSTADT (1964)
Chlamydomonas reinhardi	CsCl gradients	8×10^{-15}	CHIANG and SUEOKA (1967)
Chlamydomonas reinhardi	CsCl gradients	1×10^{-14}	SAGER and ISHIDA (1963)
Nicotiana glutinosa	CsCl gradients	4.7×10^{-15}	TEWARI and WILDMAN (1966)
Euglena gracilis	CsCl gradients	1.2×10^{-15}	EDELMAN *et al.* (1964)
Other systems:		g/unit	
Escherichia coli	Radioautography	3×10^{-15}	CAIRNS (1963)
T3 Bacteriophage	Electron microscopy	4×10^{-17}	LANG *et al.* (1967)
Animal mitochondrion	Electron microscopy	1.7×10^{-16} (single circle)	NASS (1966, 1969)
Plant mitochondrion (*Phaseolus vulgaris*)	CsCl gradients	5×10^{-16}	SUYAMA and BONNER (1966)
T4 Bacteriophage	Electron microscopy	1.7×10^{-16}	KLEINSCHMIDT *et al.* (1962)
Higher plant diploid nuclei	Feulgen spectrophotometry	varied from: 0.55×10^{-11} to 31×10^{-11} with species	MCLEISH and SUNDERLAND (1961)

magnitude smaller than those for higher plant nuclei. In terms of length of double stranded DNA, 1×10^{-16} g is equivalent to approximately 30 μ (WILKINS, 1963) with a molecular weight of about 6×10^7 daltons; this in turn can be converted to the number of nucleotides and hence amino acids coded. Thus, the range from 10^{-16} to 10^{-14} g DNA could code for from 150 to 15,000 proteins of average molecular weight 20,000. At the lower range, this calculation has significance since the limit to the number of proteins that the plastid could make is far less than the number of protein species known to be common to all plastids. Also, of course, part of the genome may code for the various RNA species. At the higher range, however, it would be theoretically possible for the plastid to code for all its proteins, but only if the DNA contained few repeating sections. Until the amount of redundancy is established the significance of plastid DNA estimates with regard to the possible autonomy of the organelle cannot be appreciated.

Estimation of plastid DNA content from CsCl gradient experiments, or other techniques involving extraction are subject to a number of errors, principally the assumption that the extraction process is complete. More direct methods of measuring DNA are available, in particular the technique developed by KLEINSCHMIDT *et al.* (1962) in which DNA lengths are recorded with the electron microscope. Such methods have been successfully adapted to the measurement of mitochondrial DNA (see for example NASS, 1969, 1966) and in the case of animal mitochondria, resulted in the discovery of its length and circularity. Application of similar spreading techniques to plastid DNA has not had such striking success. In the case of osmotically burst spinach plastids the DNA strands were seen to be intimately associated with the plastid membrane system, possibly bound to granal and intergranal membranes in places (WOODCOCK and FERNANDEZ-MORAN, 1968). Because of the obscuring effect of the membranes, accurate measurement of the length of total DNA per plastid was not possible. The observation that the DNA occurred not only as linear strands, showing the conformation typical of double-stranded DNA, but also as a meshwork undergoing sharp angular bends may be due to the constraining effect of other macromolecules in the latter case. In view of the now overwhelming agreement that plastid DNA shows the hyperchromicity expected of double-stranded DNA (Table 3), it is unlikely that the two forms represent double and single-stranded DNA respectively.

When DNA was gently extracted from purified spinach plastids and viewed with the electron microscope, it was seen to consist of predominantly linear molecules of variable length (of the order of a few microns). Similar results were obtained by UPPHADAYHA and GRUN (1968) using *Tradescantia*. The extraction of short, linear DNA molecules of non-uniform length from plastids probably reflects the physical or enzymatic breakage of longer molecules during preparation. Alternatively, it is possible that the failure to obtain molecules of uniform length may be related to the many DNA-membrane attachments at which breaks may occur during extraction. The observations (CHIBA and SUGAHARA, 1957; KIRK, 1963) that higher concentrations (2 N) of perchloric acids are needed to liberate DNA from plastids than from nuclei (0.5 N) further suggests that in plastids, the DNA is not simply free in the stroma.

Substantially larger molecules have recently been extracted from *Acetabularia* plastids [WERZ and KELLER, 1968 (1) (2)] and osmotically burst plastids from the same species liberated amounts of DNA comparable to that found in bacteria (WOODCOCK and BOGORAD, 1970; GREEN, BURTON 1970). Evidence was also presented which indicated that the chloroplasts of *Acetabularia* do not contain equal amounts of DNA; in 35% of plastids a large but variable amount was present, while in the remaining 65% any DNA present was below the limits of detection (WOODCOCK and BOGORAD, 1970).

6. Sequence Homology of Plastid DNA

To date, only two reports relating to the sequence of homology of plastid DNA has appeared. WELLS and BIRNSTIEL (1969) using the renaturation kinetics approach developed by BRITTEN and KOHNE (1968), concluded that lettuce chloroplast DNA included two fractions, one which renaturated *as though* it consisted of a number of repeating segments, each of molecular weight 3×10^6 daltons (equivalent to a length of 1.5 μ), and a second fraction with the renaturation speed expected of DNA segments of molecular weight 1.2×10^8 daltons (60 μ). Since the molecular weight as calculated from the amount of DNA per chloroplast was 2×10^9 daltons (1,000 μ), it was concluded that lettuce chloroplast DNA was extensively reiterated. STUTZ (1970) in similar studies, reported that the kinetic complexity of *Euglena gracilis* plastid DNA was approximately one thirtieth (1.8×10^8 daltons) of the analytical complexity. These data, if confirmed, are of great value in appreciating the role of the chloroplast in the cell since the potential coding capacity of the chloroplast is seen to be similar to that of a large virus, rather than a bacteria. It is to be hoped that more work in this area will soon appear.

C. Information Replication

The demonstration of the complete synthesis of a new genome by plastids has proven very difficult; *in vivo* experiments are open to the possibility of nuclear interference and *in vitro* plastids perform inadequately. However, both approaches have contributed considerable circumstantial evidence to support the hypothesis that plastids possess a DNA polymerase and are capable of replicating DNA.

CHIANG and SUEOKA (1967) using a synchronous culture of *Chlamydomonas reinhardi* were able to show that the amount of DNA per volume of culture increased during two separate phases of the life cycle; an approximate doubling prior to cell division, and a 15% increase, some 12 h before cell division. The assumption that this 15% increment involved chloroplast DNA was verified by an N^{15}/N^{14} transfer experiment [similar to that performed by MESELSON and STAHL (1958)] during which DNA was extracted and analyzed on CsCl density gradients. The chloroplast DNA band from cells grown on ^{15}N showed a sufficient density increase (about 0.02 g/cc) over that from ^{14}N cells for the two chloroplast DNAs to be resolved as separate bands. ^{15}N cells transferred to medium containing ^{14}N initially had heavy chloroplast DNA but at the time of the 15% DNA increment, a new DNA peak, intermediate in density between ^{15}N DNA and ^{14}N DNA, appeared. The appearance of this hybrid DNA indicated that a single round of semi-conservative replication had occurred. Three hours later, two chloroplast peaks were seen, one hybrid and one of ^{14}N density as a result of another round of semi-conservation replication within the chloroplast.

Although the above experiment elegantly demonstrates the replication of DNA, the possibility that the nucleus is controlling and/or performing the synthesis cannot be excluded. A second approach, which does not suffer from this complication, has been the study of DNA synthesis by isolated chloroplasts. The main difficulty here has been that although incorporation of nucleotide triphosphates into DNA may be demonstrated, it is of very short duration [about 15 min in *Euglena gracilis* chloroplasts

(Scott *et al.*, 1968)] and consequently whether true replication or merely repair is being carried out cannot be determined. However, it has been shown for *Euglena* (Scott *et al.*, 1968), *Spinacia* [Spencer and Whitfeld, 1967 (2)] and *Nicotiana* (Tewari and Wildman, 1967) that DNA is synthesized, that the reaction is DNA-dependent and that the product bands with chloroplast DNA in CsCl gradients.

D. Information Transcription

The transcription of genetic information from DNA into the various types of RNA is a fundamental stage in the processing of information in bacteria and eucaryotic cells. The RNA system is complex, involving three distinct kinds of RNA: ribosomal, transfer, and messenger. Moreover, RNA polymerase is known to be a multi-component enzyme and the details of "initiation", "termination" and template specificity are just beginning to be understood even in bacterial cells. In plastids, much of the work pertaining to transcription has been concerned with the characterization of plastid RNA species, and only recently has the RNA polymerase system itself received very much attention.

1. Plastid RNAs

RNA is readily detected in plastids with conventional staining techniques (Table 5). However, RNA stains are generally less specific than the better DNA stains and controls must be adequate. Autoradiography following the administration of RNA precursors was used to demonstrate the presence of RNA in *Zea mays* etioplasts (Jacobson *et al.*, 1963), *Euglena gracilis* chloroplasts (Sagan *et al.*, 1964), and *Acetabularia mediterranea* chloroplasts [Shephard, 1965 (1)]. Ribosome-like particles have also been seen in electron microscope preparations of plastids from many plants; in a few cases (Jacobson *et al.*, 1963; Gunning, 1965) it was further shown that the particles were not seen in tissue treated with ribonuclease.

Many chemical assays of isolated plastids have also shown RNA to be present (for references, see Kirk and Tilney-Bassett, 1967) but as with the DNA measurements on this type of preparation, the degree of contamination with RNA from other cell components has been difficult to assess.

A major advance in the study of plant RNAs came with the discovery that plastid ribosomes and ribosomal RNAs differed in sedimentation velocity from those of the cytoplasm. The first report of these differences came from Lyttleton (1962) who found two types of ribosomes in whole leaf preparations from spinach. Only the component which sedimented at the slower rate in the analytical ultracentrifuge was seen in purified chloroplast preparations. Later work has established sedimentation constants of about 70S for plastid ribosome monomers from many plants, compared with about 80S for those of the cytoplasm (see e.g., Stutz and Noll, 1967). "70S" type ribosomes are also present in mitochondria and procaryotic cells. Sedimentation constants of 23S and 16S have been established for the principal RNA components of 70S ribosomes, and 25S and 18S for the rRNA of plant 80S ribosomes (see e.g. Loening, 1969; Loening and Ingle, 1967). Early sedimentation data showed considerable variation and only recently have the "S" values quoted above been generally accepted. Improvements in technique have contributed here, especially

Table 5. *Detection of RNA in plastids—in situ studies*

Plant	Method	Control	Result	Comment	Author
Selaginella savatieri *Tradescantia fluminiensis* *Rhoeo discolor*	Pyronin stains	RNase	+ + +	Author did not consider results conclusive	CHIBA (1951)
Agapanthus umbellatus	Pyronin	Hot TCA	+		METZNER (1952)
Elodea	Pyronin	RNase	+		YOSHIDA (1962)
Selaginella savatieri *Tradescantia fluminiensis* *Rhoeo discolor*	Azur B staining	Hot TCA	+ + +	Author did not consider results conclusive	LITTAU (1958)
Chlorella vulgaris	Pyronin	RNase	+		FLAUMENHAFT *et al.* (1960)
Daucus carota (chromoplasts)	Pyronin	RNase Hot TCA	+		STRAUS (1954)
Chlorophytum comosum *Helianthus tuberosus*	Basophilia	RNase	+		SPIEKERMAN (1957)
Zea mays (etioplasts)	Azur B	RNase	+		JACOBSON *et al.* (1963)
Zea mays (etioplasts)	Electron microscopy	RNase	+	RNase-sensitive ribosome-like particles seen	JACOBSON *et al.* (1963)
Avena sativa	Electron microscopy	RNase	+	RNase-sensitive ribosome-like particles seen	GUNNING (1965)
Zea mays (etioplasts)	Radioautography following ^{3}H cytidine feeding	RNase	+		JACOBSON *et al.* (1963)
Euglena gracilis	Autoradiography following ^{3}H guanine, ^{3}H-adenine incorporation	RNase	+		SAGAN *et al.* (1964)
Acetabularia mediterranea	Radioautography following ^{3}H uridine incorporation	RNase	+		SHEPHARD (1965)

the development of adequate ribonuclease inhibitors (preventing partial digestion of the ribosomal RNA), and the adaptation of polyacrylamide gel electrophoresis with its great capacity to resolve RNAs (see, for example, LOENING, 1967). A notable exception here has been work on *Acetabularia* in which only 70S ribosome monomers, and 23S and 16S rRNAs have been reported. Possible reasons for this deviation are discussed on p. 115. A detailed account of work on plastid RNA may be found in a recent review by SMILLIE and SCOTT (1969).

Maize etioplasts contain one or more RNA polymerases. Soon after illumination of etiolated leaves the plastid RNA polymerase activity increases. Since the increase hardly occurs in plants to which chloramphenicol has been administered, it seems that some protein synthesis in the plastid may be required for the change in enzyme activity (BOGORAD, 1967, 1968). It is not known whether increased RNA polymerase activity is *required* for light-controlled etioplast maturation.

2. RNA Synthesis in Plastids

The differences between the ribosomes of the chloroplast and "cytoplasm" discussed in the preceeding section have been used to demonstrate preferential synthesis of one type of ribosome or of one pair of rRNAs. During greening of etiolated leaves, for example, ^{32}P is incorporated primarily into RNA components of chloroplast ribosomes (BOGORAD, 1967, 1968; INGLE, 1968). Such data from the *in vivo* system do not show conclusively that RNA is transcribed in plastids, since the possibility of nuclear involvement cannot be excluded. However, many successful demonstrations of RNA synthesis in isolated plastids from higher plants and from algae have been reported [KIRK, 1964; SEMAL *et al.*, 1964; SHAH and LYMAN, 1966; BERGER, 1967; SPENCER and WHITFELD, 1967 (1); BOVÉ *et al.*, 1967; TEWARI and WILDMAN, 1969; BOGORAD, 1967, 1968; BOGORAD and WOODCOCK, 1970]. In these experiments the DNA-dependence and/or the DNase-sensitivity of the reaction and [with the exception of the work on *Acetabularia* (BERGER, 1967)] the necessity for all four nucleotide triphosphates provided additional evidence that DNA-directed RNA synthesis was occurring. In two cases, the experiments were taken a step further; the products of the reactions were characterized by sucrose density centrifugation. SPENCER and WHITFELD [1967 (1)] found that the product of RNA synthesis of *Spinacia* chloroplasts had a sedimentation constant of 11S at the peak and a spread from 11S to 23S. The characterization of the product of *in vitro* RNA synthesis in *Acetabularia* chloroplasts (BERGER, 1967) is discussed on p. 116.

An elegant use of radioautography to demonstrate the synthesis of RNA by plastids *in vivo* was provided by GIBBS (1967, 1968) using *Ochromonas danica*. Very short pulses of ^{3}H-uridine were shown to be incorporated into RNA at the site of the conspicuous DNA-containing region of the plastid. Since no significant label appeared in the cytoplasm during this time, the RNA could only have been synthesized inside the plastid.

Recently, the effects of rifamycins, a class of inhibitors of transcription initiation in bacterial systems (HARTMAN *et al.*, 1967; WEHRLI *et al.*, 1968) but not in eucaryotic systems (WEHRLI *et al.*, 1968) have been studied in plastids (SURZYCKI, 1969; SURZYCKI *et al.*, 1969; BOGORAD and WOODCOCK, 1970). *In vivo* application of rifamycins causes a striking decrease in plastid rRNA synthesis both in *Zea mays* (BOGORAD and

Woodcock, 1970) and *Chlamydomonas reinhardi* (Surzycki, 1969). *In vitro* experiments suggest that the RNA polymerase of *Chlamydomonas* chloroplasts and at least some RNA polymerase activity of maize plastids is sensitive to rifamycins while the polymerase(s) of the nucleus are insensitive. Taken together, the *in vivo* and *in vitro* experiments provide evidence that the plastid genome codes for plastid rRNA.

The differences in rifamycin sensitivity and cation requirements (Bogorad and Woodcock, 1970) between the plastid and nuclear RNA polymerase system suggest possible criteria for the separation and purification of these enzymes.

3. Affinities of Plastid rRNA

Since RNA is transcribed directly from a DNA template, it follows that the base sequences of the two will be complementary, and if a single-stranded DNA is allowed to react with its RNA complement, the two molecules will pair to form a stable "hybrid". Thus, the demonstration of hybridization between a species of RNA and a particular DNA provides evidence that the DNA has the potential for directing the synthesis of that RNA. Initially introduced by Hall and Spiegelman (1961), this method has been used extensively to test for affinity between DNAs and RNAs and has been particularly valuable in detecting and measuring the number of rRNA cistrons in bacteria and in eucaryotic plants and animals (see, for example, Goodman and Rich, 1962; Gillespie and Spiegelman, 1965; Ritossa and Spiegelman, 1965; Matsudo and Siegel, 1967; Shearer and McCarthy, 1967).

An obvious application would be to test the nuclear and plastid DNAs and rRNAs for hybrid formation to find out where complementary sequences of each of the rRNA species were stored. Despite the apparent simplicity of the system, the experiments themselves are demanding both in terms of the purity and quantity of the DNA and RNA reactants and in the number of factors which must be controlled, and it is perhaps for these reasons that only two reports of such experiments have appeared to date (Scott and Smillie, 1967; Tewari and Wildman, 1968).

The methods used included refinements developed by Gillespie and Spiegelman (1965) to increase the sensitivity of hybrid detection. Typically, purified DNA is separated into single strands which are then immobilized on a filter. Radioactively labelled RNA is then added and conditions suitable for base-pairing maintained for several hours. Un-paired regions of RNA are then removed with ribonuclease (in the hybrid, RNA is protected from this enzyme), and the amount of radioactivity remaining on the filter determined.

In the case of the quality of the reactants and reagents used, the criteria which, according to Gillespie and Spiegelman (1965), are essential for the correct interpretation of hybridization results included:

1. The DNA must be completely protein- and RNase-free, and be completely denatured.

2. There must be no trace of DNA (especially labelled DNA) contaminating the RNA.

3. The RNase used to destroy un-paired RNA must be DNase free, and theRNase digestion must be carried out under conditions which favor the stability of the hybrid.

In addition, the hybrid itself, if correctly paired should give a melting profile in which the hybrid breaks down over a small temperature range (Denis, 1966; Melli and BISHOP, 1969). However, even after taking all the precautions mentioned above, it is still not possible to exclude the possibility that only a portion of each RNA molecule is correctly paried in the hybrid. The only method available of testing hybrids for complete pairing of the whole RNA molecule is by direct observation, and measurement of the length of hybrid with the electron microscope (DAVIS and HYMAN, 1970). SCOTT and SMILLIE (1967) in their hybridization studies on *Euglena* followed the procedure of GILLESPIE and SPIEGELMAN (1965), and demonstrated that chloroplast rRNA produced a stable hybrid with chloroplast DNA. As the ratio of RNA/DNA was increased, a plateau in the amount of hybrid formed was obtained showing that the reactants were suitably pure and free from nucleases. SMILLIE and SCOTT (1967) estimated that 1% of the DNA was occupied by rRNA at saturation and from the amount of DNA per plastid in *Euglena* (see p. 100) calculated that each plastid contained 20 to 45 cistrons for rRNA. No hybridization was observed between cytoplasmic rRNA and plastid DNA.

TEWARI and WILDMAN (1968) in a similar study on *Nicotiana* found that a maximum of 1.65% of plastid DNA was occupied by plastid rRNA at saturation and calculated that each plastid contained about four sites of DNA complementary to each of the two ribosomal RNA species. Hybridization also occurred between cytoplasmic rRNA and nuclear DNA to a similar extent to that found for other procaryotic and eucaryotic cells. In cross hybridization experiments, chloroplast DNA did not react appreciably with cytoplasmic rRNA, but significant hybridization (0.1%) was obtained between nuclear DNA and chloroplast rRNA. There was no competition for these sites between chloroplast and cytoplasmic rRNA; thus, each was attaching to separate regions of the nuclear DNA. TEWARI and WILDMAN interpreted this result as indicating the presence of plastid rRNA cistrons within the nucleus and calculated that each nucleus contained about 1,000 times as many sites coding for plastid rRNA as a plastid. Hybrid melting profiles were not performed in the *Euglena* or *Nicotiana* experiments.

As previously mentioned, the detection of a stable hybrid does not necessarily mean that the whole RNA molecule has been correctly matched nor is it proven that a complementary site on the DNA was active in transcribing RNA. However, the evidence from DNA/RNA hybridization is strongly suggestive that plastid rRNA synthesis is directed by the plastid genome and indeed other types of evidence also point in this direction (see p. 106). Whether we are to view the hybridization of plastid rRNA with nuclear DNA found in *Nicotiana* as evidence for an additional site of plastid rRNA synthesis, as an unused relic of the past, or as a mismatching artifact of the hybridization technique, remains to be seen.

Recently, STUTZ and RAWSON (1970) succeeded in separating the two strands of *Euglena gracilis* plastid DNA on the basis of buoyant density and found that plastid rRNA hybridized principally to the heavier strand.

E. Information Translation

1. Components of the Protein Synthetic System

The presence of plastid ribosomes and rRNA components which are physically separable from those of the cytoplasm has already been discussed (see p. 103). In addition, it has been clearly shown that the ribosomal proteins from plastid ribosomes are different from those of the chloroplast (LYTTLETON, 1968). Furthermore, most of the proteins of *Phaseolus aureus* chloroplast ribosomes migrate differently from mitochondrial ribosomal proteins of the same species on acrylamide gel electrophoresis; thus, although plastid and mitochondrial ribosomes are both approximately 70S, most if not all of their proteins are the product of separate genes (VASCONCELOS and BOGORAD, 1970). *Phaseolus aureus* cytoplasmic (approx. 80S) ribosomes contain a still differently migrating complement of proteins.

Activating enzymes for some amino acids have been detected in isolated plastid preparations [CLARK, 1958; BOVÉ and RAACKE, 1959; HENSHALL and GOODWIN, 1964 (1) (2); FRANCKI *et al.*, 1965; SISSAKIAN *et al.*, 1965]. In spinach, differences between cytoplasmic and chloroplast fractions with respect to the amount of individual amino acids activated were noted.

Evidence for the presence of tRNA in plastids is fragmentary. However, RNA fractions with the ability to accept amino acids in the presence of amino acid activating enzymes and ATP have been reported (SISSAKIAN *et al.*, 1965; SPENCER and WILDMAN, 1964). In addition, BARNETT *et al.* (1969) have shown that new types of isoleucyl-, glutamyl-, and phenylanyl-tRNA are produced after dark-grown wild-type *Euglena gracilis* cells are illuminated. The further observation that a bleached mutant of this organism does not produce these tRNAs supports the contention that the synthesis of light-induced forms is related to plastid development. These experiments do not demonstrate that the new tRNAs are formed or are localized in the plastids but the probability seems high that these are plastid-specific tRNAs.

The demonstration that N-formyl-methionyl-tRNA is involved in initiation of protein synthesis by bacterial ribosomes has stimulated interest in the problem of peptide chain initiation in chloroplasts and mitochondria. Attempts to formylate mammalian methionyl-tRNA with a variety of mammalian extracts have been unsuccessful; however, N-formyl-methionyl-tRNA has been found in yeast and rat liver mitochondria (SMITH and MARCKER, 1968) and in HeLa cell mitochondria (GALPER and DARNELL, 1969). In the case of chloroplasts, SCHWARTZ *et al.* (1967) demonstrated the presence of N-formyl-methionine in protein synthesized by a cell-free preparation of *Euglena* chloroplasts incubated with f_2 phage RNA. More recently N-formyl-^{35}S-methionyl-tRNA was recovered from an RNase digest of an incubation mixture of an extract of sonicated chloroplasts of *Phaseolus vulgaris* (French bean), tRNAs extracted from plastids, ^{35}S-methionine, and formyl-tetrahydrofolate. The implication is that N-formyl-methionyl-tRNA was formed; this contention was strengthened by the observation that treatment of cytoplasmic tRNAs in a similar manner resulted in the production of only a very small amount of formyl-methionine which the authors believed could be contamination from the chloroplasts (BURKARD *et al.*, 1969). BIANCHETTI and BOGORAD (1970) have found that chloroplasts of *Zea mays* contain formyl-tetrahydrofolate synthetase and ^{10}N-formyl-tetrahydrofolate methionyl-tRNA transformylase. During incubation of these enzymes together with ^{3}H-formate, ^{35}S-methionine and maize leaf tRNAs, a chromatographically distinct tRNA is formed which contains tritium and ^{35}S and which when digested, yields formyl methionine. These observations have been extended by SARTIRANA and BIANCHETTI (1970) who have shown that formyl-^{35}S-methionine is obtained after spinach chloroplasts incubated with ^{35}S-methionine are treated with puromycin and the radioactive peptidyl puromycin derivatives obtained chromatographically are digested with pronase. Taken together these results indicate that chloroplasts contain formyl-methionyl-tRNA, the enzymes for charging this tRNA with methionine and with formate, and that protein synthesis by chloroplast ribosomes involves formyl-methionyl-tRNA. It seems probable from the experiments of SARTIRANA and BIANCHETTI that this tRNA is involved in peptide chain initiation in plastids but none of the data demonstrates that N-formyl-methionyl-tRNA is *required* for any or all protein synthesis by these organelles.

There is no physical criterion for distinguishing mRNA and reports of "template activity" of plastid RNA fractions in the presence of cell-free bacterial protein-synthesizing systems [BRAWERMAN and EISENSTADT, 1964 (2)] should be interpreted with caution. Although it is probable that the protein synthesis measured was due to the "reading" of plastid mRNA by the bacterial system, the possibility cannot be excluded that other low molecular weight RNAs of plastid origin, such as partially degraded rRNA or tRNA, were involved.

Further indirect evidence for the presence of mRNA in plastids has come from sedimentation studies on ribosomes. Ribosome extracts from whole leaves of *Brassica pekinesis* (CLARKE *et al.*, 1964) contained material with greater sedimentation constants than the "80S" cytoplasmic and "70S" plastid monomers. On brief RNase treatment of the ribosome preparation, these heavier components disappeared, while the 80S and 70S peaks increased. This effect was probably due to the presence of polyribosomes, held together by mRNA strands. More recently, STUTZ and NOLL (1967) and BOGORAD (1967) have obtained characteristic polysome profiles from chloroplast preparations which are converted to ribosome monomers by RNase.

2. Protein Synthesis *in vitro*

One of the major objectives of studying protein synthesis by chloroplasts *in vitro*, [besides wanting to know the properties of the system *per se* in order to learn more about protein synthesis in general] is to be able to compare the characteristics of this system with those of the protein synthetic apparatus in the cytoplasm and in other subcellular components. This knowledge would help lead to an understanding of control mechanisms and integrative processes within the cell. The best way to approach these problems would probably be to separate the components of the various protein synthetic systems, to study the properties of each of the components and to examine the operation of both reconstituted homologous systems and of combinations of components from different cellular sources to see if parts are interchangeable. As already noted, it has been difficult to dissect plastid systems alone and knowledge of protein synthesis by cytoplasmic components is not appreciably greater. However, chloroplasts prepared from several species incorporate radioactive amino acids into trichloracetic acid precipitable polypeptides and these systems have been studied in some detail.

Thus far, it has not been possible to prepare plastids which will incorporate amino acids at appreciable rates for more than 30 to 60 min; in most cases, linear incorporation lasts for only a few minutes. One of the possible sources of misleading data is amino acid incorporation by contaminating bacteria and preparations which are most active for the longest period of time should be the most suspect regarding contamination (App and Jagendorf, 1964). Criteria for non-bacterial synthesis have been taken to be dependence on ATP and GTP, sensitivity to RNase, and insensitivity to respiratory poisons such as KCN (Gnanam *et al.*, 1969). Maintenance of plastid integrity may be required for sustained high activity. Electron microscopic examination of the plastids during incorporation (Margulies *et al.*, 1968) showed that typical membrane-bound plastids are converted into stromaless aggregates of unstacked membranes. Ranaletti *et al.* (1969) reported that plastid ribosomes seem to be extruded during the incorporation of amino acids. Thus, chloroplasts seem to undergo a kind of self-destruction during *in vitro* protein synthesis despite the use of iso- or hyper-tonic incubation media.

Many aspects of chloroplast isolation have been reported to influence the performance of plastids. It has recently been suggested that media containing the (allegedly) protective colloids dextran and Ficoll, originally introduced in work on tobacco chloroplasts, are essential for other species (Gnanam *et al.*, 1969). However, the observed dependence on the Honda medium may, as the authors note, be related to the preservation of the outer membrane and it seems likely that sucrose-prepared chloroplasts with intact membranes would perform as efficiently. One of the other general problems is that chloroplasts lose soluble proteins during isolation and many factors for amino acid incorporation may thus be missing. Among the reports of amino acid incorporation by plastids from many higher plants are those of Bamji and Jagendorf (1966), Parenti and Margulies (1967), Margulies *et al.* (1968), Margulies and Parenti (1968), Ranalletti *et al.* (1969), and Spencer (1965); similar reports have dealt with plastids from *Euglena* (Eisenstadt and Brawerman, 1964) and *Acetabularia* (Goffeau, 1969).

Another goal of studying protein synthesis by chloroplasts *in vitro* is to learn which RNA messages are translated by the chloroplast's apparatus for protein synthesis—i.e. to try to learn which proteins are made on plastid ribosomes. Furthermore, if chloroplast protein and RNA synthesis could be coupled under conditions where both would go on extensively *in vitro*, one might learn which proteins are coded for by chloroplast DNA. Since it might be possible (at least in principle) to do some of these things without first separating and then recombining the various components of the chloroplast system for protein synthesis, some workers have tried to identify the products of brief periods of amino acid incorporation by isolated chloroplasts.

Margulies (1970) using *Phaseolus vulgaris* was able to detect radioactivity in a protein fraction containing ribulose 1-5 diphosphate carboxylase [an enzyme comprising up to 60% of the soluble protein of plastids (Ridley *et al.*, 1965)]. However, insufficient radioactivity was present to allow the further purification of the enzyme necessary to equate enzyme and label with confidence. The failure of trypsin and chymotrypsin to solubilize the labelled protein was a further inconclusive feature of these experiments. In a similar study, Ranalletti *et al.* (1969) isolated the calcium dependent ATPase enzyme from bean plastids which had incorporated radioactive amino acids *in vitro*. They were able to show a correlation of enzyme activity with radioactivity in a column eluate, but by this time the most active fraction gave only 30 cpm above background. Chen and Wildman (1970) found no evidence that *Nicotiana tabacum* plastids were synthesizing Fraction I protein or "structural" protein *in vitro*.

Thus, we are still a long way from defining the protein synthetic capabilities of plastids by *in vitro* methods. Clearly, what is necessary at this stage is the development of a method or medium for preparing high-performance chloroplasts. In so doing, valuable information would also be gained concerning the properties of the *in vivo* environment in which synthesis is not accompanied by self-destruction.

Thus, even attempts to learn which proteins are made by the chloroplast protein system using whole chloroplast preparations, rather than fractionated and reconstituted amino acid incorporating systems have not been very successful. The problem is clear: Plastids capable of doing *in vitro* what they can do *in vivo* need to be prepared. How to achieve this or how to measure how close one is to achieving it has not yet been established. It would almost seem necessary to incubate chloroplasts in a cytoplasmic medium incapable of doing synthesis itself so that any cofactors which are exchanged with the cytoplasm can still reach the chloroplast. The amount of work required is formidable and perhaps this is a very difficult way to determine which proteins are made by the chloroplast itself.

3. Identification of the Products of Plastid Protein Synthesis by Inhibitor Studies

Although a full discussion of the topic is outside the scope of this article (for more details, see SMILLIE and SCOTT, 1969), some mention must be made of the use of inhibitors to determine the site of synthesis of particular plastid enzymes. Chloramphenicol, which blocks protein synthesis in procaryotic systems by binding to the 70S type of ribosome (VASQUEZ, 1966) has been found also to inhibit protein synthesis by isolated plastids (EISENSTADT and BRAWERMAN, 1964; SPENCER and WILDMAN, 1964; GOFFEAU and BRACHET, 1965; SISSAKIAN *et al.*, 1965; SPENCER, 1965; BAMJI and JAGENDORF, 1966; MARGULIES, 1970). In contrast, synthesis of proteins on the 80S type of ribosome is relatively insensitive to chloramphenicol. A second inhibitor with inhibitory activity complementary to chloramphenicol, is cycloheximide (ENNIS and LUBIN, 1964; KIRK and ALLEN, 1965; MORRIS, 1967; COOPER *et al.*, 1967). Thus it might be possible by treating plants *in vivo* with these drugs and monitoring enzyme levels to determine whether 80S or 70S type ribosomes were responsible for their synthesis. Two conditions must be fulfilled if evidence obtained by such methods is to be acceptable; firstly, the specificity of the inhibitors on the two protein-synthesizing systems must be demonstrated, and secondly, the inhibitors must not appreciably affect either directly of indirectly any other part of the *in vivo* system. Although the first condition has been well documented, it is clearly almost impossible satisfactorily to fulfill the second. Surely, it is over optimistic to expect that a major long-term inhibition in one cell compartment will not produce side effects in other compartments?

SMILLIE *et al.* (1967) found that the synthesis of ribulose-1,5-diphosphate carboxylase and NADP-glyceraldehyde-3-phosphate dehydrogenase in *Euglena* were strongly inhibited by chloramphenicol while somewhat stimulated by cycloheximide. However, ribulose-1,5-diphosphate carboxylase was affected both by chloramphenicol and cycloheximide in *Chlamydomonas reinhardi* (SURZYCKI *et al.*, 1970). Of two fructose-1,6-diphosphate aldolases in *Euglena*, occurring in the cytoplasm and plastids respectively, the former was inhibited only by cycloheximide, while the plastid enzyme was affected only by chloramphenicol (SMILLIE and SCOTT, 1970). The effect of the inhibitors on the synthesis of cytochrome-552, of a b-type cytochrome, and ferredoxin NADP-reductase in *Euglena* was less conclusive. Although chloramphenicol inhibited all three (Smillie *et al.*, 1963, 1967), cycloheximide also had an inhibitory effect, though at a higher concentration than needed to inhibit some cytoplasmic enzymes (Kirk and Allen, 1965; Evans *et al.*, 1967). For ferredoxin-NADP-reductase, diametrically opposite results have been obtained from *Chlamydomonas reinhardi* (Surzycki *et al.*, 1970). Hoober *et al* (1969) showed that chloramphenicol had little effect on the greening of the *Chlamydomonas reinhardi* Y-1 mutant, but the recovery rates of Photosystems I and II were reduced by 50% and 35% respectively and the membranes rarely fused. On the other hand, cycloheximide had a marked inhibition (50% to 100%) on the rate of chlorophyll formation, but did not affect Photosystems I and II (measured on a per chlorophyll basis). It was concluded that maturation of the plastid required both cytoplasmic and plastid protein synthetic systems to be active. Bonotto *et al.* (1969) observed a marked lack of specificity both of chloramphenicol and cycloheximide on the synthesis of plastid and non-plastid enzymes and other soluble and insoluble proteins in *Acetabularia*.

These disparate results could be caused by species differences, by differences in the experimental approach used, in the (highly variable) concentrations of inhibitors required, or in lack of inhibitor specificity. Thus, although there is good evidence that chloramphenicol and cycloheximide act specifically *in vitro*, there is a strong possibility that their *in vivo* effects are multiple.

Another approach to the question of which proteins are synthesized on plastid ribosomes has come from work on the *ac*-20 mutant of *Chlamydomonas reinhardi*. When these cells are grown mixotrophically on acetate, plastid ribosomes are reduced in number to one tenth of the level found in the wild type (Togasaki and Levine, 1970; Levine and Paszeiwski, 1970; Goodenough and Levine, 1970). It was found that concurrent with this reduction in plastid ribosomes, the levels of ribulose-1,5-diphosphate carboxylase, cytochrome 559, and *Q* (the fluorescence quencher of Photosystem II) were also reduced. In addition, plastid membrane organization and pyrenoid formation were affected. However, many other proteins including NADP-glyceraldehyde-3-phosphate dehydrogenase, cytochrome 553, and ferredoxin-NADP reductase (which were inhibited by chloramphenicol in *Euglena*) were unaffected. Since some plastid ribosomes were present at all times, it can only be concluded that normal levels of ribosomes are necessary for the maintenance of certain components, but not of others. However, the use of such mutants clearly has many advantages over the inhibitor approach, and the isolation of further mutants with similar deficiencies would be valuable.

F. Information Processing in Acetabularia Plastids

A great deal of information concerning DNA→RNA→protein systems in plastids has come from work on *Acetabularia*, the giant, unicellular alga. Although *Acetabularia* provides unique opportunities for studying such problems, there are a number of difficulties involved which are also specific to this organism and which should be taken into account. It, therefore, seems appropriate to consider the subject of information processing in *Acetabularia* plastids as a whole rather than dividing it among the preceding sections.

During the vegetative phase of the life cycle of *Acetabularia* which lasts a few months, the single nucleus remains in the basal rhizoid, while the stalk, which may reach 3 to 5 cm in length, contains several million chloroplasts [SHEPHARD, 1965 (2)]. Thus by simply severing the stalk from the rhizoid, a supply of plastids free from nuclear contamination is obtained. Moreover, both portions produced by such a surgical operation continue to metabolize and show marked regeneration. Nucleate halves will regenerate a complete new plant, while enucleate portions are capable of increasing in length, and in some cases can differentiate a cap (umbrella) before metabolism ceases. The discovery that interspecific grafting and nuclear implantation were possible with *Acetabularia* cells led to much valuable work concerning the interactions between the nucleus and cytoplasm (for a review, see HAEMMERLING, 1963). These studies showed that the nucleus controlled (among other things) the form of the cap, and since caps can be made by enucleate fragments, the idea of a long-lived morphogenetic messenger substance was evolved.

The search for this hypothetical substance, which is still being pursued (BRACHET, 1970) required studies of the information processing capabilities of *Acetabularia* cells, and subcellular fractions, and it was in applying standard biochemical procedures to *Acetabularia* that the problems of working with this plant became apparent. Firstly, it is not possible to use kilogram quantities of material in order to extract and purify a particular enzyme or nucleic acid; at best, one is limited to a few grams of cells (200 plants 3 cm long, weight approximately 1 g), and in enucleation experiments where the plants must be treated individually, the limitations in starting material are even more severe. Consequently, for example, much of the earlier work on the characterization of rRNA had to rely entirely on the position of small traces of radioactive *Acetabularia* material mixed with *E. coli* "carrier". Advances in technique have now made it possible to obtain enough rRNA from *Acetabularia* to give measurable optical density readings on sucrose gradients, but this nevertheless often requires working up to the limits of resolution of the method.

A second major difficulty in working with *Acetabularia* is the presence of powerful nucleases, particularly ribonucleases (SCHWEIGER, 1966; DILLARD, 1970) which are difficult to inhibit. Indeed, the use of RNA extraction methods suitable for other organisms often meets with complete failure when applied to *Acetabularia*. Even when RNA is successfully extracted, it is usually more heterogenous in size than that obtained from other sources.

Thirdly, there is the problem of obtaining and maintaining axenic cultures of *Acetabularia*. It has been shown that even cells which are bacteria-free by serological standards may harbor significant numbers of microorganisms (GREEN *et al.*, 1967). In cases where contamination must be controlled, it is therefore necessary to monitor preparations with the electron microscope.

Another factor which must be taken into account in experiments on isolated *Acetabularia* chloroplasts, is related to the cytology of the cells. In the stalk, the cytoplasm forms a dissected network of thin strands in which chloroplasts, mitochondria and other inclusions are embedded. Thus it is common in electron micrographs to see chloroplasts surrounded by an extremely thin layer of cytoplasm with the chloroplast envelope and the tonoplast almost touching. Another common sight is a chloroplast closely appressed to one or more mitochondria, with the tonoplast enclosing the group. With this type of internal morphology, it is to be expected that mild homogenization of the cells results in the formation of many "cytoplasts" (consisting of a number of organelles and a small amount of cytoplasm surrounded by tonoplast membrane) as well as free organelles. It is also to be expected that cytoplasts would not only perform much better than free chloroplasts in *in vitro* studies, but be influenced by the portion of included cytoplasm. Since small cytoplasts are indistinguishable from chloroplasts with the light microscope, electron microscopic monitoring of preparations is again called for. BRACHET (1970) considered that differences in strains, culture media, and handling techniques also contributed to the difficulty of obtaining consistent results.

In spite of these problems, work on *Acetabularia* has yielded much information on the DNA→RNA→protein system in plastids, but in assessing these contributions the limitations imposed by the organism must be borne in mind.

1. Information Storage

The advantages of using isolated *Acetabularia* plastids to test for the presence of DNA were recognized at an early stage in the development of the subject (BALTUS and BRACHET, 1963; GIBOR and IZAWA, 1963). Both reports confirmed the presence of DNA in the chloroplasts, and GIBOR and IZAWA (1963) estimated on the basis of fluorimetric analyses that each chloroplast contained about 10^{-16} g of DNA. However, the density of this DNA is still uncertain. GREEN *et al.* (1967, 1970) reported a value of 1.714 g/cc from preparations which were monitored with the electron microscope while GIBOR (1967) obtained DNA with a density of 1.695 g/cc from chloroplasts of apparently sterile plants (electron microscopic controls were not performed).

Direct visualization of *Acetabularia* chloroplast DNA has indicated that at least some chloroplasts contain very large amounts of DNA, of the same order of magnitude as bacteria (WOODCOCK and BOGORAD, 1970; GREEN, BURTON, 1970; GREEN *et al.*, 1970), thus suggesting that the estimates of GIBOR and IZAWA (1963) were too low. WERZ and KELLNER (1968) reported loops of DNA about 30 μ long (10^{-16} g) in extracts from *Acetabularia* chloroplasts, but there was no evidence that each 30 μ loop originated from a single chloroplast. The vast amount of DNA in some chloroplasts, and evidence that the quantity per chloroplast is highly variable (WOODCOCK and BOGORAD, 1970) underline the need for studies on the base sequence redundancy of this DNA.

2. Information Replication

No definitive studies on DNA replication in *Acetabularia* chloroplasts have appeared. However, the incorporation of DNA precursors by chloroplasts *in vivo* [SHEPHARD, 1965 (1)] and net DNA synthesis into chloroplast fractions from enucleate portions (HEILPORN-POHL and BRACHET, 1966) have been observed.

3. Information Transcription

As in other plants (see p. 103) much of the work on the transcription of DNA to RNA has been concerned with the identification and localization of the various RNA species involved. This task has had an added impetus in *Acetabularia* since RNA molecules are attractive candidates for the long-lived morphogenetic messenger substances. Despite this attention, the current status of our knowledge of *Acetabularia* RNAs is unsatisfactory, and is a major hindrance to further work on the transcription process itself. Probably the major cause of this state of affairs is the difficulty encountered in preparing undegraded RNA (see for example, DILLARD, 1970). Either the intracellular RNases involved are extremely resistent to the usual inhibitors or, as seems more likely, the enzymes are intimately associated with ribosomes *in vivo*, and breaking the cells stimulates an autodestructive process. In support of this latter possibility, are observations that *E. coli* ribosomes and RNAs are not degraded when mixed with *Acetabularia* homogenates (BALTUS and QUERTIER, 1966; BALTUS *et al.*, 1968).

4. Ribosomal RNAs

There is considerable agreement to date that rRNAs extracted from *Acetabularia* have sedimentation constants of about 23S and about 16S, these figures being based on co-sedimentation with *E. coli* rRNAs [BALTUS and QUERTIER, 1966; SCHWEIGER *et al.*, 1967; BERGER, 1967; BALTUS *et al.*, 1968; FARBER, 1969 (1); JANKOWSKI *et al.*, 1969; DILLARD and SCHWEIGER, 1968, 1969). These observations of only two rRNA species (compared with four species in other eucaryotes—see p. 103), suggesting that *Acetabularia* differs from other plants and animals, clearly demand careful assessment. In all cases, the analytical method used was sucrose gradient sedimentation which has limited resolving power. It is possible to resolve 23S and 25S RNAs by this method (see for example, STUTZ and NOLL, 1967) but only if the RNAs concerned form the narrow bands characteristic of undegraded RNA. In the case of the majority of studies on *Acetabularia*, the RNA bands are rather broad, suggesting a polydisperse size distribution resulting from slight degradation. Is it possible then, that *Acetabularia* cells do contain the 25S and 18S rRNAs characteristic of eucaryotic cytoplasmic ribosomes, as well as the 23S and 16S rRNAs found in procaryotes, mitochondria, and plastids? DILLARD and SCHWEIGER (1969) found that the kinetics of labelling of the 23S and 16S RNAs were consistent with two-component system, and considered the possibility that this was due to the low resolution of the sucrose density gradient analytical method. Recently, both SCHWEIGER (1970) and ourselves (WOODCOCK and BOGORAD, 1970) have examined *Acetabularia* RNA extracts using the higher resolving power of polyacrylamide gel electrophoresis (LOENING, 1967). In these experiments, the major rRNA components extracted were assigned "25S" and "18S" values on the basis of co-electrophoresis with *E. coli* rRNA. SCHWEIGER (1970) also obtained rRNA bands which travelled at the rates expected of 23S and 16S species. Despite many precautions, the effects of intracellular RNases was marked, the *Acetabularia* bands being much broader than those of the *E. coli* rRNAs. The assignment of 25S and 18S values to the *major* species of RNA isolated from Aceta-bularia cells is consistent with the results from other green plants in which 25S and 18S RNA derived from 80S ribosomes predominate over 23S and 16S RNA, derived from plastid and mitochondrial ribosomes (INGLE, 1968; SURZYCKI, 1969).

The successful resolution of *Acetabularia* rRNAs should open the way for more meaningful studies on the *synthesis* of RNA in the various cell compartments.

Several other low molecular weight RNAs have been reported in *Acetabularia*, and although some have been considered to be tRNA, no real evidence has been put forward to support this. More recently, both Farber (1969) and Janowski and Bonotto (1970) noted a stable, slowly labelled RNA fraction sedimenting at 15S which is clearly a promising candidate for the "morphogenetic messenger".

5. Ribosomes

Even more difficulty has been encountered in extracting ribosomes from *Acetabularia* than in extracting high molecular weight RNA. Baltus *et al.* (1968) obtained radioactively-labelled particles which co-sedimented with *E. coli* 70S, 50S, and 30S ribosomal particles. Similar profiles could not be obtained from anucleate fragments, nor was sufficient material extracted in any of the experiments to give optical density readings of *Acetabularia* ribosomes. In a more detailed study, Janowski *et al.* (1969) confirmed the presence of labelled 70S, 50S and 30S particles in extracts of whole cells, chloroplasts, and non-chloroplast fractions but again, the sucrose gradients were mostly analyzed by radioactivity counts rather than optical density measurements. Sufficient ribosomal material was, however, obtained from purified chloroplasts, and here the absorbance profile of the gradient showed two major peaks, at 50S and 30S positions. The failure to isolate unlabelled cytoplasmic ribosomes is puzzling since in electron micrographs, they are most conspicuous. Again, partial degradation by nucleases appears to be the most likely cause.

As in the case of the rRNA, it seems too early to conclude that all the ribosomes of *Acetabularia* are of the 70S procaryotic type. Hopefully, better methods of protecting ribosomes during extraction will be devised, although most of the standard precautions have already been tried, and it is difficult to suggest what further measures would be taken. Perhaps mild fixation with glutaraldehyde or formaldehyde prior to disruption of the cells would be worth exploring.

6. RNA Synthesis in Isolated Plastids

Although many authors have demonstrated the *in vivo* incorporation of RNA precursors into plastid fractions, only one report of incorporation *in vitro* has appeared. In this study (Berger, 1967), isolated *Acetabularia* chloroplasts synthesized RNA for a number of hours (though a plot of incorporation against time was not included). The plants used were judged to be sterile by agar-plating the medium and by light microscopy. The incorporation of uracil was stimulated by light, but inhibited about 50% by actinomycin C and by DNase. Nucleotide triphosphates were also incorporated into TCA insoluble material, but contrary to other work on isolated plastids (see p. 105) the reaction did not require all four nucleotide triphosphates to be present. When analyzed on sucrose density gradients, the RNase sensitive radioactive products of uracil uptake were found to have sedimentation constants of 23S, 16S, 9S and 4S. These experiments are remarkable for three features: the long-lived activity of the preparations, the independence of all nucleotide triphosphates, and the successful separation of the product into four distinct RNA fractions. Berger

(1967) suggested that the careful preparative procedures kept the nucleotide triphosphate pools intact and this may be the case. However, it also seems possible that a portion of the chloroplast preparations existed as "cytoplasts" (see p. 113) which may well have far greater pool sizes, activity and longevity than free chloroplasts. It is to be hoped that future experiments will include electron microscopic monitoring of plastid fractions so that careful work of this sort will not be attended by similar doubts.

7. Information Translation

Of the many components of the protein synthetic system, only ribosomes and rRNA have received detailed study in *Acetabularia*. BALTUS *et al.* (1968) observed that amounts of monomer and subunit ribosomal particle increased following a brief RNase treatment, and suggested that this was due to the breakdown of polyribosomes. However, they were unable to obtain direct evidence for the presence of these ribosome-messenger RNA aggregates.

FARBER *et al.* (1968) were able to demonstrate protein synthesis when bulk RNA extracts from *Acetabularia* were added to a cell-free protein synthesizing system from *E. coli*. However, with a heterogenous RNA preparation it is impossible to tell whether this synthesis was really a translation of the *Acetabularia* messenger. The observation that the quantity of protein synthesized decreased with the age of the plant was thought to be due to the utilization of messenger RNA, but could equally be explained by other factors such as an increase in ribonuclease activity.

Studies on net protein synthesis in *Acetabularia* have mostly been concerned with a comparison between the nucleate and anucleate condition. However, GOFFEAU and BRACHET (1965) reported a DNA-dependent protein synthesis in isolated chloroplasts and recently, GOFFEAU (1969) has taken the experiments a stage further and begun the characterization of the labelled proteins. Both studies were performed with chloroplasts isolated from enucleated plants. In the earlier experiments, incorporation of amino acids into protein lasted about 1 h, and was inhibited by chloramphenicol, puromycin and actinomycin D. DNase and RNase had little effect, and an ATP generating system was not necessary. Although bacterial contamination was monitored, and considered to be negligible (streptomycin resistance was also observed), it is clear that these results do not comply with the criteria found by other workers (APP and JAGENDORF, 1964; GNANAM *et al.*, 1969, see p. 109—110) to be necessary to dismiss microorganism contamination. In particular, the insensitivity to RNase, even after a 1 h incorporation, and the lack of dependence on added ATP require explanation. It is possible that unlike higher plants, *Acetabularia* chloroplasts retain their outer membranes, and hence RNase insensitivity during protein synthesis. The drastic effect of actinomycin D was thought to indicate that the complete process DNA→RNA→protein must be occurring inside the chloroplasts, and presumably the lack of inhibition by DNase must be attributed to lack of penetration. However, in BERGER's (1967) experiments on RNA synthesis in isolated chloroplasts of *Acetabularia*, DNase used at a much lower concentration (30 μg/ml compared with 200 μg/ml) gave about 50% inhibition. In the light of these inconsistencies, it is very difficult to assess the significance of the protein synthesis observed.

In characterizing the protein made by isolated *Acetabularia* chloroplasts GOFFEAU (1969) recognized the need for electron microscopic controls of the chloroplast preparations (which were obtained by differential centrifugation). Most of the chloroplasts retained their outer membranes and there were some mitochondria and "islands of cytoplasm". Also apparent in the published micrograph are aggregates of plastids, mitochondria and cytoplasm, possibly bounded by plasma membrane—i.e., the "cytoplasts" discussed previously (see p. 113). No microorganisms were seen. Goffeau also noted the remarkably resistant nature of the chloroplast envelope which can survive mild hypotonic conditions and low detergent concentrations without rupturing. This may partially explain the apparent impermeability of *Acetabularia* chloroplasts to enzymes, and possibly the retention of nucleotide triphosphate pools. Analysis of the protein synthesized by these preparations showed that soluble, membrane-bound, and insoluble [the "structural" component of Criddle (1966)] types were made.

It seems clear that the vast potential in *Acetabularia* for attacking many of the problems of plastid function and intracellular ecology has not yet been fully exploited, and that this has been principally due to the various features of the plant which make it unsuitable for many of the established biochemical procedures. It is probable that in this section we have been too ready to point out the inconsistencies and inadequacies of experiments, and have not done justice to the painstaking, pioneering work which has been required to bring the field to its present state. However, if the potential of *Acetabularia* is to be realized, it is essential that the ground-work be rigorously established, and we hope that our criticisms are sufficiently constructive to help towards this end.

8. Conclusions

Although many features of the natural history of chloroplasts are obscure, the data discussed here leave no doubt that at least some plastids in every cell of the plant species studied so far contain DNA and several types of RNA.

The question of distinctions between chloroplast and nuclear DNAs has been explored extensively. The conclusions seem to be: (1) if a DNA is isolated from highly purified chloroplasts, it is plastid DNA, and (2) the plastid DNA may be almost indistinguishable from the bulk of the nuclear DNA, except for its more rapid renaturation kinetics and its lack of 5-methyl cytosine (these two features do not distinguish it from mitochondrial DNA). There are simply too few criteria for distinguishing between DNAs, and even fewer ways of knowing what certain kinds of differences might mean. Some approaches which are likely to be employed in the near future are (a) analyses of the proteins which may or may not be associated with plastid vs nuclear DNAs, (b) DNA-DNA hybridization experiments or their equivalents designed to explore the extent of similarity of DNAs from various part of a cell, and (c) the use of DNA and RNA polymerases and repressors to look into regions or codons with specific operational information.

The evidence for the presence, in chloroplasts, of ribosomes containing rRNAs different from those of the cytoplasm is unequivocal and in one species the chloroplast and mitochondrial ribosomes have also been shown to be distinctly different although they are the same size. Evidence for the presence of mRNA in chloroplasts is only indirect, while that for some unique types of tRNAs is marginal. However, reinforcing data will probably appear soon.

In addition, plastids contain systems capable of forming DNA, of using a DNA template for the synthesis of RNA and of using polyribonucleotides as templates for the incorporation of amino acids into polypeptides. Despite great progress, very little is known yet about the characteristics of most of these synthetic systems. Knowledge of the fidelity of information processing and the specificity of the transcoding systems should help establish the role of these macromolecule-synthesizing complexes in plastid metabolism *in vivo*. Eventually, it should be possible to resolve questions such as: Does the DNA polymerase system of the plastid replicate only plastid DNA or would it copy with equal fidelity DNA from some other part of the cell, which might be transported, somehow, into the plastid? Is there any DNA replication restriction in the plastid and, if so, what is the mechanism? How specific is the chloroplast RNA polymerase (or polymerases) and is only chloroplast DNA recognized and copied faithfully? If plastid RNA polymerase encountered some foreign DNA in a plastid would a faithful copy be made? Are special recognition sites required on the DNA for an RNA polymerase to initiate transcoding? Again, how specific is the protein synthetic machinery? Does it recognize only certain initiations codons? Can it use only tRNAs from a chloroplast or is its specificity mixed? Are there some mRNAs made in the nucleus with initiation codons which permit reading by the chloroplast polypeptide synthesizing system?

Many of the questions posed will probably be at least partially resolved in a few years and may bring us to an understanding of how eucaryotic cells are integrated. Judging from the study of protein synthesis by bacterial ribosomes *in vitro*, is it possible that conditions such as ion concentration which prevail within the plastid may impose certain specificity limitations on enzymes of DNA, RNA, and protein synthesis, and that these limiting factors may not be recognized from experiments *in vitro* for some time.

It would be pleasant if we could write now that "chloroplasts possess systems which replicate chloroplast DNA, transcribe the information of this DNA into RNA, and translate information in the plastid mRNA into specific proteins which remain in the chloroplast or are exported to the mitochondria or cytoplasm. Instead, as already mentioned, we are limited to the statement that "plastids contain systems capable of forming DNA, of using a DNA template for the synthesis of RNA and of using polyribonucleotides as templates for the incorporation of amino acids into polypeptides". At the present time, we have virtually no conclusive evidence which compels us to believe that the amino acid sequence of any single protein is dictated by information stored in the chloroplast or that such information was transmitted according to the idealized sequence stated early in the previous sentence. Conversely, there is little compelling evidence that any chloroplast protein originates from information stored in the nucleus and is made on cytoplasmic or on chloroplast ribosomes.

The techniques developed by BRITTEN and KOHNE (1968) and used by WELLS and BIRNSTIEL (1969) and STUTZ (1970) give a measure of gene redundancy in a genome; data obtained by these methods should help in establishing an upper limit for the number of genes in plastid DNA. It is likely that a number of reports of this sort will be forthcoming. It will be interesting to see how widely these values will differ from species to species, and to try to relate the number of possible gene products to the number of proteins known to be localized in plastids.

The next problem will be to determine which specific proteins are formed according to directions in the plastid genome—some investigations of this sort have already been undertaken and have been discussed here. Much work on plastids is pursued with the idea that general principles of division of activities among cellular components are sure to emerge. That is, if a given protein is made in a plastid from plastid information in one species this will also be the case in all other species. While it would be orderly and pleasing for this to be the case, there is no basis for this view. The unit of evolution with regard to distribution of intracellular functions has been the organism and the cell. It seems reasonable that some information needed somewhere in the cell could reside equally well in (e.g.) the nucleus-cytoplasm or in the mitochondria or in the chloroplasts and thus, different species might divide these functions differently. A single plastid-localized enzyme might be coded for by nuclear DNA in one species, and by plastid DNA in a second (and mitochondrial DNA in a third). Recognition of these differences will probably come only after we understand how to discover with relative ease the sites of information storage, and the sites of synthesis of compounds which occur in the chloroplasts and other subcellular elements.

The potential value of *Acetabularia* as an experimental organism for studying chloroplast-cytoplasmic-nuclear relationships has been recognized for a long time but its present limitations are just becoming apparent as new kinds of biochemical investigations are being attempted. There is no reason to doubt that the technical problems, once revealed, will be resolved. Very many profitable experiments can easily be visualized.

The question of the origin of plastids (and mitochondria) continues to be popular. After more than 80 years there are still two proposals: (1) Plastids were originally entirely autonomous elements and arose as symbionts from photosynthetic bacteria or blue-green algae or, (2) mitochondria and chloroplasts (as well as nuclei) arose by segregation of parts of a primitive procaryote and its genome. In many respects—such as the possession of smaller ribosomes, the sensitivity of protein and RNA synthesizing systems to inhibitors which affect some bacteria and blue-green algae but do not inhibit nuclear-cytoplasmic systems, and the absence of massive histone-protein coverings on their DNA—chloroplasts (and mitochondria) resemble procaryotes more than they do the nuclear-cytoplasmic complex of eucaryotes. This has encouraged supporters of the "symbiont hypothesis" in their views. This may well be the way things went long ago but it seems as difficult to reach a final positive decision as to do the experiment to show that blue-green algae and bacteria arose as escapees from an intracellular community. Regardless of prejudices about how organelles originated, most students of plastids and mitochondria believe that it would be useful to succeed at this experiment—i.e. to get plastids and mitochondria to escape into culture. We agree.

G. Literature

App, A. A., Jagendorf, A. T.: C^{14} amino acid incorporation by spinach chloroplast preparations. Plant Physiol. **39**, 772—779 (1964).

Baltus, E., Brachet, J.: Presence of deoxyribonucleic acid in the chloroplasts of *Acetabularia mediterranea*. Biochim. biophys. Acta (Amst.) **76**, 490—492 (1963).

— EDSRTÖM, J. E., JANOWSKI, M., HANOCQ-QUERTIER, J., TENCER, R., BRACHET, J.: Base composition and metabolism of various RNA fractions in *Acetabularia mediterranea.* Proc. nat. Acad. Sci. (Wash.) **59**, 406—413 (1968).

— QUERTIER, J.: A method for the extraction and characterization of RNA from subcellular fractions of *Acetabularia.* Biochim. biophys. Acta (Amst.) **119**, 192—194 (1966).

BAMJI, M. S., JAGENDORF, A. T.: Amino acid incorporation by wheat chloroplasts. Plant Physiol. **41**, 764—770 (1966).

BARD, S. A., GORDON, M. P.: Studies on spinach chloroplast and nuclear DNA using large-scale tissue preparations. Plant Physiol. **44**, 377—384 (1969).

BARNETT, W. E., PENNINGTON, C. J., JR., FAIRFIELDS, S. A.: Induction of *Euglena* transfer RNA's by light. Proc. nat. Acad. Sci. (Wash.) **63**, 1261—1268 (1969).

BAXTER, R., KIRK, J. T. O.: Base composition of DNA from chloroplasts and nuclei of *Phaseolus vulgaris.* Nature (Lond.) **222**, 272—273 (1969).

BELL, P. R.: Interaction of nucleus and cytoplasm during oogenesis in *Pteridium aquilinum.* Proc. roy. Soc. **B 153**, 412—432 (1961).

— MÜHLETHALER, K.: Evidence for the presence of deoxyribonucleic acid in the organelles of the egg cells of *Pteridium aquilinum.* J. molec. Biol. **8**, 853—862 (1964).

BERGER, S.: RNA synthesis in *Acetabularia.* II. RNA synthesis in isolated chloroplasts. Protoplasma (Wien) **64**, 13—25 (1967).

BERIDZE, T. G., ODINTSOVA, M. S., SISSAKYAN, N. M.: Distribution of DNA components of bean leaves among cellular fractions. J. molec. Biol. **1**, 122—133 (1967).

BIANCHETTI, R., BOGORAD, L.: (1970) (unpublished observations).

BIGGINS, J., PARK, R. B.: Nucleic acid content of chloroplasts of spinach isolated by a non-aqueous technique. Nature (Lond.) **203**, 425—426 (1964).

BISALPUTRA, T., BISALPUTRA, A. A.: The occurrence of DNA fibrils in chloroplasts of *Laurencia spectabilis.* J. Ultrastruct. Res. **17**, 14—22 (1967).

— — The ultrastructure of chloroplasts of a brown alga *Sphacelaria* sp. I. Plastid DNA configuration—the chloroplast genophore. J. Ultrastruct. Res. **29**, 151—170 (1969).

BOGORAD, L.: Aspects of chloroplast assembly. Organizational biosynthesis, p. 395—418 (H. J. VOGEL, J. O. LAMPEN, V. BRYSON, Eds.). London: Academic Press 1967.

— Control mechanisms in developmental processes. 26th Symposium Society of Developmental Biology. Developmental Biology Suppl. 1, p. 1—31. London: Academic Press 1968.

— WOODCOCK, C. L. F.: Rifamycins: variable effects on chlorophyll formation and the inhibition of plastid RNA. Symposium on autonomy and biogenesis of mitochondria and chloroplasts (N. K. BOARDMAN, A. W. LINNANE, R. M. SMILLIE, Eds.). North Holland Publishing Company 1970 (in press).

BONOTTO, S., GOFFEAU, A., JANOWSKI, M. VAN DEN DRIESSCHE, T., BRACHET, J.: Effects of various inhibitors of protein synthesis on *Acetabularia mediterranea.* Biochim. biophys. Acta (Amst.) **174**, 704—712 (1969).

BOVÉ, J. M., BOVÉ, C., RONDOT, M. J., MOREL, G.: Chloroplasts and virus RNA synthesis. Biochemistry of chloroplasts, Vol. II, p. 329 (T. W. GOODWIN, Ed.). London: Academic Press 1967.

— RAACKE, I. D.: Amino acid activating enzymes in isolated chloroplasts from spinach leaves. Arch. Biochem. **85**, 521—531 (1959).

BRACHET, J.: Concluding remarks. The Biology of *Acetabularia*, p. 273 (J. BRACHET, S. BONOTTO, Eds.). London: Academic Press 1970.

BRAWERMAN, G., EISENSTADT, J. M.: (1) Deoxyribonucleic acid from the chloroplasts of *Euglena gracilis.* Biochim. biophys. Acta (Amst.) **91**, 477—485 (1964).

— — (2) Template and ribosomal ribonucleic acids associated with the chloroplasts and cytoplasm of *Euglena gracilis.* J. molec. Biol. **10**, 403—411 (1964).

BRITTEN, R. J., KOHNE, D. E.: Repeated sequences in DNA. Science **161**, 529—540 (1968).

BURKARD, G., ECHLANCHAR, B., WEIL, J. H.: Presence of N-formyl-methionyl-tRNA in bean chloroplasts. FEBS Letters **4**, 571—574 (1967).

CAIRNS, J.: The bacterial chromosome and its manner of replication as seen by autoradiography. J. molec. Biol. **6**, 208—213 (1963).

CHEN, J. L., WILDMAN, S. G.: Free and membrane-bound ribosomes and nature of the products formed by isolated tobacco chloroplasts incubated for protein synthesis. Biochim. biophys. Acta (Amst.) **209**, 207—217 (1970).

CHIBA, Y.: Cytochemical studies on chloroplasts. I. Cytologic demonstration of nucleic acids in chloroplasts. Cytologia (Tokyo) **16**, 259—264 (1951).

— SUGAHARA, K.: The nucleic acid content of chloroplasts isolated from spinach and tobacco leaves. Arch. Biochem. **71**, 367—376 (1957).

CHIANG, K. S., SUEOKA, N.: Replication of chloroplast DNA in *Chlamydomonas reinhardi* during vegetative cell cycle: its mode and regulation. Proc. nat. Acad. Sci. (Wash.) **57**, 1506—1513 (1967).

CHUN, E. H. L., VAUGHAN, M. H., RICH, A.: The isolation and characterization of DNA associated with chloroplast preparations. J. molec. Biol. **7**, 130—141 (1963).

CLARK, J. M., JR.: Amino acid activation in plant tissues. J. biol. Chem. **233**, 421—424 (1958).

CLARK, M. F., MATTHEWS, R. E. F., RALPH, R. K.: Ribosomes and polyribosomes in *Brassica pekinensis*. Biochim. biophys. Acta (Amst.) **91**, 289—304 (1964).

COOPER, D., BANTHORPE, D. V., WILKIE, D.: Modified ribosomes conferring resistance to cycloheximide in mutants of *Saccharomyces cerevisiae*. J. molec. Biol. **26**, 347—350 (1967).

COOPER, W. D., LORING, H. S.: The ribonucleic acid composition and phosphorus distribution of chloroplasts from normal and diseased turkish tobacco plants. J. biol. Chem. **228**, 813—822 (1966).

CRIDDLE, R. S.: Protein and lipo-protein organization in the chloroplast. Biochemistry of chloroplasts, Vol. I, p. 203 (T. W. GOODWIN, Ed.). London: Academic Press 1966.

DAVIS, R. W., HYMAN, R. W.: Cold Spr. Harb. Symp. quant. Biol. 1970 (in press).

DENIS, H.: Gene expression in amphibian development. I. Validity of the method used: interspecific and intraspecific hybridization between nucleic acids. Properties of messenger RNA synthesized by developing embryos. J. molec. Biol. **22**, 269—283 (1966).

DILLARD, W. L.: RNA synthesis in *Acetabularia*. The biology of *Acetabularia*, p. 17 (J. BRACHET, S. BONOTTO, Eds.). London: Academic Press 1970.

— SCHWEIGER, H. G.: Kinetics of RNA synthesis in *Acetabularia*. Biochim. biophys. Acta (Amst.) **169**, 561—563 (1968).

— — RNA synthesis in *Acetabularia*. III. Kinetics of RNA synthesis in nucleate and enucleate cells. Protoplasma (Wien) **67**, 87—100 (1969).

EDELMAN, M., COWAN, C. A., EPSTEIN, H. T., SCHIFF, J. A.: Studies of chloroplast development in *Euglena*. VIII. Chloroplast-associated DNA. Proc. nat. Acad. Sci. (Wash.) **52**, 1214—1219 (1964).

EISENSTADT, J. M., BRAWERMAN, G.: The protein synthesizing systems from the cytoplasm and chloroplasts of *Euglena gracilis*. J. molec. Biol. **10**, 392—402 (1964).

ENNIS, H. L., LUBIN, M.: Cycloheximide: aspects of inhibition of protein synthesis in mammalian cells. Science **146**, 1474—1476 (1964).

EVANS, W. R., WALENGA, R., JOHNSON, C.: Effect of cycloheximide on chloroplast development in *Euglena*. Fourth Annual Report, Charles F. Kettering Research Laboratories, p. 99. Ampersand Press 1967.

FARBER, F. E.: (1) Studies on RNA metabolism in *Acetabularia mediterranea*. I. The isolation of RNA and labelling studies on RNA of whole plants and plant fragments. Biochim. biophys. Acta (Amst.) **174**, 1—11 (1969).

— (2) Studies on RNA metabolism in *Acetabularia mediterranea*. II. The localization and stability of RNA species: effects on RNA metabolism of dark periods and actinomycin D. Biochim. biophys. Acta (Amst.) **174**, 12—17 (1969).

— CAPE, M., DELROY, M., BRACHET, J.: The *in vitro* translocation of *Acetabularia mediterranea* RNA. Proc. nat. Acad. Sci. (Wash.) **61**, 843—846 (1968).

FLAUMENHAFT, E., CONRAD, S. M., KATZ, J. J.: Nucleic acids in some deuterated green algae. Science **132**, 892—894 (1960).

FRANCKI, R. I. B., BOARDMAN, N. K., WILDMAN, S. G.: Protein synthesis by cell-free extracts from tobacco leaves. I. Amino acid incorporating activity of chloroplasts in relation to their structure. Biochemistry **4**, 865—872 (1965).

GALPER, J. P., DARNELL, J. E.: The presence of N-formylmethionyl-tRNA in HeLa cell mitochondria. Biochem. biophys. Res. Commun. **34**, 205—214 (1969).
GIBBS, S. P.: Synthesis of chloroplast RNA at the site of chloroplast DNA. Biochem. biophys. Res. Commun. **28**, 653—657 (1967).
— Autoradiographic evidence of the *in situ* synthesis of chloroplast and mitochondria RNA. J. Cell Sci. **3**, 327—340 (1968).
GIBOR, A.: DNA synthesis in chloroplasts. Biochemistry of chloroplasts, Vol. II, p. 321 (T. W. GOODWIN, Ed.). London: Academic Press 1967.
— GRANICK, S.: Plastids and mitochondria: inheritable systems. Science **145**, 890—897 (1964).
— IZAWA, M.: The DNA content of the chloroplasts of *Acetabularia*. Proc. nat. Acad. Sci. (Wash.) **50**, 1164—1169 (1963).
GILLESPIE, D., SPIEGELMAN, S.: A quantitative assay for DNA-RNA hybrids with DNA immobilized on a membrane. J. molec. Biol. **12**, 829—842 (1965).
GIVNER, J., MORIBER, L. G.: An attempt to demonstrate DNA in the chloroplastids of *Euglena gracilis* using fluorescence microscopy. J. Cell Biol. **23**, 36A (1964).
GNANAM, A., JAGENDORF, A. T., RANALLETTI, M. L.: Chloroplasts and bacterial amino acid incorporation: a further comment. Biochim. biophys. Acta (Amst.) **186**, 205—213 (1969).
GOFFEAU, A.: Incorporation of amino acids into the soluble and membrane-bound proteins of chloroplasts. Biochim. biophys. Acta (Amst.) **174**, 340—350 (1969).
— BRACHET, J.: DNA-dependent incorporation of amino acids into the proteins of chloroplasts isolated from anucleate *Acetabularia* fragments. Biochim. biophys. Acta (Amst.) **95**, 302—313 (1965).
GOODENOUGH, U. W., LEVINE, R. P.: Chloroplast structure and function in *ac*-20 a mutant strain of *Chlamydomonas reinhardi*. III. Chloroplast ribosomes and membrane organization. J. Cell Biol. **44**, 547—562 (1970).
GOODMAN, H. M., RICH, A.: Formation of a DNA-soluble RNA hybrid and its relation to the origin, evolution, and degeneracy of soluble RNA. Proc. nat. Acad. Sci. (Wash.) **48**, 2101—2109 (1962).
GRANICK, S., GIBOR, A.: The DNA of chloroplasts, mitochondria and centrioles. Progr. Nucleic Acid Res. **6**, 143—186 (1967).
GREEN, B. R., BURTON, H.: *Acetabularia* chloroplast DNA. Electron microscopic visualization. Science **168**, 981—982 (1970).
— BURTON, H., HEILPORN, V., LIMBOSCH, S.: The cytoplasmic DNAs of *Acetabularia mediterranea:* their structure, and biological properties. In: The biology of *Acetabularia*, p. 35 (J. BRACHET, S. BONNOTTO, Eds.). London and New York: Academic Press 1970.
— GORDON, M. P.: The satellite DNAs of some higher plants. Biochim. biophys. Acta (Amst.) **145**, 378—390 (1967).
— HEILPORN, V., LIMBOSCH, S., BOULOUKHERE, M., BRACHET, J.: The cytoplasmic DNAs of *Acetabularia mediterranea*. Proc. nat. Acad. Sci. (Wash.) **58**, 1351—1358 (1967).
GUNNING, B. E. S.: The fine structure of chloroplast stroma following aldehyde-osmium tetroxide fixation. J. Cell Biol. **24**, 79—93 (1965).
HAEMMERLING, J.: Nucleo-cytoplasmic interactions in *Acetabularia* and other cells. Ann. Rev. Plant Physiol. **14**, 65—89 (1963).
HALL, B. D., SPIEGELMAN, S.: Sequence complementarity of T_2 DNA and T_2 specific RNA. Proc. nat. Acad. Sci. (Wash.) **47**, 137—146 (1961).
HARTMAN, G., HONIKEL, K. O., KNÜSEL, F., NÜESCH, J.: The specific inhibition of the DNA-directed RNA synthesis by rifamycin. Biochim. biophys. Acta (Amst.) **145**, 843—844 (1967).
HEILPORN-POHL, V., BRACHET, J.: Net DNA synthesis in anucleate fragments of *Acetabularia mediterranea*. Biochim. biophys. Acta (Amst.) **119**, 429—431 (1966).
HENSHALL, J. D., GOODWIN, T. W.: Amino-acid activating enzymes in germinating pea seedlings. Phytochemistry **3**, 677—691 (1964).
HERRMANN, R. G., KOWALLIK, K. V.: Selective presentation of DNA regions and membranes in chloroplasts and mitochondria. J. Cell Biol. **45**, 198—202 (1970).
HOLDEN, M.: The fractionation and enzymic breakdown of some phosphorus compounds in leaf tissue. Biochem. J. **51**, 433—442 (1952).

Hoober, J. K., Siekewitz, P., Palade, G. E.: Formation of chloroplast membranes in *Chlamydomonas reinhardi* Y-1. J. biol. Chem. **244**, 2621—2631 (1969).

Ingle, J.: The effect of light and inhibitors on chloroplast and cytoplasmic RNA synthesis. Plant Physiol. **43**, 1850—1854 (1968).

Iwamura, T., Kuwashima, S.: Two DNA species in chloroplasts of *Chlorella*. Biochim. biophys. Acta (Amst.) **174**, 330—339 (1969).

Jacobson, A. B., Swift, H., Bogorad, L.: Cytochemical studies concerning the occurrence and distribution of RNA in plastids of *Zea mays*. J. Cell Biol. **17**, 557—570 (1963).

Jagendorf, A. T., Wildman, S. G.: The proteins of green leaves. VI. Centrifugal fractionation of tobacco leaf homogenates and some properties of isolated chloroplasts. Plant Physiol. **29**, 270—279 (1954).

Janowski, M., Bonotto, S.: A stable RNA species in *Acetabularia mediterranea*. The biology of *Acetabularia* (J. Brachet, S. Bonotto, Eds.). London: Academic Press 1970.

— — Bolоukhere, M.: Ribosomes of *Acetabularia mediterranea*. Biochim. biophys. Acta (Amst.) **174**, 525—535 (1969).

Kern, H.: Über das Vorkommen von Nucleinsäuren in isolierten Chloroplasten. Protoplasma (Wien) **50**, 505—543 (1959).

Kirk, J. T. O.: The deoxyribonucleic acid of broad bean chloroplasts. Biochim. biophys. Acta (Amst.) **76**, 417—424 (1963).

— DNA-dependent RNA synthesis in chloroplast preparations. Biochem. biophys. Res. Commun. **14**, 393—397 (1964).

— Will the real chloroplast DNA please stand up? Symp. on autonomy and biogenesis of chloroplasts and mitochondria (N. K. Boardman, A. W. Linnane, R. M. Smillie, Eds.). North Holland Publishing Co. 1970 (in press).

— Allen, R. L.: Dependence of chloroplast pigment synthesis on protein-synthesis: effect of actidione. Biochem. biophys. Res. Commun. **21**, 523—530 (1965).

— Tilney-Bassett, R. A. E.: Chapter X, 302. The plastids. W. H. Freeman and Co. 1967.

Kislev, N., Swift, H., Bogorad, L.: DNA from chloroplasts and mitochondria of swiss chard. J. Cell Biol. **25**, 327—333 (1965).

Kleinschmidt, A. K., Lang, D., Jacherts, D., Zahn, R. K.: Darstellung und Längenmessungen des gesamten Desoxyribonucleinsäureinhaltes von T_2-Phagen. Biochim. biophys. Acta (Amst.) **61**, 857—864 (1962).

Kung, S. D., Williams, J. P.: Chloroplast DNA from broad bean. Biochim. biophys. Acta (Amst.) **195**, 434—445 (1969).

Lang, D., Bujard, H., Wolff, B., Russell, D.: Electron microscopy of size and shape of viral DNA in solutions of different ionic strengths. J. molec. Biol. **23**, 163—181 (1967).

Leff, J., Mandel, M., Epstein, H. T., Schiff, J. A.: DNA satellites from cells of green and aplastidic algae. Biochem. biophys. Res. Commun. **13**, 126—130 (1963).

Levine, R. P., Paszewski, A.: Chloroplast structure and function in ac-20 a mutant strain of *Chlamydomonas reinhardi*. II. Photosynthesic electron transport. J. Cell Biol. **44**, 540—546 (1970).

Lima-da-Faria, A., Moses, M. J.: Labelling of *Zea mays* chloroplasts with H^3-thymidine. Hereditas (Lund) **52**, 367—378 (1965).

Littau, V. C.: A cytochemical study of the chloroplasts of some higher plants. Amer. J. Bot. **45**, 45—53 (1958).

Loening, U. E.: The fraction of high molecular weight ribonucleic acid by polyacrylamide gel electrophoresis. Biochem. J. **102**, 251—257 (1967).

— Molecular weights of ribosomal RNA in relation to evolution. J. molec. Biol. **38**, 355—365 (1969).

— Ingle, J.: Diversity of RNA components in green plant tissue. Nature (Lond.) **215**, 363—367 (1967).

Lyttleton, J. W.: Isolation of ribosomes from spinach chloroplasts. Exp. Cell Res. **26**, 312—317 (1962).

— Protein constituents of plant ribosomes. Biochim. biophys. Acta (Amst.) **154**, 145—149 (1968).

Margulies, M. M.: *In vitro* protein synthesis by plastids of *Phaseolus vulgaris*. V. Incorporation of ^{14}C leucine into a protein fraction containing ribulose-1,5-diphosphate carboxylase. Plant Physiol. **46,** 136—141 (1970).

— Parenti, F.: *In vitro* protein synthesis by plastids of *Phaseolus vulgaris*. III. Formation of lamellar and soluble chloroplast proteins. Plant Physiol. **43**, 504—514 (1968).

— Gantt, E., Parenti, F.: *In vitro* protein synthesis by plastids of *Phaseolus vulgaris*. II. The possible relation between RNase intensive amino-acid incorporation and the presence of intact chloroplasts. Plant Physiol. **43**, 495—503 (1968).

Matsuda, K., Siegel, A.: Hybridization of plant ribosomal RNA to DNA: the isolation of a DNA component rich in ribosomal RNA cistons. Proc. nat. Acad. Sci. (Wash.) **58**, 673—680 (1967).

McClendon, J. H.: The intracellular localization of enzymes in tobacco leaves. I. Identification of components of the homogenate. Amer. J. Bot. **39**, 275—282 (1952).

McLeisch, J., Sunderland, N.: Measurements of deoxyribonucleic acid (DNA) in higher-plants by feulgen photometry and chemical methods. Exp. Cell Res. **24**, 527—540 (1961).

Melli, M., Bishop, J. O.: Hybridization between rat liver DNA and complementary RNA. J. molec. Biol. **40**, 117—136 (1969).

Meselson, M., Stahl, F. W.: The replication of DNA in *E. coli*. Proc. nat. Acad. Sci. (Wash.) **44**, 671—683 (1958).

Metzner, H.: Cytochemische Untersuchungen über das Vorkommen von Nucleinsäuren in Chloroplasten. Biol. Zbl. **71**, 257—272 (1952).

Morris, I.: The effect of cycloheximide (actidione) on protein and nucleic acid synthesis by *Chlorella*. J. exp. Botany **18**, 54—64 (1967).

Nass, N. M. K.: The circularity of mitochondrial DNA. Proc. nat. Acad. Sci. (Wash.) **56**, 1215—1222 (1966).

— Mitochondrial DNA: advances, problems and goals. Science **165**, 25—35 (1969).

Orth, G. M., Cornwall, D. G.: The isolation and composition of chloroplasts and etiolated plastids from corn seedlings. Biochim. biophys. Acta (Amst.) **71**, 734—736 (1963).

Pavulans, J.: DNA content of several angiosperm female nuclei. Protoplasma (Wien) **34**, 22—29 (1940).

Pollard, C. J.: The deoxyribonucleic acid content of purified spinach chloroplasts. Arch. Biochem. **105**, 114—119 (1964).

Puiseux-Dao, S., Gibello, D., Hoursiango-Neubrun, D.: Techniques du mise en évidence du DNA dans les plastes. Compt. rend. **D 265**, 406—414 (1967).

Randall, J., Disbrey, C.: Evidence for the presence of DNA at basal body sites in *Tetrahymena pyriformis*. Proc. roy. Soc. **B 162**, 473—491 (1965).

Ranalletti, M. L., Gnanam, A., Jagendorf, A. T.: Amino acid incorporation by isolated chloroplasts. Biochim. biophys. Acta (Amst.) **186**, 192—204 (1969).

Ray, D. S., Hanawalt, P. C.: Properties of the satellite DNA associated with chloroplasts in *Euglena gracilis*. J. molec. Biol. **9**, 812—824 (1964).

Richards, O. C.: Hybridization of *Euglena gracilis* chloroplast and nuclear DNA. Proc. nat. Acad. Sci. (Wash.) **57**, 156—163 (1967).

Ridley, S. M., Thornber, J. P., Bailey, J. L.: A study of spinach beet chloroplasts with particular reference to Fraction I protein. Biochim. biophys. Acta (Amst.) **140**, 62—79 (1967).

Ris, H., Plaut, W.: The ultrastructure of DNA-containing areas in the chloroplast of *Chlamydomonas*. J. Cell Biol. **13**, 383—391 (1962).

Ritossa, F. M., Spiegelman, S.: Localization of DNA complementary to ribosomal RNA in the nucleolus organizer region of *Drosphilia melanogaster*. Proc. nat. Acad. Sci. (Wash.) **53**, 737—745 (1965).

Ruppel, H. G.: Über Nucleinsäuren in Chloroplasten von *Allium porrum* und *Antirrhinum majus*. Biochim. biophys. Acta (Amst.) **80**, 63—72 (1964).

— Nucleic acids in chloroplasts. I. Characterization of the DNA and RNA from *Antirrhinum majus*. Z. Naturforsch. **22 b**, 1068—1076 (1967).

— van Wyck, D.: Über die Desoxyribonucleinsäure in Chloroplasten von *Antirrhinum majus*. Z. Pflanzenphysiol. **53**, 32—38 (1965).

SAGAN, L., BEN-SHAUL, Y., SCHIFF, J. A., EPSTEIN, H. T.: Radiographic localization of DNA in the chloroplasts of *Euglena*. J. Cell Biol. **23**, 81a—82a (1964).

SAGER, R., ISHIDA, M. R.: Chloroplast DNA in *Chlamydomonas*. Proc. nat. Acad. Sci. (Wash.) **50**, 725—730 (1963).

SARTIRANA, M. L., BIANCHETTI, R.: N-Formylmethionyl-tRNA and its involvement in polypeptide chain initiation in plants. Symposio Sintesi Paroteice Controllo, Palanza, Italy 1969 (in press).

SCHILDKRAUT, C. L., MARMUR, J., DOTY, P.: Determination of the base composition of DNA from its buoyant density in CsCl. J. molec. Biol. **4**, 430—443 (1962).

SCHWARTZ, J. H., MEYER, R., EISENSTADT, J. M., BRAWERMAN, G.: Involvement of N-formylmethionine in initiation of protein synthesis in cell-free extracts of *Euglena gracilis*. J. molec. Biol. **25**, 571—574 (1967).

SCHWEIGER, H. G.: Ribonuclease-Aktivität in *Acetabularia*. Planta **68**, 247—255 (1966).

— DILLARD, W. L., GIBOR, A., BERGER, S.: RNA synthesis in *Acetabularia*. I. RNA synthesis in enucleated cells. Protoplasma (Wien) **64**, 1—12 (1967).

SCOTT, N. S., SHAH, V. C., SMILLIE, R. M.: Synthesis of chloroplast DNA in isolated chloroplasts. J. Cell Biol. **38**, 151—157 (1968).

— SMILLIE, R. M.: Evidence for the direction of chloroplast ribosomal RNA synthesis by chloroplast DNA. Biochem. biophys. Res. Commun. **28**, 598—603 (1967).

SEMAL, J., SPENCER, D., KIM, Y. T., WILDMAN, S. G.: Properties of a ribonucleic acid synthesizing system in cell-free extracts of tobacco leaves. Biochim. biophys. Acta (Amst.) **91**, 205—216 (1964).

SHEARER, R. J., MCCARTHY, B. J.: Evidence for ribonucleic acid molecules restricted to the cell nucleus. Biochem. J. **6**, 283—289 (1967).

SHEPHARD, D. C.: (1) An autoradiographic comparison of the effects of enucleation and actinomycin D on the incorporation of nucleic acid and protein precursors by *Acetabularia* chloroplasts. Biochim. biophys. Acta (Amst.) **108**, 635—643 (1965).

— (2) Chloroplast multiplication and growth in the unicellular alga *Acetabularia mediterranea*. Exp. Cell Res. **37**, 93—110 (1965).

SHIPP, W. S., KIERAS, F. J., HASELKORN, R.: DNA associated with tobacco chloroplasts. Proc. nat. Acad. Sci. (Wash.) **54**, 207—213 (1965).

SISSAKIAN, N. M., FILIPPOVITCH, I. I., SVETAILLO, E. N., ALEVY, K. A.: On the protein-synthesizing system of chloroplasts. Biochim. biophys. Acta (Amst.) **95**, 474—485 (1965).

SMILLIE, R. M., EVANS, W. R., LYMAN, H.: Metabolic events during the formation of a photosynthetic from a nonphotosynthetic cell. Brookhaven Symposia in Biol. **16**, 89—108 (1963).

— GRAHAM, D., DWYER, M. R., GRIEVE, A., TOBIN, N. F.: Evidence for the synthesis *in vivo* of proteins of the Calvin cycle and of the photosynthetic electron transfer pathway on chloroplast ribosomes. Biochem. biophys. Res. Commun. **28**, 604—610 (1967).

— SCOTT, N. S.: Organelle biogenesis: The chloroplast. Progress in subcellular and molecular biology, Vol. I, p. 136—202 (F. E. HAHN, F.E. SPRINGER, T. T. PUCK, K. WALLENFELS, Eds.). Berlin-Heidelberg-New York: Springer 1969.

SMITH, A. E., MARCKER, K. A.: N-formylmethionyl-tRNA in mitochondria from yeast and rat liver. J. molec. Biol. **38**, 241—243 (1968).

SPIEKERMAN, R.: Cytochemische Untersuchungen zum Nachweis von Nucleinsäuren in Protoplastiden. Protoplasma (Wien) **48**, 303—324 (1957).

SPENCER, D.: Protein synthesis by isolated spinach chloroplasts. Arch. Biochem. **111**, 381—390 (1965).

— WILDMAN, S. G.: The incorporation of amino acids into protein by cell-free extracts from tobacco leaves. Biochemistry **3**, 954—966 (1964).

— WHITFELD, P. R.: (1) Ribonucleic acid synthesizing activity of spinach chloroplasts and nuclei. Arch. Biochem. **212**, 336—345 (1967).

— — (2) DNA synthesis in isolated chloroplasts. Biochem. biophys. Res. Commun. **28**, 538—542 (1967).

STEFFENSEN, D. M., SHERIDAN, W. F.: Incorporation of ^{3}H thymidine into chloroplast DNA of marine algae. J. Cell Biol. **25**, 619—626 (1965).

STOCKING, C. R., GIFFORD, E. M.: Incorporation of thymidine into chloroplasts of *Spirogyra*. Biochem. biophys. Res. Commun. **1**, 159—164 (1959).

STRAUS, W.: Properties of isolated carrot chloroplasts. Exp. Cell Res. **6**, 392—402 (1964).

STUTZ, E.: The kinetic complexity of *Euglena gracilis* chloroplasts DNA. FEBS Letters **8**, 25—28 (1970).

— NOLL, H.: Characterization of cytoplasmic and chloroplast polysomes in plants: Evidence for 3 classes of ribosomal RNA in nature. Proc. nat. Acad. Sci. (Wash.) **57**, 774—781 (1967).

— RAWSON, J. R.: Separation and characterization of *Euglena gracilis* chloroplast single-strand DNA. Biochim. biophys. Acta (Amst.) **209**, 16—23 (1970).

SURZYCKI, S. J.: Genetic function of the chloroplast of *Chlamydomonas reinhardi*. Effect of rifampin on chloroplast DNA-dependent RNA polymerase. Proc. nat. Acad. Sci. (Wash.) **63**, 1327—1334 (1969).

— GOODENOUGH, U. W., LEVINE, R. P., ARMSTRONG, J. J.: Nuclear and chloroplast control of chloroplast structure and function in *Chlamydomonas reinhardi*. Symposia Soc. exp. Biol. **24**, 13—35 (1970).

SUYAMA, Y., BONNER, W. D., JR.: DNA from plant mitochondria. Plant Physiol. **41**, 383—388 (1966).

SHAH, V. C., LYMAN, H.: DNA-dependent RNA synthesis in chloroplasts of *Euglena gracilis*. J. Cell Biol. **29**, 174—176 (1966).

TEWARI, K. K., WILDMAN, S. G.: Chloroplast DNA from tobacco leaves. Science **153**, 1269—1271 (1966).

— VÖTSCH, W., MAHLER, H. R., MACKLER, B.: Biochemical correlates of respiratory deficiency. VI. Mitochondrial DNA. J. molec. Biol. **20**, 453—481 (1966).

— WILDMAN, S. G.: DNA polymerase in isolated tobacco chloroplasts and nature of the polymerised product. Proc. nat. Acad Sci. (Wash.) **58**, 689—696 (1967).

— — Function of chloroplast DNA. I. Hybridization studies involving nuclear and chloroplast DNA with RNA from cytoplasmic (80S) and chloroplast (70S) ribosomes. Proc. nat. Acad. Sci. (Wash.) **59**, 569—576 (1968).

— — Function of chloroplast DNA. II. Studies on DNA-dependent RNA polymerase activity of tobacco chloroplasts. Biochim. biophys. Acta (Amst.) **186**, 358—372 (1969).

TOGASAKI, R. K., LEVINE, R. P.: Chloroplast structure and function in *ac*-20 a mutant strain of *Chlamydomonas reinhardi*. I. CO_2 fixation and ribulose-1,5-diphosphate carboxylase synthesis. J. Cell Biol. **44**, 531—539 (1970).

UPPHADAYHA, K. C., GRUN, P.: Structure and form of chloroplast DNA. J. Cell Biol. **39**, 138a (1968).

VASCONCELOS, A. C. L., BOGORAD, L.: Proteins of cytoplasmic, chloroplast, and mitochondrial ribosomes of some plants. Biochim. biophys. Acta (Amst.) **228**, 492—502 (1970).

VASQUEZ, D.: Mode of action of chloramphenicol and related antibiotics. Symp. Soc. Gen. Microbiol. **16**, 169—171 (1966).

VOSA, C., KIRK, J. T. O.: (1963) (unpublished).

WEHRLI, W., NÜESCH, J., KNÜSEL, F., STAEHLIN, M.: Action of rifamycins on RNA polymerase. Biochim. biophys. Acta (Amst.) **157**, 215—217 (1968).

WELLS, R., BIRNSTEIL, M.: A rapidly renaturing DNA component associated with chloroplast preparations. Biochem. J. **105**, 53p (1967).

— — Kinetic complicity of chloroplastal deoxyribonucleic acid and mitochondrial nucleic acid from higher plants. Biochem. J. **112**, 777—786 (1969).

WERZ, G., KELLNER, G.: (1) Molecular characteristics of chloroplast DNA of *Acetabularia* cells. J. Ultrastruc. Res. **24**, 109—115 (1968).

— — (2) Isolierung und elektronenmikroskopische Charakterisierung von Desoxyribonucleinsäure aus Chloroplasten kernloser *Acetabularia*. Z. Naturforsch. **23 b**, 1018 (1968).

WHITFELD, P. R., SPENCER, D.: Buoyant density of tobacco and spinach chloroplast DNA. Biochem. biophys. Acta (Amst.) **157**, 333—343 (1968).

WILKINS, M. H. F.: Molecular configuration of nucleic acids. Science **140**, 941—950 (1963).

WOLLGIEHN, R., MOTHES, K.: Über die Inkorporation von ^{3}H-Thymidin in die Chloroplasten-DNS von *Nicotiana rustica*. Exp. Cell Res. **35**, 52—57 (1964).

WOLSTENHOLME, D. R., GROSS, N. J.: The form and size of mitochondria DNA of the red bean *Phaseolus vulgaris*. Proc. nat. Acad. Sci. (Wash.) **61**, 245—252 (1968).

Woodcock, C. L. F.: (1970) (unpublished observations).
— Bogorad, L.: Evidence for variation in the quantity of DNA among plastids of *Acetabularia.* J. Cell Biol. **44**, 361—375 (1969).
— — On the isolation and characterization of ribosomal RNA from *Acetabularia.* (1970) Biochim. biophys. Acta (Amst.) **224**, 639—643 (1970).
— Fernandez-Moran, H.: Electron microscopy of DNA conformations in spinach chloroplasts. J. molec. Biol. **31**, 627—631 (1968).
Yoshida, Y.: Nuclear control of chloroplast activity in *Elodea* leaf cells. Protoplasma (Wien) **54**, 476—492 (1962).

Lipids of Chloroplasts

ANDREW A. BENSON

Chlorophyll is a hydrophobic substance. In order that it be oriented in the lamellae for optimum efficiency of photon capture and transfer of excitation energy to adjacent molecules, chlorophyll must associate closely with the organized structure of its environment. With neither covalent nor ionic bonds its position is somehow firmly fixed in the lamellar molecular architecture. Although chlorophyll is soluble in many organic solvents, these frequently fail to extract it from the chloroplasts. Its association with lamellar protein presents a problem basic to the understanding of membrane function.

Chloroplast lamellae, like many functional membrane systems in Nature, are composed of high density lipoprotein — half lipid and half protein (LICHTENTHALER and PARK, 1963). The ratio in higher plants is very nearly one-to-one while some bacterial membranes may be only 20% lipid. The lipids include the chlorophylls, carotenoids, sterols, quinones, and, in major amounts, the phospholipids and glycolipids. The last, having both hydrophobic long chain fatty esters and hydrophilic sugar or ionic groups, seek locations at the boundaries between hydrophilic and hydrophobic regions in the chloroplast. They resemble the detergents but differ in that they always have two fatty hydrocarbon chains. When isolated by solvent extraction or when dislocated from their positions in membranes they exhibit a strong tendency to aggregate like other surfactant molecules. These micelles often present a lamellar appearance resembling myelin in electron micrographs of chloroplasts perturbed by solvents, heat, or detergents.

Although chloroplast lamellae may resemble myelin in dimensions and appearance in the electron microscope there is no resemblance at the molecular level. The lipids of the lamellae are intimately associated with specific protein molecules while those of myelin and "myelin figure" artifacts, seen in plant cells, are strictly the result of the tendency for association of the surfactant lipids; the longer and the more saturated the hydrocarbon chains the better they aggregate in myelin structures. The fatty chains in myelin of mammalian brain have saturated C_{18}, C_{24}, up to C_{36} chains. The hydrocarbon chains of lipids in chloroplasts are usually polyunsaturated C_{16} or C_{18} fatty esters.

Lipids occur in isolated microglobules as well as intimately associated with lamellar protein. The globules, occurring in chloroplasts of both higher plants and algae, apparently do not participate directly in photosynthetic function; they will be discussed in a subsequent section.

Polyisoprenoids are important lipid components of the chloroplast; the phytol of the chlorophylls is a C_{20} alcohol with one double bond and comprises one third the total weight of chlorophyll (Fig. 1). Phytol's four methyl side chains result from condensation of isopentenylpyrophosphate by enzyme systems in the chloroplast stroma. Similar condensations produce the C_{40}-C_{55} polyisoprene side chains of plastoquinone and other quinones (see GOODWIN, p. 256). The pigments and surfactant lipids are 65—70% hydrocarbon by weight. Hence, on a dry weight basis, chloroplast lamellae are one third hydrocarbon. Arrangement of this hydrophobic material in a system surrounded by water remains a perplexing problem of membrane structure.

That most of the lipids of chloroplasts or other cell membrane systems are amphipathic or surfactant suggests that they may form micelles. These geometrical aggregates, characteristic of soaps and natural surfactants, have been considered as a basis for membrane s ructure. The lipid bilayer or "unit membrane" model for cell membranes is a stable structure but fails to meet the requirements of selectivity,

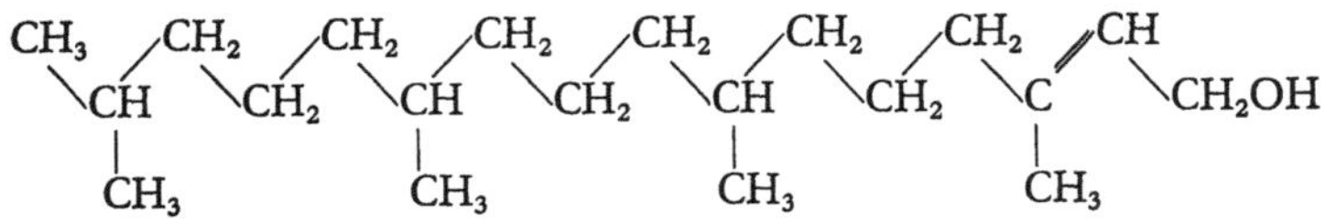

Fig. 1. Phytol, $C_{20}H_{40}O$

adaptability, and metabolic control characteristic of functional membranes. Whether a fraction of the lipid of the chloroplasts is "stored" in lamellar or spherical micelles to be drawn upon as needed by adjacent lipoprotein must be considered.

A. Phospholipids of Chloroplasts

The best known membrane lipids of the surfactant type are the phospholipids (Fig. 2). In chloroplasts these are predominantly glycerol phosphatides. Phosphatidyl choline (lecithin) and phosphatidyl ethanolamine (cephalin), the major phosphatides of animal tissues, occur widely in plant membrane systems. Mitochondria of cauliflower, avocado, and sweet potato contain major amounts of phosphatidyl choline as do mammalian mitochondria. The chloroplast outer membrane has not yet been investigated in this respect and may resemble plant mitochondria (BIALE, YANG and BENSON, 1966; DOUCE *et al.*, 1968) in lipid composition. Mitochondria and ion-transporting membranes, in general, appear to utilize choline lipids. Chloroplast lamellae, however, contain very little of these zwitterionic phosphatides. The vesiculated periphery of bundle sheath chloroplasts from C_4-pathway plants (see p. 181) such as sugar cane and *Amaranthus* (LAETSCH, 1969) exhibits structure similar to that of certain mitochondria. The relation of chloroplast outer membrane to mitochondrial structure remains to be established. Since mitochondria invariably contain much phosphatidyl choline (lecithin) while chloroplast lamellae have almost none, the content of this lipid in an isolated cell fraction can be indicative of its organelle source. The cinematographic observations of WILDMAN, HONGLADAROM, and HONDA (1962) and the electron microscopic studies of THOMSON (1969) in which chloroplast outer membranes appear to form and to fuse with mitochondria may yet find analytical and enzymatic verification.

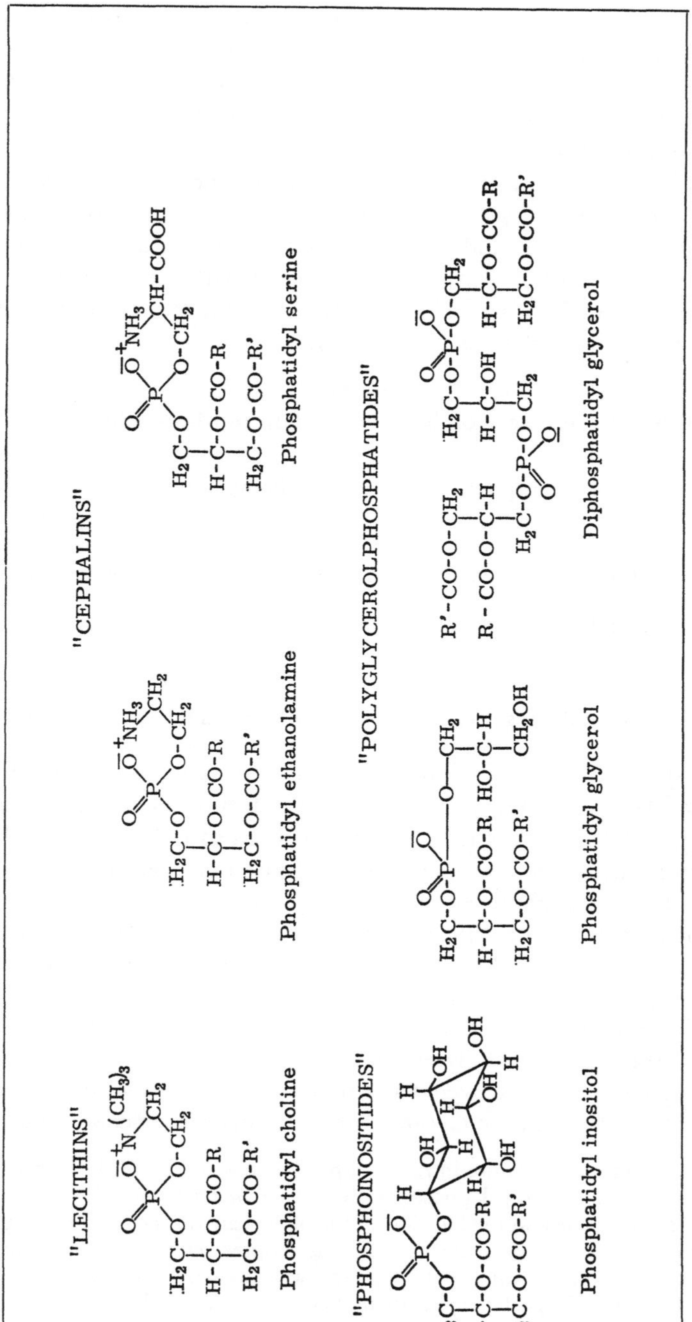

Fig. 2. The glycerolphosphatides found in plants

Phospholipid composition of lamellar lipoprotein is unique in that phosphatidyl glycerol, a derivative of glycerol phosphate, is the predominant phosphorus-containing lipid. In *Anacystis nidulans* it makes up over 90% of the phospholipid (ALLEN, HIRAYAMA, and GOOD 1966). The bacteria possess comparable or even greater concentrations of phosphatidyl glycerol.

Phosphatidyl glycerol is a metabolic product of the chloroplast (BENSON, 1958). Its formation requires ATP, cytidine triphosphate (CTP) D-glycerol-1-phosphoric acid[1], fatty acids, coenzyme A and ACP (acyl carrier protein, a functional protein analog of coenzyme A). Its biosynthesis is integrated with three functions of the chloroplast, ATP production, fatty acid biosynthesis, and carbohydrate biosynthesis. These may be schematically described as follows:

$$\text{Dihydroxyacetone phosphate} \xrightleftharpoons{\text{DPNH}} \text{D-Glycerol-1-phosphate} \quad (1)$$

$$\text{D-Glycerol-1-phosphate} \xrightarrow{\text{2 Acyl—ACP}} \text{Phosphatidic acid (2,3-diacyl-D-glycerol-1-phosphate)} \quad (2)$$

$$\text{Phosphatidic acid + CTP} \rightleftharpoons \text{Cytidine diphosphodiglyceride + Pyrophosphate} \quad (3)$$

$$\text{CDP-diglyceride + D-Glycerol-1-phosphate} \rightarrow \text{Phosphatidyl-L-glycerol-3-phosphate + CMP} \quad (4)$$

$$\text{Phosphatidyl glycerol phosphate} \xrightarrow{\text{Phosphatase}} \text{Phosphatidyl glycerol + Pi} \quad (5)$$

Enzymatic synthesis of CDP-diglyceride was observed in mitochondria and microsomes by BAHL *et al.* (1970); no evidence for formation of its precursor in chloroplast preparations could be demonstrated. It remains to be determined which of the above intermediates are actually formed in chloroplasts, which are transferred, and how such a transfer might be effected (see p. 228).

In this sequence it is seen that the configuration of the natural glycerophosphate appears to be altered from "D-" to "L-". Asymmetry of glycerophosphate is the result of both the mechanism of triose phosphate reduction (step 1) and of the specificity of glycerol kinase, an enzyme which selects only one of the two primary —OH groups of the symmetrical glycerol molecule (see footnote).

Phosphatidylglycerol was discovered as a result of identification of its deacylated derivative, diglycerophosphate, in *Chlorella* where it can occur in concentrations ten

[1] Conventional lipid nomenclature introduced by E. BAER and H. O. L. FISCHER refers to glycerophosphate as L-α-glycerol phosphate by virtue of its potential oxidizability to L-glyceraldehyde-3-phosphate and the location of the phosphate on the end or "α" position of glycerol. Carbohydrate nomenclature used in this chapter refers to assignment of the glycerol with the —OH group on the right in the well-known Emil Fischer projection as "D"-and placement of the important substituent, phosphate, in the "1-position" at the top.

$$\begin{array}{c} H_2COPO_3H_2 \\ | \\ H—C—OH \\ | \\ H_2COH \end{array} \quad \text{"Natural" D-glycerol-1-phosphoric acid}$$

Contemporary asymmetric atom terminology has been adopted by which the above compound is termed *sn*-glycerol-3-phosphoric acid.

times as great as those of the phosphorylated intermediates of sugar synthesis. It is not yet known whether it is actually a component of the chloroplast, where phosphatidylglycerol is concentrated or whether it is a cytoplasmic phosphate ester.

The relationship between phosphatidylglycerol and photosynthetic activity suggests that this lipid represents a pool of "reduced" triose phosphate of considerable concentration (0.006 M in *Chlorella*). Its effect on ^{14}C-labeling patterns of hexoses in $^{14}CO_2$ fixation is of concern (BENSON, 1958; PRASAD and BENSON, 1969).

Chloroplasts of *Tetragonia*, spinach, and lettuce have been analyzed for phospholipid. The lamellar lipoprotein of chloroplasts contains phosphatidylglycerol as the major phospholipid. Very little diphosphatidylglycerol, resulting from further phosphatidylation of phosphatidylglycerol has been observed in chloroplasts. In

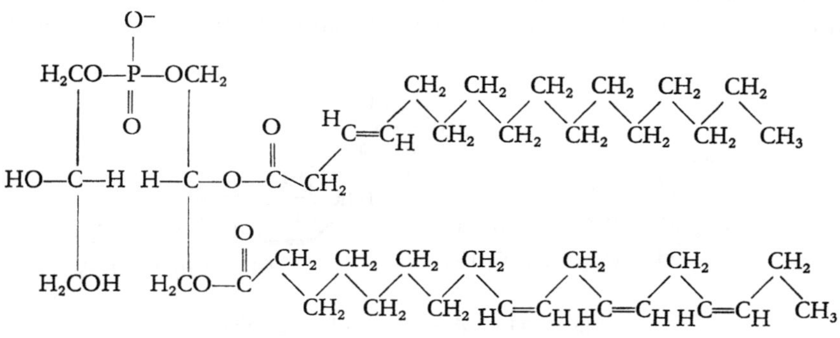

Fig. 3. The major phosphatidylglycerol of spinach chloroplasts. (After F. HAVERKATE and L. L. M. VAN DEENEN, 1965)

mammalian mitochondria and in bacterial membranes diphosphatidyl glycerol, cardiolipin, is a characteristic lipid component. Its formation may be mediated by phospholipase D whose phosphatidyl-enzyme intermediate has been found capable of phosphatidylating phosphatidyl glycerol to form diphosphatidyl glycerol (cardiolipin) (STANACEV and STUHNE-SEKALEC, 1970). It is the transphosphatidylase action of this enzyme which may lead to formation of phosphatidyl ethanol, etc. during alcohol extractions (YANG, FREER, and BENSON, 1967). The di-anionic lipid, cardiolipin, is an important component of membranes. Its two negative charges enhance the cation binding potential at its sites in the membrane.

The fatty acid composition of chloroplast phosphatidylglycerol is unique. HAVERKATE and VAN DEENEN (1965) separated the several phosphatidylglycerols of spinach leaves by thin layer chromatography on silver nitrate-impregnated silica gel. Half (47%) of the phosphatidylglycerol was found in one component, 2-*trans*-Δ^3-hexadecenoyl-3-linolenoyl-D-glycerol-1-phosphoryl-L-glycerol (Fig. 3). This is the only lipid in the green leaf where *trans*-Δ^3-hexadecenoic esters have been recognized. The nature of chloroplast acyl lipids and their fatty acids is reviewed by NICHOLS and JAMES (1968).

B. Glycolipids of Chloroplasts

The glycolipids comprise four-fifths of the lipids of spinach lamellar lipoprotein (Wintermans, 1960). Of these lipids 90% are monogalactosyl diglycerides and digalactosyl diglycerides while sulfolipid, in amounts varying with sulfate nutrition of the plant, makes up the rest. The structures of these three lipids and their predominant fatty acyl components are shown in Fig. 4.

It is seen that two of these are neutral amphipathic lipids with highly water-soluble carbohydrates bound to diglycerides. The galactolipids also occur in mammalian brain and are β-linked as are the cerebrosides. The galactosyl diglycerides result from galactosylation of appropriate diglycerides by a reaction shown to occur with UDP-galactose (UDP-Gal) in chloroplasts (Neufeld and Hall, 1964). The α-galactosyl linkage in UDP-Gal appears to produce the β-linked product by a simple displacement reaction. The subsequent α-galactosylation at C-6 has been shown (Mudd, 1967) to proceed at a different site and by an independent mechanism. Incorporation of

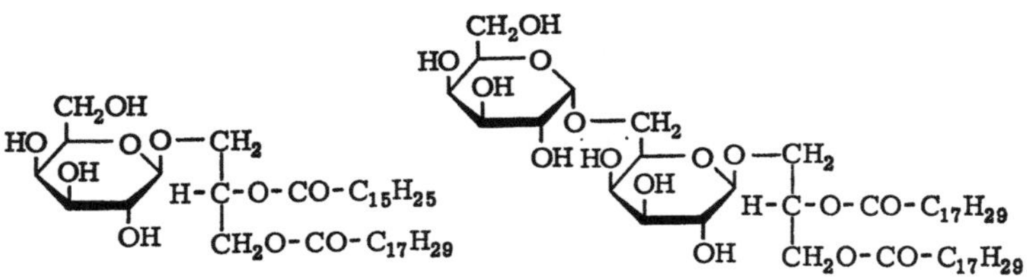

Fig. 4. The galactolipids of green plants

glucose-^{14}C by chloroplasts yields predominantly digalactolipid (Ongun and Mudd, 1968) while $^{14}CO_2$ fixation by *Chlorella* (Ferrari and Benson, 1961) produced predominantly monogalactolipid during 15 sec of photosynthesis:

$$\text{Diglyceride} \xrightarrow{\text{UDP-Gal}} \text{Monogalactosyl diglyceride} \xrightarrow{\text{UDP-Gal}} \text{Digalactosyl diglyceride.} \quad (6)$$

The reactions of (6), therefore proceed at different sites in the chloroplast and are surprisingly independent (see p. 228).

Independence of mono- and digalactolipid syntheses may be presumed from the striking difference in their fatty ester content. While both are among the most unsaturated lipids in Nature, only the monogalactolipid contains the shorter chain fatty ester, as much as 25% of hexadecatrienoic (16:3) acid. *Euglena gracilis* galactolipids (Rosenberg, Gouaux and Milch, 1966) differ even more strikingly and include hexadecatetraenoic (16:4) acid in only the monogalactolipid. The several species of polyunsaturared galactolipids of spinach have been separated by Nichols and Moorhouse (1969) using TLC on silver nitrate-impregnated silicic acid.

C. α-Linolenic Esters in Lipids of Oxygen-Producing Chloroplasts

The most unique aspect of the galactolipids of chloroplasts is their remarkably high α-linolenate (18:3) content (Table 1). As much as 95% of α-linolenate has been reported in spinach digalactosyldiglyceride (Fig. 5). Polyunsaturated lipids are difficult to prevent from oxidizing during isolation. It is surprising, indeed, that they accumulate and remain stable in the chloroplast where one finds high oxygen diffusion pressure, free radicals, light, and sensitizing pigment molecules. One would expect these unsaturated lipids to catch fire or at least to polymerize to a hard resin. On the

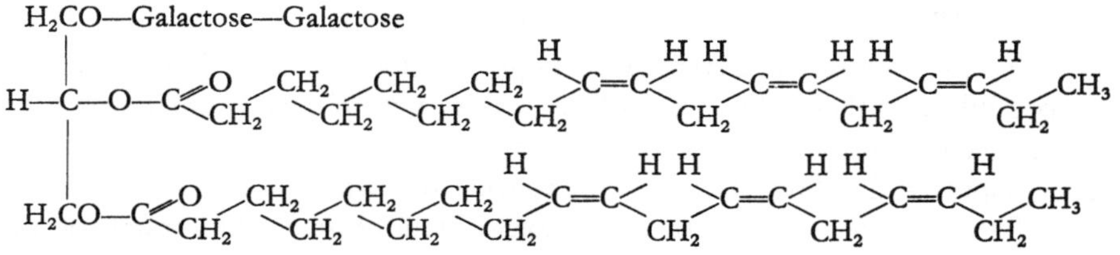

Fig. 5. Digalactosyl dilinolenin, a major chloroplast lipid

contrary, chloroplast lipids appear to be quite stable and the turnover of fatty acid components is slower than that of their carbohydrate moieties.

A direct relationship between α-linolenate content and oxygen-producing capability was recognized by ERWIN and BLOCH (1963). Except in the rare case of *Anacystis nidulans* where polyunsatured acids are not formed, linolenate production and

Table 1. *Monogalactosyl diglyceride fatty acid composition, weight per cent*

Fatty acid (carbons : double bonds)	*Euglena gracilis*[a]	Spinach lamellae[b]	*Anacystis nidulans*[b]	Alfalfa leaves[c]
16:0	4.2		36	2.7
16:1				
16:3	16.8	25	28	trace
16:4	17.8			trace
18:1			13	trace
18:2	8.1	2		1.7
18:3	34.8	72		95.0

[a] VAN DEENEN and HAVERKATE (1966).
[b] ALLEN, HIRAYAMA and GOOD (1966).
[c] O'BRIEN and BENSON (1964).

photosynthetic capability develop simultaneously. Two relationships are possible: a) galactolipids containing α-linolenate are involved in oxygen production or required for proper conformation of the oxygen-producing system; b) oxygen is a requirement for oxidative desaturation of linoleic acid (18:2) and its production facilitates synthesis of α-linolenate. Mutants of the alga, *Scenedesmus obliquus*, incapable of oxygen production, synthesize α-linolenic acid. Either aerobic culture of the organism permits linolenate synthesis as suggested above (b) or part of the oxygen-producing structure may be intact while the genetic block may be restricted to another aspect of the photosynthetic oxygen production. WINTERMANS *et al.* (1969) have related photochemical activities to the extent of enzymatic transacylation and deacylation in isolated chloroplasts. In sodium chloride solutions monogalactolipase was observed. In sucrose or mannitol media the digalactolipid was degraded by transacylation reactions. Both the Hill reaction and NADP reduction in Photosystem I were found sensitive to chloroplast galactolipid degradation.

Absence of α-linolenate in *Anacystis* is unique but not restrictive (Table 1). It is entirely possible that these blue green algae possess a membrane protein structure designed to accommodate less saturated hydrocarbon chains. Further, the recognition that galactolipids and α-linolenate are components of plant mitochondria (BIALE, YANG and BENSON, 1966) suggests that these major lipids of green chloroplasts occur and function more widely than initially suspected. It appears that there is a remarkable affinity of polyunsaturated fatty acids for many membrane proteins. It has been suggested (BENSON, 1966) that this affinity is a result of overlap of pi orbitals of the olefinic bonds with those of the N⋯C⋯O pi systems of the protein chains. The ease with which the linolenic acid chain may assume a helical configuration and the similar spacing of pi orbital systems in the α-helix structure of membrane protein is consistent with this proposal.

D. The Plant Sulfolipid

The major soluble anionic sugar derivatives in the chloroplast are the plant sulfolipid, Fig. 6 (O'BRIEN and BENSON, 1964), and its deacylation product α-D-sulfoquinovosyl glycerol, Fig. 7 (BENSON, 1963). The biosynthesis of these compounds, is dealt with on p. 230.

The sulfolipid has a charge and fatty acid composition similar to that of phosphatidyl inositol or phosphatidyl glycerol. It does not hydrolyze, however, with snake venom phospholipase A. It possibly binds calcium and magnesium less specifically than do the anionic phospholipids. Its normally sluggish turnover does not indicate metabolic involvement in carbohydrate biosynthesis. Sulfolipid stabilized the chloroplast CF_1 coupling factor against cold inactivation. The sulfolipid was considerably more effective than galactolipids or phospholipid. Mitochondrial F_1, on the other hand, required phospholipid but sulfolipid had no particular stabilization properties (LIVNE and RACKER, 1969).

In spite of its high concentration in green plants and apparent concentration in chloroplasts (SHIBUYA, MARUO and BENSON, 1965) the relation of chloroplast metabolism to sulfonic acid biosynthesis and degradation is not yet understood. The sulfonic acid group of the sulfosugar, cysteinolic acid, and the sulfopropanediol apparently is not derived from cysteic acid, the well known sulfonic acid. Formation of sulfopropanediol from labeled sulfoquinovose by chloroplasts implicates this compound in degradative metabolism of the sulfosugar (R. F. LEE, personal communication) (cf. Fig. 8). Its formation may be the result of reactions related to photosynthetic activity. In most cases the sulfuryl group of 3-phosphoadenylyl sulfate, PAPS, is transferred to an hydroxyl oxygen acceptor by a sulfokinase to yield sulfate ester, much like the formation of phosphate esters from ATP. In the chloroplast, however, a transfer to a carbon acceptor with consequent synthesis of a sulfonic acid, R-SO_3H, must occur. The sulfonic acids being even stronger than sulfuric acid, are dissociated at all biological pH conditions and probably even in the hydrophobic environment of the lamellar lipoprotein where aqueous dissociation constants do not apply. The sulfo

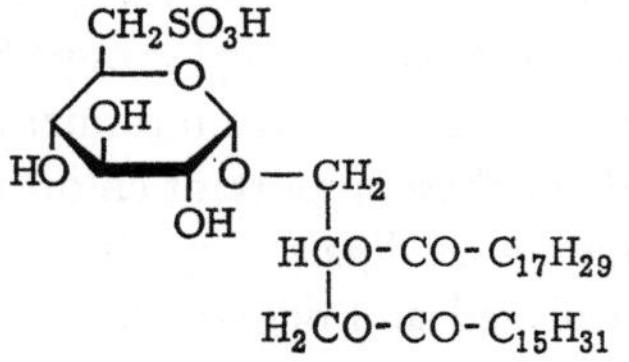

Fig. 6. The plant sulfolipid

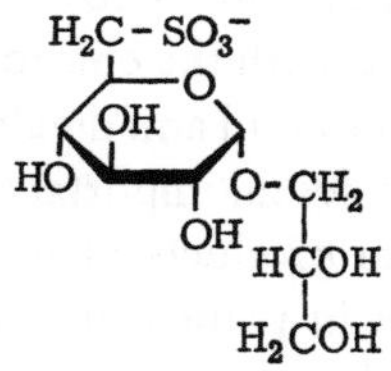

Fig. 7. 6-sulfo-α-D-quinovopyranosyl-(1→1′)-D-glycerol, deacylation product of the plant sulfolipid

group is formed *in vitro* when sulfite reacts with enolic compounds like phosphopyruvic acid (LEHMANN and BENSON, 1964). Whether the reduction product of PAPS is involved in this kind of reaction is not yet established.

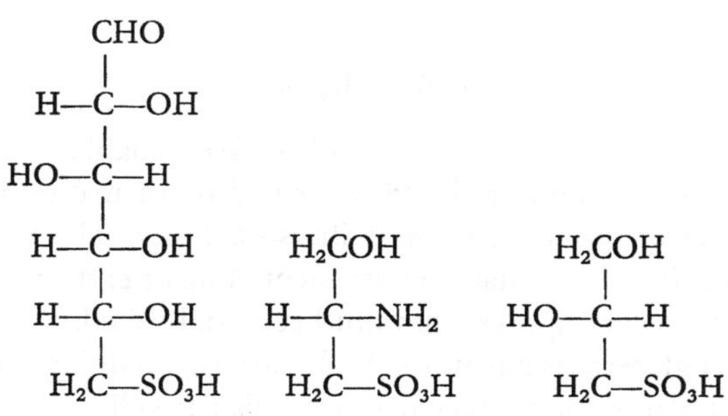

Fig. 8. Sulfoquinovose, cysteinolic acid, and sulfopropanediol

E. Deacylation of Chloroplast Lipids

The fatty acyl groups of the lipids may be removed easily *in vitro* by base-catalyzed transesterification in methanol. In 0.1 N KOH in anhydrous methanol at room temperature, the fatty acids are removed in 15 to 30 min. Similar reactions are catalyzed *in vivo* by phospholipases and glycolipases. These degradations diminish chloroplast activities after isolation or sonic disruption. McCARTY and JAGENDORF

(1965) recognized the decay of electron transport activity in isolated spinach chloroplasts and demonstrated that it was indeed the result of galactolipase activity released during disruption of the cells.

1. Galactolipase

Sastry and Kates (1964) selected runner bean (*Phaseolus multiflorus*) leaves for isolation of ^{14}C-labeled chloroplast lipids. To their surprise they found little or no galactolipid. They had discovered a unique source of the lipolytic enzyme, galactolipase, associated with the chloroplasts which removes the two fatty acyl groups from both monogalactolipid and digalactolipid. The enzyme was purified from runner bean leaves and provides an important tool for *in vitro* study of these important chloroplast lipids. The final products of galactolipase action, mono- or digalactosylglycerol do not accumulate in plant cells; they are further hydrolyzed and degraded. The related α-galactosylglycerols, floridoside and isofloridoside, of red algae, on the other hand, are metabolically stable carbohydrate reservoirs in the plant.

Chloroplast function is impaired by the free fatty acids liberated by galactolipase action during isolation or incubation of bean chloroplasts (Clendenning and Gorham, 1950; Bamberger and Park, 1966). Fortunately the enzyme activity is negligible in most chloroplast preparations. A transacylase activity, leading to 40% 6-O-acyl, 30% 4-O-acyl, and 6% 3-O-acyl-galactosyldiglyceride during chloroplast isolation was recognized by Heinz (1965). Enzymatic transacylations, transphosphatidylations, transphosphorylations, and transglycosylations are typical in disrupted plant cell systems and must be anticipated in research work. Gressel and Avron (1965) had studied similar effects of phospholipases upon photophosphorylation in chloroplasts.

2. Sulfolipase

Disrupted *Scenedesmus* cells possessed a sulfolipase capable of removing the two fatty acyl groups from the sulfolipid with intermediate formation of a lysosulfolipid with a single fatty ester group (Yagi and Benson, 1962). The final product, sulfoquinovosyl glycerol (Fig. 7) is a stable constituent of plant extracts. Whether it leaves the chloroplast after *in vivo* lipolysis has not been established.

Sulfoquinovosyl glycerol is a major soluble sulfur-containing component of *Chlorella* and many other plants. Its resistance to metabolism in sulfur-sufficient cells and dramatic conversion of its sulfur to protein was revealed by the work of the S. Miyachis (1966). In sulfur-deficient media both sulfolipid and sulfoquinovosyl glycerol reservoirs were rapidly and extensively depleted with conversion of the labeled sulfur to insoluble (protein) components (Fig. 9a and b).

3. Phospholipases

Very little if any lysophospholipid occurs in chloroplast or whole cell extracts, indicating that phospholipase A activity *in vivo* is low. Deacylated derivatives of phospholipids do however accumulate in plant cells and algae. Diglycerophosphate is by far the most frequently occurring and predominant of these. Its concentration in *Chlorella* often exceeds those of hexose phosphates or any other phosphate esters by a factor of ten or more. On illumination its concentration decreases with concomitant formation of phosphatidyl glycerol. It is not yet established that this re-acylation or re-synthesis proceeds within the chloroplast but since phosphatidyl glycerol is largely a chloroplast lipid one may presume that it contains diglycerophosphate.

Similar deacylation reactions occur with the phosphatidyl choline and phosphatidyl ethanolamine and glycerophosphoryl choline and glycerophosphoryl ethanolamine may accumulate in the cell. These are broken by a phosphodiesterase to yield glycerophosphate and the corresponding base. The enzyme from the bacterium *Serratia*

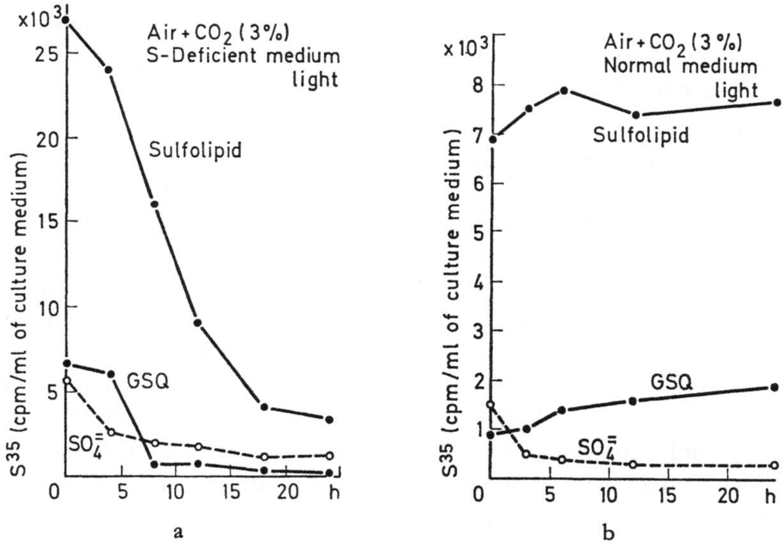

Fig. 9a and b. Metabolism of ^{35}S-labeled sulfolipid and glyceryl sulfoquinovoside in *Chlorella ellipsoidea* (S. Miyachi and S. Miyachi, 1966)

plymuthica was shown to react more readily with glycerophosphoryl esters of choline and ethanolamine than with diglycerophosphate (Prasad and Benson, 1969). The role of diglycerophosphate in chloroplast metabolism has not yet been clearly established. Its importance as a potential reservoir for "reduced" triose phosphate in the chloroplast is apparent from the dependence of its concentration upon photosynthesis and has been discussed in a previous section.

F. "Osmiophilic" Lipid Globules of Chloroplasts

Lipid globules accumulate in regions of active lipid biosynthesis. They decrease during the transformation of prolamellar bodies into the lamellar system of the grana and increase during periods of intense growth in strong light (Bailey and Whyborn, 1963). Electron microscopy of osmium-stained chloroplast sections usually reveals black globules of diameter varying from thylakoid thickness to that of several grana (100 to 5000 Å). These lipid components of the stroma exhibit neither external membranes nor internal structure (LICHTENTHALER and PEVELING, 1966). They appear to function as extra-thylakoidal pools of lipid from which the lipoprotein of the lamellar membrane is assembled (SPREY and LICHTENTHALER, 1966; LICHTENTHALER, 1966). In etiolated barley the prolamellar bodies possess aggregates of osmiophilic globules. Their number decreases as grana are produced.

The osmiophilic globules of chloroplasts have been concentrated by flotation centrifugation (GREENWOOD, LEECH and WILLIAMS, 1963). On the basis of their absorption spectra it is clear that they contain plastoquinone and carotenoids whose synthesis is light-independent. They contain only 0.06% of the leaf's chlorophylls but have 75% of its phytoene, no β-carotene, a galactolipid composition like that of lamellae, and 40 to 50% of the chloroplast's plastoquinone. Their composition is indicative of their serving as galactolipid reservoirs during chloroplast development (LICHTENTHALER and PEVELING, 1966; SPREY and LICHTENTHALER, 1966).

In the degeneration of senescent or damaged chloroplasts, the number of osmiophilic globules again increases. Such dispersal of lipoprotein components is clearly characteristic of the more rapid damage to chloroplasts by solvents or detergents which result in intracellular development of myelin figures from lipid removed from the lamellae.

In the exceptionally regular grana of *Zea mays* chloroplasts small osmiophilic globules occur at the edges of the thylakoids. One may surmise that these lipid pools serve in assembly as well as in degradation of lipoprotein structures. The lipid components of lipoprotein membrane surfaces may accumulate beyond the capacity of the protein to hold them. Then, as in the case of excess lipid or film pressure on a monolayer of lipid, the lipids aggregate in globules or micelles. The fact that lipid compositions of the globules differ from that of the lamellar lipoprotein suggests a degree of specificity of the protein for a limited number and type of lipid molecules; that is, there appears to be a specificity for association of lipid and protein in the membrane lipoprotein.

G. Assembly of Chloroplast Lipoprotein

Current concepts of membrane structure involve hydrophobic association of amphipathic lipids such as those of the chloroplast with membrane structural proteins (BENSON, 1966). These have been studied by WEBER (1962) and CRIDDLE (1966), and found to have much in common with mitochondrial structural protein; they have no disulfide cross links, 21—28,000 molecular weight, high content of hydrophobic amino acids and glutamic and aspartic acids, and low content of the basic amino acids. The lamellar lipoprotein, "quantasomes", of PARK and PON (1963) contains such proteins in association with chlorophyll and the other lipid components of the thylakoid (BAILEY, THORNBER and WHYBORN, 1966). The structural protein has been isolated by displacement of the lipid components by detergents followed by extraction of the detergent. It has also been prepared by solution of the protein in formic acid, an effective solvent for these low molecular weight proteins containing 20 to 30% of glutamic and aspartic acid residues. These proteins have a strong tendency to aggregate when free of lipid and therefore their properties are not easily examined.

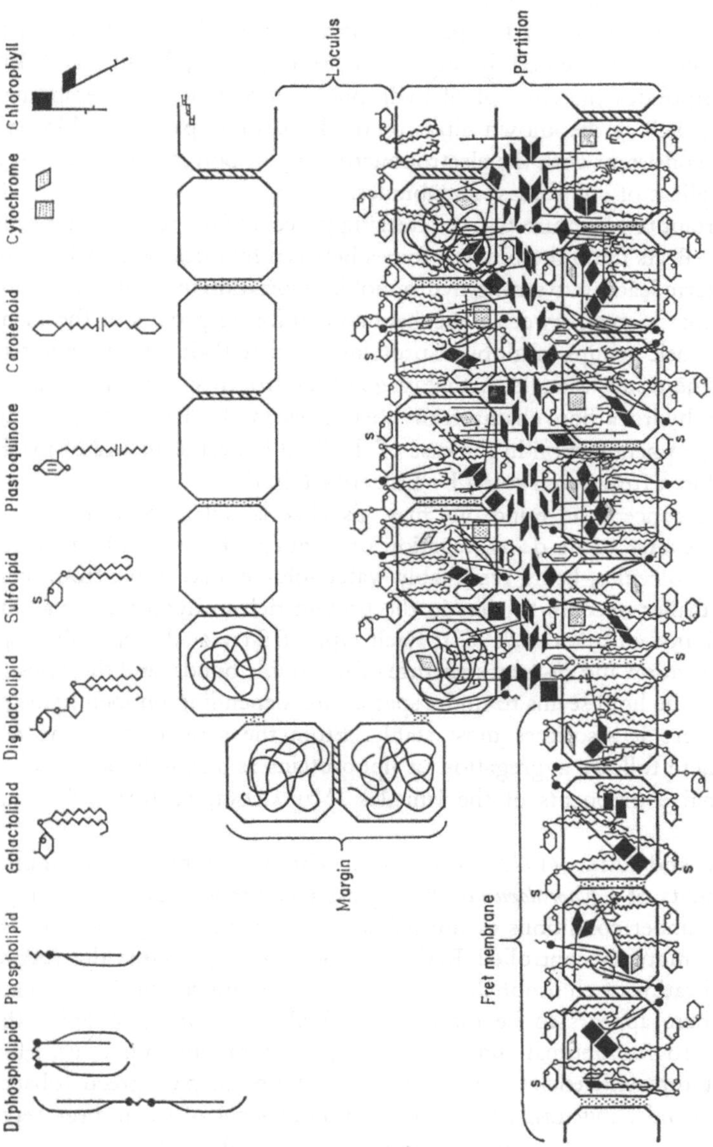

Fig. 10. Proposed schematic model for lamellar lipoprotein organization (WEIER and BENSON, 1966)

Aggregation of lipoprotein subunits, of course, is an attribute essential for formation and repair of membranes.

In the presence of chlorophyll or lipids, lamellar protein may accept a limited number of hydrophobic molecules to form lipoprotein of reproducible composition. From the known hydrophobic nature of protein internal structure, it was deduced that the hydrocarbon chains of chlorophyll, glycolipids, phospholipids and terpenoids should associate with the interior of the protein molecules rather than be exposed to the external water phase. The hydrophilic moieties of the lipids then cover or surround the protein surface. A model for thylakoid molecular arrangement is presented in Fig. 10 (Benson, 1964; Weier and Benson, 1966). It incorporates the concept of hydrophobic association of membrane lipids and protein as well as the known ultrastructural relationships in the chloroplast. It appears to be consistent with the electronmicroscopy of both stained sections and of freeze-etch replicas of chloroplast membranes.

It is important to recognize that the freezing process of the freeze-etch technology has profound effects upon the binding forces between hydrocarbon chains and lipid. The entropy term leading to strong hydrophobic association in a liquid water medium disappears. Ionic forces between phosphate and sulfonate groups of the membrane lipids and the basic groups of globular proteins associated with them are not affected by freezing. The result, then, is that cleavage of the frozen specimen can most easily "uproot" the hydrocarbon chains from their sites in the lamellar lipoprotein. A related cleavage phenomenon in the case of the lipid bilayer structure of myelin was demonstrated in a freeze-etch study by Branton (1967).

With these concepts in mind one may visualize lamellar lipoprotein formation *in vivo* as the reverse of its de-aggregation or cleavage. Upon synthesis of lamellar protein, lipid molecules, being reasonably water-soluble, may move from their lipid globules or smaller micelles to sites in the protein rich in hydrophobic amino acid side chains (Kirk and Allen, 1965). Each type of fatty acid, depending upon its chain length location and *cis-trans* structure of its double bonds, and their position on the glycerol of the lipid seems to find a special site, depending on membrane protein sequences, where it associates most stably. From the symmetry of the resultant lipoprotein there follows aggregation of lipoproteins as a mosaic of photosynthetically functional components of the lamellae (Murakami, Katoh and Takamiya, 1965).

Goldberg and Ohad (1970), observed synchronous synthesis of pigment and lipid components of *Chlamydomonas* chloroplast membranes during greening. They conclude that under conditions of normal active development synthesis of lamellar components is mutually controlled. Ratios of mono- and digalactosyldiglycerides and sulfolipid and ratios of chlorophylls *a* and *b* were constant to within less than 10%. In general, there appears to be more chlorophyll and more glycolipids than are actually required for optimal function of the photosynthetic apparatus. However, excellent quantum efficiencies have been demonstrated in pale green tobacco mutants (Schmid and Gaffron, 1968). The sulfolipid level of sugar beet leaves can vary over a considerable range without apparent effect on lamellar function (Wintermanns, 1960).

H. Reconstitution of Chloroplast Membrane Lipoprotein

The process of lamellar development involves coordination and control of both lipid and protein synthesis. The chloroplast prolamellar body is a regular array of lipoprotein which may be termed a crystal lattice. Its structure has been revealed by many workers, notably Gunning (1965) and Wehrmeyer (1965). It appears that reorganization of the tetrahedral subunits of the prolamellar body must require synthesis of new lipid or protein components essential for lamellar development. It is possible that some of the quasicrystalline bodies observed in electron micrographs of stained sections (PRICE and THOMSON, 1967), are in fact aggregates of incomplete membrane lipoprotein molecules. Similarly there are the pro-mitochondrial crystals in *Hydra* from which mitochondrial membranes appear to develop (DAVIES, 1967). One may envisage a process of lipoprotein conformational alteration involving association of essential lipid components with the lipoprotein subunits.

As a model for study of lipid association in functional lipoprotein the lamellar membrane of chloroplasts offers advantages. Its lipid components are few and relatively simple in structure. Reconstitution of its natural organization may be followed by assay of its electron transport capabilities. The reagent, being the photon, is readily added without problems of diffusion or transport. The extensive studies of light-induced electron transport in chloroplasts provide bases for estimation of the degree of their reconstitution from separated lipid and protein.

Lipids are removed from lamellar lipoprotein preparations by cold 70% acetone extraction. The extracted protein is red as a result of rearrangement of its β-carotene following removal of chlorophyll and many lipids (NISHIMURA and TAKAMATSU, 1957). The association of the pigment with a site in the hydrophobic interior of the protein leads to a shift in absorption maximum from 480 nm to 538 nm. This is an example of specificity of hydrophobic association of a hydrocarbon chain with a protein to form a lipoprotein (JI and BENSON, 1968). Extraction with 100% acetone or with hexane readily removed β-carotene from the red lipoprotein. This characteristic shift in absorption spectrum is a specific indicator of structure and was used in developing techniques for re-introducing β-carotene, chlorophyll and the amphipathic lipids into the extracted lamellar membrane structural protein. DEROCHE and COSTES (1969) have observed multiple absorption bands for several carotenoids. It is clear that these hydrocarbon components of the chloroplast interact in specific ways which lead to individual absorption characteristics.

Since membrane proteins are, in general, not cross-linked by S-S bonds and are low molecular weight proteins (23—28,000) they are resistant to irreversible denaturation by aqueous methanol. JI, HESS, and BENSON (1968) suspended the extracted protein in methanol in the presence of β-carotene. Upon gradual dialysis against water the red lipoprotein, λmax 538 nm, was reconstituted. The same technique was used in determining the number of lipids or pigment molecules which associate with each molecule of membrane protein (molecular weight, 23—29,000). Chlorophyll was introduced into the protein in methanol solution. The amount bound, determined colorimetrically, showed that the protein saturated with association of 30 molecules of chlorophyll. Table 2 shows that 29 to 36 hydrocarbon chains associate with each protein molecule. These figures are comparable to the ratio of lipids to protein in the original lamellar membrane. The amount of detergent necessary for stabilization of membrane components varies but recent analyses by ALLEN and GOOD (1971), indicate that the number of detergent hydrocarbon chains associated with a lipoprotein is almost exactly equal to the number of chloroplast lipid and chlorophyll molecules displaced in the preparation. These include the detergent-solubilized preparations, TSF-1 and TSF-2 (VERNON *et al.*, 1966) which exhibit Photosystems I and II activities. An exception, the diphytyl ether analog of phosphatidylglycerol-3-phosphate of the wall membrane of *Halobacterium cutirubrum* reached saturation when six molecules of the lipid were associated with the chloroplast lamellar protein. Apparently the lamellar protein was not designed to associate with this bacterial lipid.

Table 2. *Reassociation of chloroplast lamellar lipoprotein* (Ji and Benson, 1968)

Lipid	Moles of lipid bound per mole of lamellar protein[a]	Moles of hydrocarbon chains per mole of lamellar protein	Hydrophilic group	Hydrophobic group
Palmitic acid	36	36	Carboxylic acid	Palmitic acid
Chlorophyll	30	30	Carbonyl	Phytol, porphyrin ring
Monogalactosyl diglyceride	16	32	Galactose	Linolenic acid
Phosphatidyl diglyceride	15	30	Glycerophosphate	Linolenic acid
Digalactosyl diglyceride	14·5	29	Digalactose	Linolenic acid
PGP[b]	5·7	11·4	Glycerol 1,3-diphosphate	Dihydro-phytyl

[a] Lipid chains bound to lamellar protein at saturation.

[b] PGP, the diphytyl ether analog of phosphatidyl glycerol-3-phosphate occurring in membranes of *Halobacterium cutirubrum*. The low affinity is identical to the association ratio of this lipid in the bacteria *in vivo*.

Benson *et al.* (1969, 1970), demonstrated reconstitution of the P 700 chlorophyll lipoprotein, Photosystem I, from 80% acetone extracted chloroplasts. The lipids and pigments were restored by suspending the protein in 80% acetone and addition of water to a 65% acetone concentration during 16 h at − 10°. Photoreduction of methyl red by the reconstituted lipoprotein was observed using a 2,6-dichlorophenol indophenol-ascorbic acid couple. The reconstituted chlorophyll protein exhibited 4 to 6 times the rate of photoreduction of methyl red or TPN than chlorophyll alone when suspended in 2% Triton X-100. A two-fold stimulation was demonstrated upon addition of plastocyanin and crude ferredoxin. These results indicated restoration of electron flow through Photosystem I. Maximal activity of the reconstituted system was observed at 2% Triton concentration; this equaled that of the original chloroplast preparation in a Triton concentration of 0.2%.

A different approach to introduction of lipids and pigments into chloroform-extracted lamellar protein was taken by Shibuya and Maruo (1966) (Shibuya, Honda, and Maruo, 1965). They suspended protein and sonicated lipid in buffer and stored the suspension at − 35° for 5 days. The membrane protein, no longer constricted by presence of a surrounding water phase was apparently free to relax and accept the lipid chains as they slowly diffused. The recombined preparations possessed up to 70% of the Hill activity (Photosystem II) characteristic of sonicated 14S particles (quantasomes) which had been stored under identical conditions.

The recombination of Photosystems I and II lipoproteins isolated by detergent (Triton X-100) dispersal and density gradient centrifugation has been reported by Huzishige *et al.*, 1969. The aggregates in Triton solution did not aggregate but formed 1:1 dimers with photoactivity, electron transport, and oxygen production characteristic of photosynthesis.

The two procedures for introducing lipids and pigments into the solvent-extracted membrane structural proteins are related in that they both remove the effects of the aqueous medium surrounding the protein to reduce its internal hydrophobic association ("hydrophobic bonding"). The lipids and pigments then may diffuse into the interior of the protein whereupon thawing or gradual addition of water to the non-aqueous suspending medium may again lock the hydrophobic lipid and pigment hydrocarbon chains in the resulting lipoprotein.

Non-biological substances also associate with chloroplast lamellar lipoprotein. DDT and its biodegradation product, DDE, are absorbed by isolated swelled chloroplasts with 50% inhibition of electron transport rates at 5×10^{-6} M DDT (GEE and BOWES, 1970; BENSON *et al.*, 1970). Certain weed-killing oils, DCMU, industrial pollutants, and bacterial components can also affect membrane function in plants. The successful competition by these substances for the lipid-specific sites in lamellar lipoprotein presents a constant problem to the optimization of photosynthesis (BENSON *et al.*, 1970).

The photoreactions, electron transport steps, and enzyme reactions which may be observed in chloroplast preparations provide a variety of quantitative criteria for evaluating the extent of chloroplast integrity. They may reveal aspects of lipoprotein and membrane structure important for all areas of biology.

I. Literature

ALLEN, C. F., HIRAYAMA, O., GOOD, P.: Lipid composition in photosynthetic systems. In: Biochemistry of chloroplasts, vol. I, pp.195—200 (T. W. GOODWIN, Ed.). London-New York: Academic Press 1966.

— GOOD, P.: Lipid composition of P-700 enriched Triton subchloroplast fractions. Biochem. biophys. Res. Commun. (1971) (in press).

BAHL, J., GUILLOT-SALOMON, T., DOUCE, R.: Synthese enzymatique du cytidine diphosphate diglyceride dans les végétaux supérieurs. Physiol. Végétale (1970) (in press)

BAILEY, J. L., WHYBORN, A. G.: The osmiophilic globules of chloroplast. II.Globules of the spinach-beet chloroplast. Biochim. biophys. Acta (Amst.) **78**, 163—174 (1963).

— THORNBER, J. P., WHYBORN, A. G.: The chemical nature of chloroplast lamellae. In: Biochemistry of chloroplasts, vol. I, pp. 243—255. (T. W. GOODWIN, Ed.). London-New York: Academic Press 1966.

BAMBERGER, E. S., PARK, R. B.:: Effect of hydrolytic enzymes on the photosynthetic efficiency and morphology of chloroplasts. Plant Physiol. **41**, 1591—1600 (1966).

BENSON, A. A.: Dynamic interrelationships of enzyme systems in hexose synthesis *in vivo*. In: Proc. Int. Symp. Enz. Chem., pp. 189—191. Tokyo and Kyoto 1957 (K. ICHIHARA, Ed.). Tokyo: Maruzen Press 1958.

— The plant sulfolipid. Advanc. Lipid Res. **1**, 387—394 (1963).

— Plant membrane lipids. Ann. Rev. Plant Physiol. **15**, 1—16 (1964).

— Metabolism of chloroplast lipids and its relation to photosynthetic carbohydrate synthesis. In: Abstr. 10th Int. Bot. Cong. Edinburgh, 1964, pp. 160—161. Edinburgh: T. and A. Constable, Ltd. 1964.

— On the orientation of lipids in chloroplast and cell membranes. J. Amer. Oil Chemists' Soc. **43**, 265—270 (1966).

— GEE, R. W., BOWES, G. W.: Lipid-protein interactions and chloroplast membrane function. Biophys. Soc. Abst. **193 a** (1970).

— — — Inhibition of electron transport by p,p'-DDT and p,p'-DDE. Fed. Proc. **29**, 880 (1970).

— — JI, T.H., BOWES, G. W.: Lipid-protein interaction in chloroplast lamellar membrane as bases for reconstitution and biosynthesis. In: Autonomy and biogenesis of mito-

chondria and chloroplasts (N. K. BOARDMAN, A. W. LINNANE, R. M. SMILLIE, Eds.). Amsterdam: North Holland Publishing Company 1970.
BENSON, A. A., JI, T. H., GEE, R. W.: Lipid binding and reconstitution of membrane lipoprotein systems. Biophys. J. **9**, A-36 (1969).
— — — Molecular organization of productive chloroplast lipoprotein structures. In: Proc. Int. Symposium on productivity of photosynthetic systems (A. A. NICHIPOROVICH, Ed.). Moscow: Academy of Sciences, USSR 1970 (in press).
BIALE, J. G., YANG, S. F., BENSON, A. A.: Lipids in plant mitochondria. Fed. Proc. **25**, 405 (1966).
BRANTON, D.: Fracture faces of frozen myelin. Exp. Cell Res. **45**, 703—707 (1967).
BUSBY, W. F.: Sulfopropanediol and cysteinolic acid in the diatom. Biochim. biophys. Acta (Amst.) **121**, 160—161 (1966).
CLENDENNING, K. A., GORHAM, P. R.: Photochemical activity of isolated chloroplasts in relation to their source and previous history. Canad. J. Res. C, **28**, 114—139 (1950).
CRIDDLE, R. S.: Protein and lipoprotein organization in the chloroplast. In: Biochemistry of chloroplasts, vol. II, pp. 203—231 (T. W. GOODWIN, Ed.). London-New York: Academic Press 1966.
DAVIS, L. E.: Intramitochondrial crystals in *Hydra*. J. Ultrastruct. Res. **21**, 125—133 (1967).
DEROCHE, M.-E., COSTES, C.: Heterogeneity of carotenoids in chloroplasts. In: Progr. in photosynthesis research, vol. II, pp. 681—693 (H. METZNER, Ed.). Tübingen: Intern. Union Biol. Sci. 1969.
DOUCE, R., GUILLOT-SALOMON, T., LANCE, C., SIGNOL, M.: Etude comparée de la composition en phospholipides de mitochondries et de chloroplastes isoles de quelques tissus végétaux. Bull. Soc. franc. physiol. végét. **14**, 351—373 (1968).
ERWIN, J., BLOCH, K.: Polyunsaturated fatty acids in some photosynthetic microorganisms. Biochem. Z. **338**, 496—511 (1963).
— — Biosynthesis of unsaturated fatty acids in microorganisms. Science **143**, 1006—1012 (1964).
FERRARI, R. A., BENSON, A. A.: The paths of carbon in photosynthesis of the lipids. Arch. Biochem. **93**, 185—192 (1961).
GEE, R. W., BOWES, G. W.: Inhibition of electron transport by p,p'-DDT and p,p'-DDE. Fed. Proc. **29**, 880 (1970).
GOLDBERG, I., OHAD, I.: Biogenesis of chloroplast membrane. IV. Lipid and pigment changes during synthesis of chloroplast membranes in a mutant of *Chlamydomonas reinhardii* y-1. J. Cell Biol. **44**, 563—571 (1970).
— — Biogenesis of chloroplast membranes. V. A radioautographic study of membrane growth in a mutant of *Chlamydomonas reinhardii* y-1. J. Cell Biol. **44**, 572—591 (1970).
GREENWOOD, A. D., LEECH, R. M., WILLIAMS, J. P.: The osmiophilic globules of chloroplasts. I. Osmiophilic globules as a normal component of chloroplasts and their isolation and composition in *Vicia faba* L. Biochim. biophys. Acta (Amst.) **78**, 148—162 (1963).
GUNNING, B. E. S.: The greening process in plastids. 1. The structure of the prolamellar body. Protoplasma (Wien) **60**, 111—130 (1965).
HAVERKATE, F., VAN DEENEN, L. L. M.: Isolation and chemical characterization of phosphatidyl glycerol from spinach leaves. Biochim. biophys. Acta (Amst.) **106**, 78—92 (1965).
HEINZ, E.: Fermentative Acylierung von Monogalaktosyldiglycerid bei der Isolierung von Chloroplasten. Z. Naturforsch. **20b**, 83 (1965).
HUZISHIGE, H. H., USIYAMA, T., KIKUTI, T., AZI, T.: Purification and properties of the photoactive particle of Photosystem II. Plant and Cell Physiol. **10**, 441—456 (1969).
JI, T. H., BENSON, A. A.: Association of lipids and proteins in chloroplast lamellar membrane. Biochim. biophys. Acta (Amst.) **150**, 686—693 (1968).
— HESS, J. L., BENSON, A. A.: Studies on chloroplast membrane structure. I. Association of pigments with chloroplast lamellar protein. Biochim. biophys. Acta (Amst.) **150**, 676—685 (1968).
KIRK, J. T. O., ALLEN, R. L.: Dependence of chloroplast pigment synthesis on protein synthesis. Biochem. biophys. Res. Commun. **21**, 523—530 (1965).

LAETSCH, W. N.: Relationships between chloroplast structure and photosynthetic carbon fixation pathways. Sci. Progr. **57**, 323—351 (1969).
— Specialized chloroplast structure of plants exhibiting the dicarboxylic acid pathway of photosynthetic CO_2 fixation. In: Progr. in photosynthesis research, vol. I, pp. 36—46 (H. METZNER, Ed.). Tübingen: Intern. Union Biol. Sci. 1969.
LEHMANN, J., BENSON, A. A.: The plant sulfolipid. IX. Sulfosugar syntheses from methyl hexoseenides. J. Amer. chem. Soc. **86**, 4469—4472 (1964).
LICHTENTHALER, H. K. O.: Plastoglobuli und Plastidenstruktur. Ber. dtsch. bot. Ges. **79**, 82—88 (1966).
— PARK, R. B.: Chemical composition of chloroplast lamellae from spinach. Nature (Lond.) **198**, 1070—1072 (1963).
— PEVELING, E.: Osmiophile Lipideinschlüsse in den Chloroplasten und in Cytoplasma von *Hoya carnosa* R. Br. Naturwissenschaften **53**, 534—535 (1966).
LIVNE, A., RACKER, E.: Partial resolution of the enzymes catalyzing photophosphorylation. V. Interaction of coupling factor 1 from chloroplasts with ribonucleic acid and lipids. J. biol. Chem. **244**, 1332—1338 (1969).
MCCARTY, R. E., JAGENDORF, A. T.: Chloroplast damage due to enzymatic hydrolysis of endogenous lipids. Plant Physiol. **40**, 725—735 (1965).
MUDD, J. B.: Fat metabolism in plants. Ann. Rev. Plant Physiol. **18**, 229—252 (1967).
MURAKAMI, S., KATOH, S., TAKAMIYA, A.: Lamellar structure in chloroplasts and its function. In: Intracellular and membranous structure. Proc. Int. Symp. Cellular Chem., pp. 173—189 (S. SENO, E. V. COWDRY, Eds.). Oktsu 1965.
MIYACHI, S., MIYACHI, S.: Sulfolipid metabolism in *Chlorella*. Plant Physiol. **41**, 479—486 (1966).
NEUFELD, E. F., HALL, C. W.: Formation of galactolipids by chloroplasts. Biochem. biophys. Res. Commun. **14**, 503—508 (1964).
NICHOLS, B. W., MOORHOUSE, R.: The separation, structure and metabolism of monogalactosyl diglyceride species in *Chlorella vulgaris*. Lipids **4**, 311—316 (1969).
— JAMES, A. T.: Acyl lipids and fatty acids of photosynthetic tissue. In: Progr. in Phytochemistry, Vol. 1, pp. 1—48 (L. REINHOLD, Y. LIWSHITZ, Eds.). London: Interscience Publishers 1968.
NISHIMURA, M., TAKAMATSU, K.: A carotene-protein complex isolated from green leaves. Nature (Lond.) **180**, 699—700 (1957).
O'BRIEN, J. S., BENSON, A. A.: Isolation and fatty acid composition of the plant sulfolipid and galactolipids. J. Lipid Res. **5**, 432—436 (1964).
ONGUN, A., MUDD, J. B.: Biosynthesis of galactolipids in plants. J. biol. Chem. **243**, 1558—1566 (1968).
PARK, R. B., PON, N.: Correlation of structure with function in *Spinacea oleracea* chloroplasts. J. molec. Biol. **3**, 1—10 (1961).
PRASAD, R., BENSON, A. A.: Enzymatic hydrolysis of glycerophosphoryl esters. II. Cleavage of D-glycerol-1-phosphoryl-L-glycerol. Biochim. biophys. Acta (Amst.) **187**, 269—271 (1969).
PRICE, J. L., THOMSON, W. W.: Occurrence of a crystalline inclusion in the chloroplasts of Macadamia leaves. Nature (Lond.) **214**, 1148—1149 (1967).
ROSENBERG, A., GOUAUX, J., MILCH, P.: Monogalactosyl and digalactosyl diglycerides from heterotrophic, hetero-autotrophic, and photobiotic *Euglena gracilis*. J. Lipid Res. **7**, 733—738 (1966).
SASTRY, P. S., KATES, M.: Hydrolysis of monogalactosyl and digalactosyl diglycerides by specific enzymes in runner bean leaves. Biochemistry **3**, 1280—1287 (1964).
SCHMID, G. H., GAFFRON, H.: Photosynthetic units. J. gen. Physiol. **52**, 212—239 (1968).
SPREY, B., LICHTENTHALER, H.: Zur Frage der Beziehungen zwischen Plastoglobuli und Thylakoidgenese in Gerstenkeimlingen. Z. Naturforsch. **21b**, 697—699 (1966).
SHIBUYA, I., HONDA, H., MARUO, B.: Reconstitution of quantasomes. III. Studies on the factors affecting the reappearance of the activity. J. Japan. Biochem. Soc. **37**, 561 (1965).
— MARUO, B.: Preparations, characterization, and reconstruction of quantasomes. Shishitsu Seikagaku Kenkyu (Biochemistry of lipids). Japanese Conferences on the Biochemistry of Lipids, Tokyo 1965, pp. 129—131.

SHIBUYA, I., MARUO, B., BENSON, A. A.: Sulfolipid localization in lamellar lipoprotein. Plant Physiol. **40**, 1251—1256 (1965).

STANACEV, N. Z., STUHNE-SEKALEC, L.: On the mechanism of enzymatic phosphatidylation. Biosynthesis of cardiolipin catalyzed by phospholipase D. Biochim. biophys. Acta (Amst.) (1970) (in press).

THOMSON, W. W.: Ultrastructural studies of ripening oranges. In: Proc. First Int. Citrus Sympos., 1969, pp. 1163—1169.

VAN DEENEN, L. L. M., HAVERKATE, F.: Chemical characterization of phosphatidylglycerol from photosynthetic tissues. In: Biochemistry of chloroplasts, vol. I, pp. 117—131 (T. W. GOODWIN, Ed.). London-New York: Academic Press 1966.

VERNON, L. P., SHAW, E. R., KE, B.: A photochemically active particle derived from chloroplasts by the action of the Triton X-100. J. biol. Chem. **241**, 4101—4109 (1966).

WEBER, P.: Über lamellare Strukturproteide aus Chloroplasten verschiedener Pflanzen. Z. Naturforsch. **17b**, 683—688 (1962).

WEHRMEYER, W.: Zur Kristallgitterstruktur der sogenannten Prolamellarkörper in Proplastiden etiolierter Bohnen. Z. Naturforsch. **20b**, 1270—1296 (1965).

WEIER, T. E., BENSON, A. A.: The molecular nature of chloroplast membranes. In: Biochemistry of chloroplasts, Vol. I, pp. 91—113 (T. W. GOODWIN, Ed.). London-New York: Academic Press 1966.

WILDMAN, S. G., HONGLADAROM, T., HONDA, S. I.: Chloroplasts and mitochondria in living plant cells: Cinephotomicrographic studies. Science **138**, 434—436 (1962).

WINTERMANS, J. F. G. M.: Concentrations of phosphatides and glycolipids in leaves and chloroplasts. Biochim. biophys. Acta (Amst.) **44**, 49—54 (1960).

— HELMSING, P. J., POLMAN, B. J. J., VAN GISBERGEN, J., COLLARD, J.: Galactolipid transformations and photochemical activities of spinach chloroplasts. Biochim. biophys. Acta (Amst.) **189**, 95—105 (1969).

YAGI, T., BENSON, A. A.: Plant sulfolipid. V. Lysosulfolipid formation. Biochim. biophys. Acta (Amst.) **57**, 601—603 (1962).

YANG, S. F., FREER, S., BENSON, A. A.: Transphosphatidylation by phospholipase D. J. biol. Chem. **242**, 477—484 (1967).

Biochemistry of Photophosphorylation

MORDHAY AVRON

A. Historical Introduction

The realization that adenosine triphosphate may serve as a chemical-storage form of the light energy absorbed by photosynthesizing cells was first clearly formulated by RUBENS (1943). When the path of carbon in photosynthesis was unraveled by CALVIN, BASSHAM, BENSON and collaborators, it became clear that the energy requirements of the cycle can best be met by a supply of ATP and NADPH. Nevertheless, early attempts to demonstrate a light-dependent ATP synthesis *in vivo* in whole cells or *in vitro* in isolated chloroplasts, resulted in failure.

The first clear demonstration of a light-induced ATP formation was reported in 1954 with chromatophores isolated from a photosynthetic bacterium by FRENKEL (1954) and with chloroplasts from leaves by ARNON, ALLEN and WHATLEY (1954). This was followed by reports from many laboratories, which demonstrated that this process of photophosphorylation occurred in chloroplasts isolated from a wide variety of higher plant leaves, in chloroplast fragments isolated from algae and in chromatophores isolated from several photosynthetic bacteria. In recent years evidence is accumulating which clearly indicates the important role of photophosphorylation *in vivo*, not only to supply the ATP required to run the photosynthetic CO_2 cycle, but as a general energy-pool for many other processes of the cell, such as ion-uptake and conversion of glucose to starch.

B. Electron Flow Patterns

Two basic types of light induced electron flow patterns, termed "non-cyclic" and "cyclic", have been described in isolated chloroplasts.

In non-cyclic electron flow, the flow of electrons is initiated by the absorption of a light quantum by a photosynthetic pigment (chlorophyll or an accessory pigment). Part of the energy of the absorbed quantum is preserved by using this excitation energy via a photosystem of the chloroplast in promoting an endergonic oxidation reduction reaction. If the quantum in question was channelled into the reaction center of the photosystem termed Photosystem II, the reaction promoted will be the oxidation of Z and the reduction of Q (see Fig. 1). Since Q has a potential of around $E_0' =$ 0.0 V (CRAMER and BUTLER, 1969) and Z of around $E_0' = 0.8$ V, the reaction is endergonic and part of the energy of the quantum has to be utilized to allow the reaction to proceed. The oxidized Z produced oxidizes water through an unknown number of intermediates to yield free oxygen and hydrogen ions. Reduced Q, via a number of intermediates reduces P_{700} in a series of dark enzymic exergonic reactions. Part of the energy stored in reduced Q is used during the electron transfer to perform the endergonic reaction of the formation of ATP from ADP and inorganic phosphate. The reduced P_{700} is oxidized by X in a step that requires the investment of another light quantum, this time absorbed by another Photosystem, termed Photosystem I. The requirement for the absorption of a light quantum is again due to the large difference in potential $E'_0 = 0.4$ V for P_{700} (Kok, 1965) to $E'_0 = -0.6$ V for X (Zweig and Avron 1965; Kok, Rurainski and Owens, 1965). The reduced X produced reduces NADP via two additional enzymes. Thus, the net result of the absorption of four quanta, two in each photosystem, is a transfer of two electrons through the chain, producing one atom of oxygen, one molecule of NADPH and one molecule of ATP. This non-cyclic photophosphorylation produces both the NADPH and ATP necessary to drive the CO_2 fixation cycle.

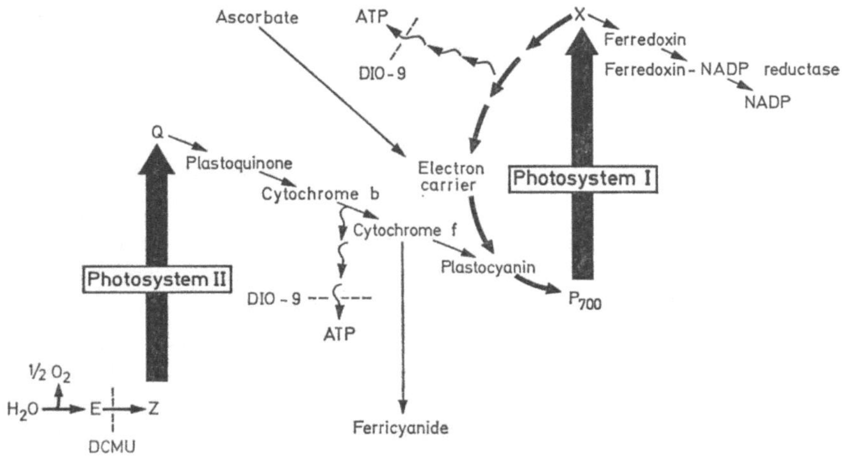

Fig. 1. A scheme of photoinduced electron transport patterns, coupled sites of phosphorylation, and of the effect of inhibitors. See text for details

The evidence which support the generally accepted sequence of reactions just described is voluminous. A few of the major types of evidence which led to its formulation will now be described. First, let us consider what is the evidence that required the formulation of two rather than one light absorption step. The first suggestion can be traced back to the observation by Emerson and Lewis (1943) who noticed that photosynthesis was rather inefficient when illumination was limited to wavelengths longer than 680 mμ ("far-red light") despite the fact that the chlorophyll *in vivo* still absorbs a considerable fraction of the light in this region (Fig. 2). This, so called "red-drop phenomenon", can be explained in view of present knowledge, as due to the accumulation of most of the "long-wavelength-absorbing-chlorophyll" only in Photosystem I. Since equal excitation of both photosystems is required for photosynthesis to proceed (production of both ATP and NADPH—see Fig. 1), it is clear why the efficiency (i.e., the amount of NADPH produced by one absorbed quantum, in this case) considerably decreases when one photosystem absorbed most of the exciting light.

This interpretation of the red drop phenomenon was strongly supported by the observations of HOCH and MARTIN (1963) in isolated chloroplasts. They could show that as expected non-cyclic electron transport to NADP in chloroplasts did indeed show the red-drop phenomenon. However, if in place of water, ascorbate and dichlorophenolindophenol (DPIP) were used as electron donors, the efficiency of electron transfer to NADP increased, rather than decreased, when illumination was

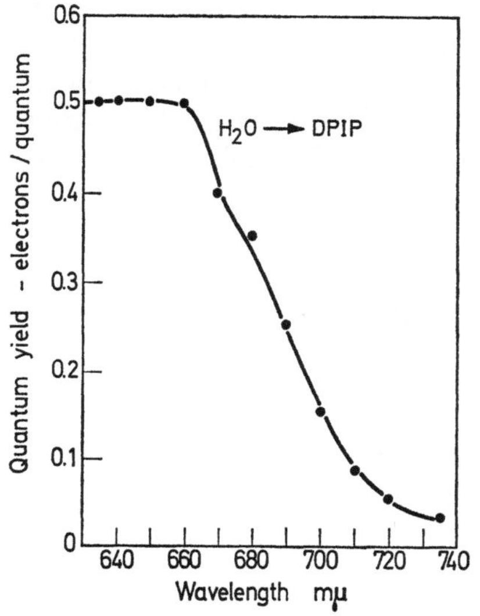

Fig. 2

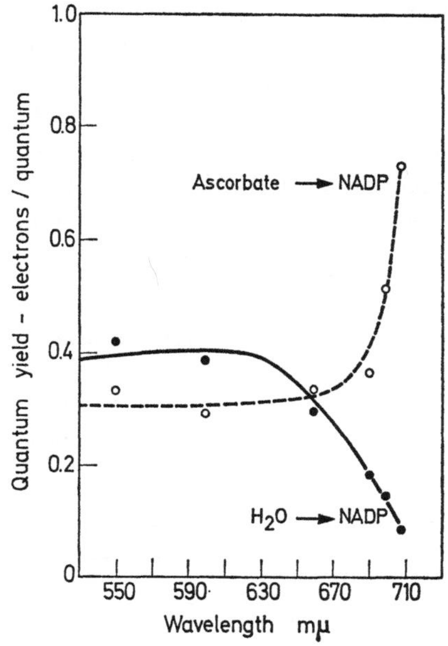

Fig. 3

Fig. 2. The "red drop" in quantum efficiency in the photoreduction of dichlorophenolindophenol by isolated chloroplasts. (After SAUER and KELLY, 1965)

Fig. 3. The quantum yield of the photoreduction of NADP with water or ascorbate as electron donors in isolated chloroplasts. (After SAUER and BIGGINS, 1965)

limited to "far-red" light (Fig. 3). Since it was known from previous work that ascorbate and DPIP donated their electrons to the chain at a point much further along the chain than water (see Fig. 1) it was clear that this transfer of electrons was sensitized only by Photosystem I, and so strongly supported the conclusion that the "long-wavelength-absorbing-chlorophyll" was located mostly in Photosystem I.

The low efficiency of photosynthesis and of the non-cyclic electron-flow from water to NADP, when illumination was limited to "far-red" light, could be considerably improved, if simultaneously some shorter wavelength light (e.g. red) was provided. This effect, termed the "enhancement" phenomenon is normally measured by determining whether or not the rate of photosynthesis or of NADPH production under *simultaneous* illumination with red and far-red light, is higher than the sum of the rates whed illumination is carried separately with the same intensity of red and far-red light. This can be expressed mathematically:

$$\text{Rate}_{(\text{Red} + \text{Far-Red})} > \text{Rate}_{(\text{Red})} + \text{Rate}_{(\text{Far-Red})}$$

or, as more commonly defined:

$$\text{Enhancement} = \frac{\text{Rate}_{(\text{Red} + \text{Far-Red})} - \text{Rate}_{(\text{Red})}}{\text{Rate}_{(\text{Far-Red})}} \, .$$

Enhancement values larger than one in the latter equation indicate a cooperation between the two photosystems. Thus, the low efficiency of the far-red light due to its being absorbed mostly by Photosystem I, can be corrected if one supplies the system simultaneously with red-light which is mostly absorbed by Photosystem II. Recently, it was shown that significant enhancement could be observed in the reduction of NADP by water, but not in that of ferricyanide by water (Govindjee, Govindjee and Hoch, 1964; Avron and Ben Hayyim, 1969; Table 1). Thus, both

Table 1. *Enhancement in NADP and ferricyanide photoreduction by chloroplasts (After* Avron *and* Ben-Hayyim, *1969)*

Electron acceptor	Enhancement
NADP	2.5 ± 0.1
Ferricyanide	0.9 ± 0.1

photosystems must act during the photoreduction of NADP, but only Photosystem II is active in the photoreduction of ferricyanide (see Fig. 1).

Finally, utilizing sensitive spectrophotometric techniques it was shown by Duysens in 1962 with algae (Duysens and Amesz, 1962) and by Avron and Chance in 1966 with chloroplasts (Avron and Chance, 1966) that the cytochrome which is bound within the chloroplast membrane behaves as if it is located in between two different photosystems (see Fig. 1). It was oxidized when the chloroplasts were illuminated with far-red light and reduced when illumination was limited to red-light (Fig. 4).

All these observations, amply confirmed and extended in other laboratories with many different types of cells, provided the basis for the existence and the order of action of the two photosystem as pictured in the figure.

The participation and the order of the other components indicated in Fig. 1 have a varied amount of experimental support. Thus, for ferredoxin and the flavoprotein, ferredoxin-NADP reductase, the evidence is overwhelming (see Avron, 1967). These two enzymes have been obtained in a highly purified form. The addition of both enzymes to depleted chloroplasts is required only when the photoreduction of NADP is involved. When added cytochrome *c* or ferrihemoglobin serve as the electron acceptor only the addition of ferredoxin is required. The reduction of ferredoxin itself does not require the addition of the flavoprotein, nor is it inhibited by the addition of an antibody to the flavoprotein, which fully inhibits NADP reduction (San Pietro and Black, 1965). On the other hand, the reduction of the flavoprotein is linearly dependent upon the amount of ferredoxin added, and the oxidation of reduced ferredoxin by NADP is totally inhibited by the addition of the antibody (Avron and Chance, 1966).

The existence of a component "X" between Photosystem I and ferredoxin has only indirect evidence to support it. Illuminated chloroplasts have been shown to be capable of reducing compounds with a redox potential considerably lower than that of ferredoxin down to at least − 0.6 V (Zweig and Avron, 1965; Kok, Rurainski and Owens, 1965). Also it was shown that after washing out of chloroplasts all detectable ferredoxin, they could still photooxidize cytochrome *f* via Photosystem I,

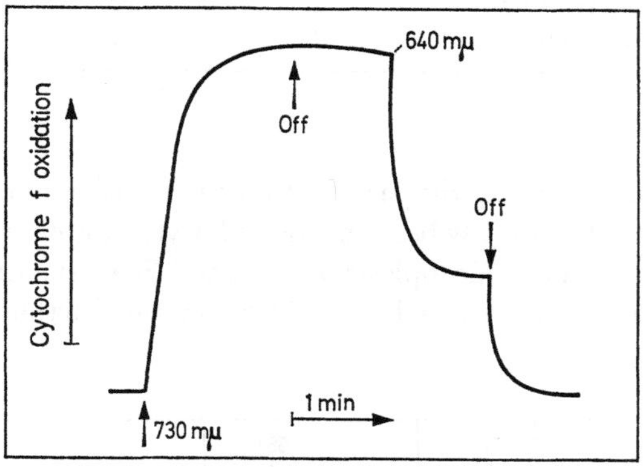

Fig. 4. The oxidation of cytochrome *f* by far-red light and its reduction by red light in isolated chloroplasts. (After Avron and Chance, 1966)

and therefore must have an electron acceptor other than ferredoxin (Chance *et al.*, 1965). Very recently a factor has been isolated and termed "ferredoxin reducing substance" (FRS) which seems to fullfill the function hitherto assigned to "X" (Yocum and San Pietro, 1969).

P_{700} is a chlorophyll-like compound, which was clearly shown by Kok and co-workers to serve as the initial electron donor to Photosystem I (Kok, 1965). Changes in its oxidation-reduction state have been observed by marked changes in absorption peaking around 700 mμ, which accompany them. Thus, upon illumination of Photosystem I (illumination with far-red light) the absorption at 700 mμ of chloroplasts decreases markedly, indicating the oxidation of P_{700}. This photooxidation still proceeds at liquid nitrogen temperature, and when activated by a short intense flash has been shown to be the fastest reaction observable. Therefore, it is generally assumed that P_{700} occupies the position of the closest donor of electrons to Photosystem I.

Plastocyanin is a copper containing protein specifically located in chloroplasts. It was first isolated and characterized by Katoh and co-workers, and was shown to be required in the transfer of electrons from ascorbate to NADP, but not from water to ferricyanide (Katoh and San Pietro, 1966). The exact position of plastocyanin in the electron transfer sequence rests mostly on evidence derived from mutants of the alga *Chlamydomonas* by LEVINE and co-workers (1969). A cytochrome-*f*-less mutant was shown to catalyze the photoinduced transfer of electrons from ascorbate to NADP, while a plastocyanin-less mutant was unable to catalyze this reaction (GORMAN and LEVINE, 1965, Table 2).

Table 2. *Photoreduction of NADP from water or ascorbate by several mutant strains of Chlamydomonas. (After* GORMAN *and* LEVINE, *1965)*

Chloroplast fragments isolated from	$H_2O \rightarrow$ NADP Relative rates	Ascorbate $\rightarrow$ NADP
Wild type	100	100
Cytochrome *f*-less mutant	0	53
Plastocyanin-less mutant	0	5

The evidence which indicates the participation and position of cytochrome *f* in the electron transfer chain has already been mentioned. Cytochromes *f* and *b* were shown to be components of isolated chloroplasts by HILL and SCARISBRICK (1951). Recently, fractionation of chloroplasts was achieved into two subfractions, one capable of

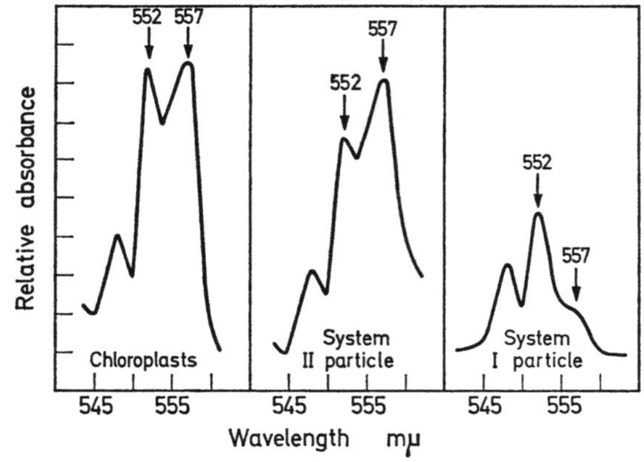

Fig. 5. The distribution of cytochromes between Photosystem II and Photosystem I enriched particles. Recorded are difference spectra between reduced (by ascorbate) and oxidized (by ferricyanide) particles, measured at 77 °K. The peak at 557 mµ corresponds to a b-type cytochrome, and that at 552 to cytochrome *f*. (After BOARDMAN and ANDERSON, 1967)

catalyzing Photosystem I dependent reactions only, such as the photoinduced transfer of electrons from ascorbate to NADP, and the other capable of catalyzing only Photosystem II dependent reactions, such as the photoinduced transfer of electrons from water to ferricyanide. BOARDMAN and ANDERSON (1967) found that a cytochrome *b* was located in the Photosystem II particle, while cytochrome *f* was located in the Photosystem I particle (Fig. 5).

Plastoquinone was shown to occur in chloroplasts in large amounts by CRANE and co-workers (LESTER and CRANE, 1959). Chloroplasts from which plastoquinone was removed are incapable of catalyzing the transfer of electrons from water to

Table 3. *The effect of extraction of plastoquinone on the photoreduction of NADP from water or ascorbate. (After* ARNON *and* HORTON, *1963)*

Electron donor	Treatment of chloroplasts		
	None	Extracted	Extracted + Plastoquinone
	Relative rates		
Water	100	14	82
Ascorbate	100	92	—

ferricyanide, but can still catalyze the ascorbate to NADP reaction (ARNON and HORTON, 1963; TREBST, 1963, Table 3). Several types of observations have indicated the existence of a large pool of electron accepting capacity close to Photosystem II in isolated chloroplasts. Plastoquinone, is the only known compound present in sufficient amounts (about one tenth of the total chlorophyll) to fulfill this function, and therefore its position is generally believed to be as indicated in Fig. 1.

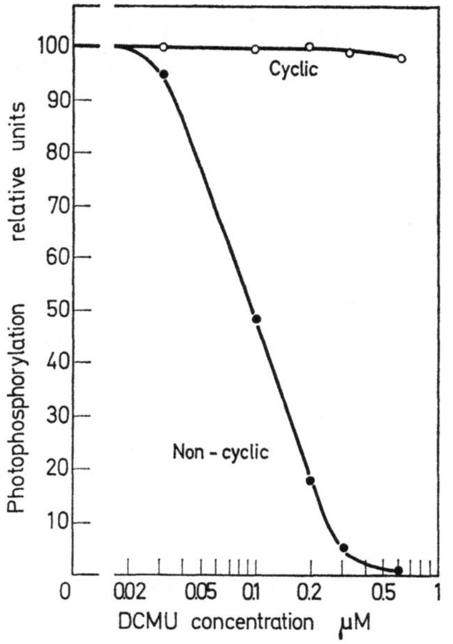

Fig. 6. The differential effect of DCMU on cyclic and non-cyclic photophosphorylation. In the experiment presented the cyclic photophosphorylation was that catalyzed by phenazine methosulphate, and the non-cyclic that coupled to the reduction of ferricyanide (unpublished experiment from the author's laboratory)

Q, Z, and E, are—as their names indicate—hypothetical intermediates. They were postulated to explain observations on transients in fluorescence yield, and in oxygen evolution during the first few seconds of illumination. The evidence supporting their existence and position will not be reviewed here (see Avron, 1967).

We have described the non-cyclic electron flow patterns which isolated chloroplasts can carry out. One from water to NADP, the second from ascorbate to NADP, and the third from water to ferricyanide. In addition, chloroplasts were early shown to catalyze a photoinduced cyclic electron flow. Upon the addition of a small amount of an electron carrier, such as phenazine methosulphate, or pyocyanine, or large amounts of ferredoxin, chloroplasts catalyze a light dependent phosphorylation which is independent of any net electron transfer, and is not inhibited by many of the inhibitors of non-cyclic electron transfer from water, such as DCMU (Fig. 6). This cyclic photophosphorylation was shown to be sensitized by Photosystem I only, and is therefore thought to proceed via the path indicated by the heavy arrows in Fig. 1. No intermediate electron carriers have been clearly indicated, but some evidence exists for the participation of a *b*-type cytochrome, different from the one associated with the non-cyclic electron flow (Cramer and Butler, 1967; Hind and Olson, 1966).

C. Coupling of Phosphorylation to Electron Flow

As already mentioned, the photoinduced electron transfer in chloroplasts is associated with concomitant ATP formation. The intimate relation and dependence of the phosphorylation on the electron transfer, is indicated by the ratio of about 1 ATP formed for each 2 electrons which traverse the chain from water to ferricyanide or to NADP, and by the enhanced rate of electron flow upon concomitant phosphorylation (Arnon, Whatley and Allen, 1958; Avron, Krogmann and Jagendorf, 1958).

The fact that about half of the electron transfer rate can proceed in the absence of concomitant phosphorylation, may be interpreted as indicating either poor coupling, or the existence of an equal number of phosphorylating and non-phosphorylating chains. Evidence supporting the second alternative was provided by Good and coworkers, who showed that under a variety of conditions which change the observed overall ATP/2 electrons ratio, a constant "corrected" ATP/2 electrons of 2 can be obtained if one considers only the electrons transported in the "phosphorylating chain" (Good, Izawa and Hind, 1966; Fig. 7).

The site or sites of ATP formation in chloroplasts are a subject of considerable current interest and controversy. A site of phosphorylation preceding cytochrome *f* in the electron transport has been clearly established by Avron and Chance, who showed that the addition of ADP to illuminated chloroplasts which accelerated the rate of electron flow brought about a reduction of cytochrome *f* (Avron and Chance, 1966, Fig. 8). The exact position of this site which precedes cytochrome *f* has not been established. If one accepts the suggestion, supported among others by the evidence on subfractionation mentioned earlier, that ferricyanide is reduced via cytochrome *f*, then the well established observation that the ATP/2 electrons ratio for the non-cyclic electron flow from water to ferricyanide or to NADP is the same, excludes another site for ATP formation between cytochrome *f* and NADP. This conclusion is in agreement with the contention of several laboratories (though objected to by others) that the electron transfer from ascorbate to NADP is devoid of a phosphorylating step. The difficulties in interpreting the data arise from the fact that all of the electrons carriers added to promote electron flow from ascorbate to NADP, also catalyze cyclic photophosphorylation. It is therefore extremely

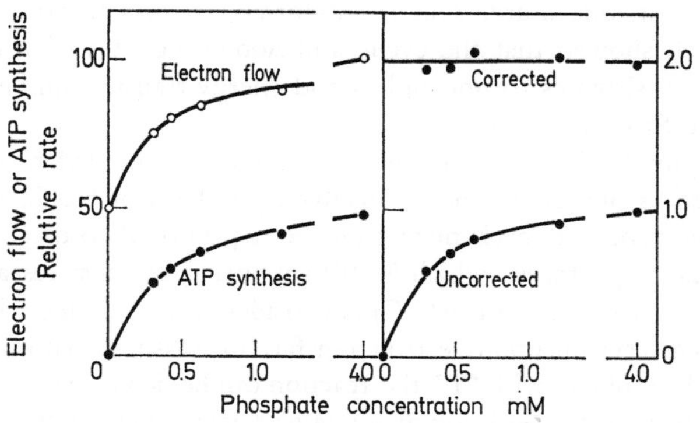

Fig. 7. The stimulation of the rate of electron flow by concomitant phosphorylation. The suspension contained a complete non-cyclic (ferricyanide) phosphorylating medium, except for the phosphate concentration which was varied, as indicated. [After GOOD, IZAWA, and HIND (1966), and a personal communication from Drs. IZAWA and GOOD]

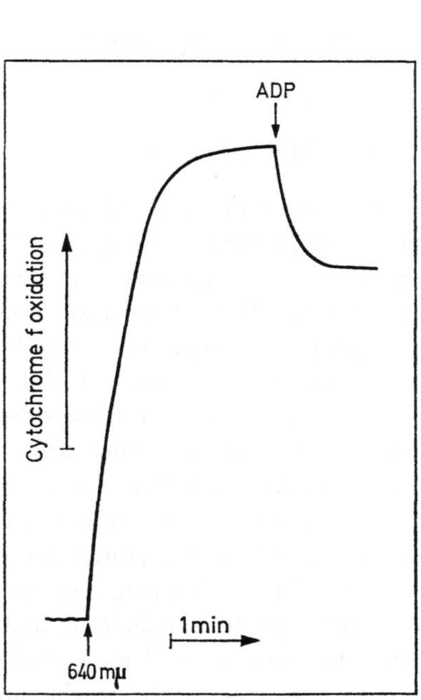

Fig. 8. The reduction of cytochrome *f* on ADP addition. The suspension contained, before the illumination with 640 mμ light, a complete phosphorylating medium, lacking only ADP. (After AVRON and CHANCE, 1966)

difficult to ascertain whether the phosphorylation which has been observed to occur concurrently with the electron transfer from ascorbate to NADP, does or does not originate from a simultaneously occurring cyclic phosphorylation (Avron and Neumann, 1968).

If we accept the evidence which indicates the absence of a site of phosphorylation between cytochrome *f* and NADP, we are forced to suggest that the site of ATP formation in cyclic phosphorylation is located at a different position and within the cycle, as indicated in the figure. This is supported, by observations from several laboratories which showed that the cyclic and non-cyclic sites of ATP formation possess different sensitivities to uncouplers and energy-transfer inhibitors (Arnon, 1965; Avron and Shavit, 1965).

The intimate mechanism whereby ATP formation is coupled to electron flow can be considered as one of the major unsolved problems of biochemical research today. Two basic types of mechanism have been proposed to-date. In the more classical "chemical" hypothesis, which is still supported by most researchers in the field, a component of the chloroplasts (I) is considered to combine with a member of the electron transfer chain, upon electron transfer. If we assume that it is the oxidized component which combines with "I", the reaction can be depicted as seen in Eq. (1) below. This electron-carrier $\sim$ I complex retains a major part of the energy which would have been dissipated in its absence during the electron transfer step. Next, the electron carrier is released from the complex, leaving a "non-phosphorylating high-energy-intermediate" (designated here $\sim$ I). The energy stored in this $\sim$ I is used to drive the endergonic synthesis of ATP from ADP and inorganic phosphate (Slater, 1967).

$$AH_2 + B \rightarrow BH_2 + A{\sim}I \quad (1)$$

$$A{\sim}I \rightarrow A + {\sim}I \quad (2)$$

$$ {\sim}I + ADP + Pi \rightarrow ATP + I \quad (3)$$

In the second, "chemiosmotic" hypothesis, no chemical high-energy intermediate involving an electron carrier is postulated. It is suggested that the electron carriers are arranged in such a way that at each coupling site there is an electron transfer from a reduced electron-carrier in which protons are bound, to one where these protons are released. Thus, appropriate couples for a coupling site could be a quinone and a cytochrome, or a flavin enzyme and plastocyanin, but not two cytochromes nor a quinone and a flavin enzyme. The essence of this hypothesis lies in its assumption that the protons are released in a directional manner into only one side of the membrane bound electron carrier. Therefore, electron transfer results in the accumulation of protons and positive charge, on only one side of the membrane. In these "proton gradient" and "membrane potential" lie the stored energy. This energy is then utilized by a direction-sensitive ATP synthesizing enzyme complex, which can use the energy released by returning the protons to their original position and so restoring proton and charge neutrality (Jagendorf and Uribe, 1967).

Several observations are in agreement with the chemiosmotic hypothesis, such as the observations of electron-transport induced proton movements in isolated chloroplasts, and ATP formation by an acid-base transition which will be discussed below. However, several recent observations are in strong disagreement with the theory, such as the observation made in Avron's laboratory that phosphorylation increases, rather than decreases, the extent of proton accumulation in chloroplasts (Karlish and Avron, 1967, 1968). Only further experimentation will decide which of the two presently expounded theories or what adaptations of these may be close to the actual mechanism.

D. Phosphorylation Inhibitors

Three distinct ways of inhibiting ATP formation have been described. These may be termed electron transport inhibition, uncoupling, and energy transport inhibition (Good, Izawa and Hind, 1966; Losada and Arnon, 1963, Fig. 9). An

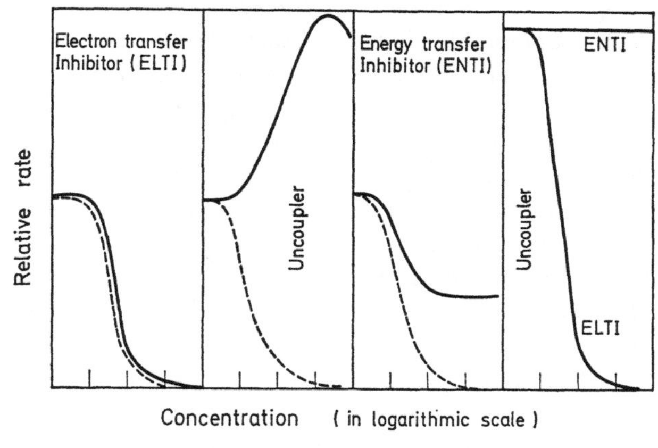

Fig. 9. An idealized drawing of the effect of varying concentrations of a typical electron transfer inhibitor, uncoupler or energy transfer inhibitor, on the electron transfer rate (solid lines) and ATP formation (dashed lines) in non-cyclic photophosphorylation. The right-hand figure depicts the effect of an energy transfer inhibitor, on the rate of electron transfer in the presence of a constant optimal uncoupler concentration. (After Avron and Shavit, 1965; Avron and Jagendorf, 1959; Izawa, Winget, and Good, 1966)

agent which inhibits electron-transport will obviously inhibit the phosphorylation associated with it. Among the common inhibitors which are considered to belong to this class are: DCMU [(3, (3,4-dichlorophenyl)-1, 1-dimethyl urea)], ortho-phenanthroline and NQNO (2-n-nonyl-4-hydroxyquinoline-N-oxide). A proposed site of action for DCMU is indicated in Fig. 1. This type of inhibitor reduces the rate of electron flow and of ATP formation to a similar degree, and the inhibition is unaffected by the further addition of an uncoupler (see below).

Uncouplers are thought to act by promoting the breakdown of a high-energy intermediate or "state". Thus, according to the chemical theory any compound which accelerates the conversion of $A \sim I$ to $A + I$ and $\sim I$ to I in Eqs. (1, 2, 3) above, is an uncoupler. According to the chemiosmotic theory also compounds which increase the permeability of the chloroplast membrane to protons (and so tend to decrease the proton gradient established during electron transport) are uncouplers. Uncouplers inhibit ATP formation without inhibiting electron transport. If the utilization of the high energy intermediate or state by the uncoupler is faster than it is during ATP formation an acceleration of electron flow may be seen. Under conditions where the rate of utilization of the high energy state limits the overall flow of electrons (such as in the absence of ADP, inorganic phosphate or Mg^{2+}) uncouplers do markedly stimulate the rate of electron flow. The simplest uncoupler described is arsenate which was shown to compete with phosphate for the active site on the ATP synthesizing machinery (AVRON and JAGENDORF, 1959). Since arsenate diester-bonds are easily hydrolyzed, the addition of arsenate accelerates the breakdown of the high energy state without ATP formation and so fulfills all the requirements of an uncoupler. It is of interest to note that the acceleration of electron flow, in the absence of phosphate, by arsenate was shown to require both Mg^{2+} and ADP. Thus the enzymic complex which catalyzes Reaction (3) must require ADP in order to be able to combine with phosphate to form a diester linkage. Other common uncouplers of phosphorylation are: ammonium ions, atebrin, derivatives of carbonyl-cyanide phenylhydrazones, and several antibiotics such as gramicidin. Since all of these, except arsenate, relieve the inhibition of electron transport imposed by energy transfer inhibitors (see below), they are considered to act at a point closer to the electron transport chain than the site of phosphate or arsenate action.

Energy transfer inhibitors act similarly to electron-transfer inhibitors in that they inhibit both electron transfer and ATP formation, although the latter somewhat more severely. However, their distinctive feature, is the release of their inhibition of electron flow (but not of ATP formation) by the further addition of an appropriate uncoupler (Fig. 9). Thus, they may be considered to act by preventing Reaction (3) above from taking place, and so inhibiting electron flow by removing the normal path for the utilization of $\sim$ I. Addition of an uncoupler, which provides an alternative path for the conversion of $\sim$ I to I, should relieve such inhibition of electron flow. Only two compounds have been described to date which act as energy transfer inhibitors in photophosphorylation: the glucoside phloridzin and the antibiotic Dio-9 (IZAWA, WINGET and GOOD, 1966; MCCARTY, GUILLORY and RACKER, 1965, see Fig. 1).

E. Partial Reactions Requiring a High Energy State

Several reactions have been described which utilize only a part of the ATP synthesizing machinery. These reactions are extremely useful in attempts to understand both the mechanism of coupling of energy transport to electron transport, and the sequence of the energy-transport reactions.

1. Adenosine Triphosphatase

Reversal of the reactions leading to ATP synthesis, should result in the hydrolysis of ATP or ATPase activity. Since this is an exergonic reaction no input of energy is required and the reaction should proceed spontaneously. In contrast to mitochondria, chloroplasts, however, do not significantly hydrolyze ATP, unless previously treated in one of several ways which activate their latent ATPase. Three procedures have been described, each of which activates an ATPase with different characteristics.

Treatment of chloroplasts with trypsin releases a calcium dependent ATPase, which has been separated and purified from chloroplasts (VAMBUTAS and RACKER, 1965). Treatment of the chloroplasts with ethylenediamine tetraacetic acid (EDTA), releases a protein from the chloroplasts, and completely inhibits their ability to catalyze photophosphorylation. However, they remain fully capable to catalyze light-induced electron flow. The released protein was shown to (a) restore the ability of the chloroplasts to phosphorylate upon its readdition to the chloroplasts (AVRON, 1963, Table 4) and (b) to be converted to the calcium dependent ATPase upon trypsin treatment (McCARTY and RACKER, 1966). An antibody to the purified EDTA-released protein (termed coupling-factor, since it recouples the non-phosphorylating EDTA-treated chloroplasts) inhibits photophosphorylation and the ATPase activity (McCARTY and RACKER, 1966). Thus, it is thought that this isolated protein may catalyze the last reaction in the ATP synthesizing sequence of chloroplasts.

Treatment of chloroplasts with sulfhydryl reagents induces a "light-triggered" ATPase (PETRACK *et al.* 1965). This ATPase becomes activated by a short period

Table 4. *Loss and reconstitution of cyclc phosphorylation by the chloroplasts coupling factor. (After* AVRON, *1963)*

Preparation	Photophosphorylating activity	
	No coupling factor	Coupling factor
	Relative units	
Chloroplast fragments	100	112
Chloroplast fragments treated with EDTA	1	39

of illumination, and thereafter no longer requires light and will hydrolyze ATP for 30 min in the dark at an unabated rate (HOCH and MARTIN, 1963). The light requirement seems to be related to a requirement for the establishment of a high-energy-state, which once having been established in the light, can be maintained in the dark by ATP. Thus, the "light-triggered state" decays in a minute or so if the addition of ATP is delayed after turning the light off. However, if ATP is added immediately upon turning the light off, no decay is apparent for at least 30 min. The reaction is a very useful one for studying the mechanism of ATP formation since it has a temporal separation of the "light-activating" stage and the "ATP hydrolysis" stage. It was shown, for example, that some uncouplers like atebrin, and all energy transfer inhibitors, affect only the "ATP hydrolysis" stage and not the "light activating stage". Other uncouplers like NH_4Cl inhibited both the light and dark stages, and all electron transfer inhibitors, like DCMU, inhibited only the "light-activating" stage (PETRACK *et al.* 1965; CARMELI and AVRON, 1967).

Finally, a "light-dependent" ATPase is induced in chloroplasts by the addition of calcium. This ATPase ceases to hydrolyze ATP as soon as the light is turned off (BENNUN and AVRON, 1965, Fig. 10).

All three ATPases are fully inhibited by the removal of the coupling-factor, and seem therefore to represent different variations of a reversal of the latter stages of ATP synthesis in photophosphorylation (McCarty and Racker, 1966).

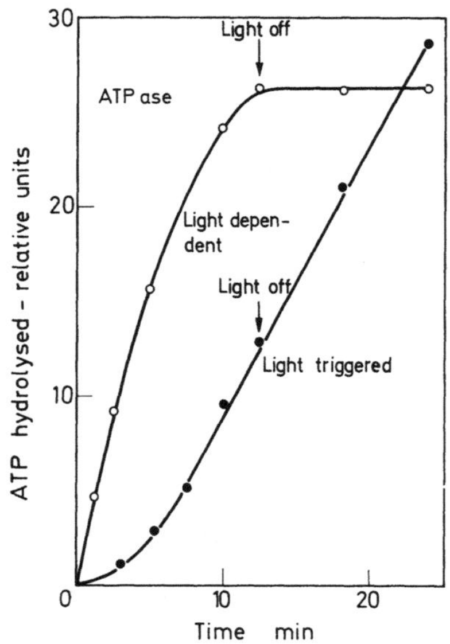

Fig. 10. Light dependent and light-triggered adenosine triphosphatase in isolated chloroplasts. (After Bennun and Avron, 1965)

2. ATP-Pi Exchange

When chloroplasts activated to catalyze the light-triggered-ATPase were placed in a solution containing ^{32}P-labeled inorganic phosphate, they were found to incorporate some of the radioactive phosphate into ATP (Carmeli and Avron, 1967). Thus, after light triggering, these chloroplasts both hydrolyze ATP and catalyze ATP-Pi exchange reaction. This "light-triggered ATP-Pi exchange" reaction was shown to possess very similar characteristics to the light-triggered ATPase. It responded to electron-transfer inhibitors, energy transfer inhibitors, and uncouplers in the manner described above for the light-triggered ATPase. However, as can be expected, arsenate did not inhibit the ATPase reaction, but fully inhibited the ATP-Pi exchange. It may be expected from Reaction (3) that a treatment which induced an ATP-Pi exchange reaction should also induce an ADP-ATP exchange reaction. This, however, was found not to be the case. One possible explanation may be that ADP is so strongly bound to the exchange enzyme that no exchange between it and the medium ADP is possible.

3. Two-Stage Photophosphorylation

ATP formation during photophosphorylation can also be performed in two stages: a light-stage and a dark-stage. Chloroplasts which are placed in the light, at pH 6 in the absence of phosphorylating reagents (magnesium, ADP and inorganic phosphate), are able to form ATP in the dark when subsequently transferred to a solution containing the phosphorylating reagents at pH 8 (HIND and JAGENDORF, 1964). Thus, a high energy intermediate (termed "Xe") is thought to be formed in the light, the energy of which is subsequently utilized to form ATP in the dark. "Xe" has a very short half-life of about 2 sec at pH 8, and 30 sec at pH 6. It amounts to about $^1/_5$ of the total chlorophyll in the chloroplast and therefore is generally considered to represent a "pool" of a high-energy state rather than an intermediate of the $\sim$ I type (JAGENDORF and URIBE, 1967). The latter would be expected to exist in concentrations similar to those of electron transport components such as cytochromes, which are present at a ratio of $^1/_{100}$ to $^1/_{500}$ that of the chlorophyll.

In two-stage photophosphorylation like in the light triggered ATPase and ATP-Pi reactions, one can test whether phosphorylation inhibitors inhibit the light or the dark stage of the reaction. Here again, energy transfer inhibitors and atebrin inhibited only the dark stage, NH_4Cl and carbonyl-cyanide phenylhydrazone derivatives inhibited both the light and dark stages, and electron transfer inhibitors only the light-stage (HIND and JAGENDORF, 1963; GROMET-ELHANAN and AVRON, 1965, Table 5).

Table 5. *The effect of inhibitors and uncouplers on the separate light and dark stages of ATP formation during two-stage photophosphorylation. (After* GROMET-ELHANAN *and* AVRON, *1965)*

Compound added	ATP formed with added compound	
	In light stage	In dark stage
	Relative units	
None	100	100
DCMU — 2 μM	14	90
Atebrin — 50 μM	100	0
NH_4Cl — 0.5 mM	70	30

4. Light-Induced Proton Gradient

When unbuffered chloroplasts are illuminated in the presence of electron acceptors or electron transport carriers, the pH of the medium increases, while the internal pH of the chloroplasts decreases. This light-induced proton-uptake was first observed by JAGENDORF and NEUMANN (1964) who showed that it had many properties in common with "Xe". It was maximal around pH 6, was inhibited by uncouplers and electron transport inhibitors, and amounted under optimal conditions to about 1 proton taken into the chloroplast per molecule of chlorophyll.

This reaction is of particular interest in terms of the chemiosmotic theory discussed earlier. It is a prediction of this theory that the energy source of ATP formation is created and maintained in the form of a gradient of protons (and possibly also of charge) across membranes. The properties of the light induced proton uptake are therefore in agreement with the chemiosmotic theory. However, recent studies of

Karlish and Avron (1967 and 1968) indicated that conditions under which ATP formation took place, not only did not decrease the light induced proton uptake, as predicted by the chemiosmotic theory, but considerably enhanced it. From this and other effects they suggest that the function of the light-induced proton pumping was to promote co-transport of the negatively charged complex of ADP, phosphate and magnesium into the interlamellar spaces, where it may be converted into ATP.

5. ATP Formation by an Acid-Base Transition

If indeed, as proposed by the chemiosmotic theory, proton gradients can be converted into ATP, it was thought possible that ATP could be formed by transferring chloroplasts from a low pH (high internal proton concentration) to a high pH (low external proton concentration) in the dark. Indeed, Jagendorf and collaborators clearly showed that ATP can be formed under such circumstances, provided a weak acid such as succinate was present in the acid stage (Jagendorf and Uribe, 1966). Again the properties of this reaction were in general agreement with the proton uptake and "Xe" data. The reaction was inhibited by the usual uncouplers of photophosphorylation, required a pH differential of at least 3.5 pH units between the acid and the basic stages, and under optimal conditions 1 μmole of ATP was synthesized per 7 μmoles of chlorophyll (Jagendorf and Uribe, 1967, Table 6).

Table 6. *ATP formation by an acid-base transition in the dark. (After* Jagendorf *and* Uribe, *1967)*

pH of acid stage	pH of basic stage	ATP formed mμmoles/μmole chl.
3.8	8.4	190
5.2	8.4	40
7.0	8.4	3
3.8	6.6	5

6. Light-Induced Conformational Changes

Conditions which induce proton-uptake into chloroplasts also induce a change in conformation which can be followed spectrophotometrically or by direct volume measurements (Packer and Crofts, 1967). As with all other energy dependent processes, the reaction is eliminated by uncouplers. It has a pH optimum around 6, and is dependent upon coupled electron transport. Many varieties of light induced conformational changes have been described, and it is clear that several different kinds of changes exist which lead to a conformational change. Some seem to be a direct result of the low pH produced in the interlamellar space by the light-induced proton uptake, while others seem to be more closely related to uptake or release of salts which results in water movement in or out of the interlamellar spaces.

F. Photophosphorylation *in vivo*

The existence and function of non-cyclic and cyclic photophosphorylation *in vivo* has been a subject of considerable controversy, cyclic photophosphorylation in particular has been suggested to be a test-tube artifact, since "the" endogenous electron transport mediator has not been found.

Several workers, however, provided rather convincing evidence that both non-cyclic and cyclic photophosphorylation exist *in vivo* and play a major role as energy suppliers for a multitude of cell activities over and above their function in the photosynthetic process, *per se*. Thus, glucose conversion into starch was shown to be strictly light dependent under anaerobic conditions (MACLACHLAN and PORTER, 1959). Active ion uptake in the roots was shown to be highly stimulated in the light, even in the total absence of CO_2, indicating that cyclic-photophosphorylation can provide sufficient energy to drive the process (MARRE, *et al.* 1966). Similarly, glucose uptake, acetate uptake, and potassium uptake into algal cells was clearly shown to proceed under conditions, where cyclic photophosphorylation provided the only path for ATP formation (WIESSNER, 1966).

It should, however, be borne in mind that the "*in vivo* cyclic photophosphorylation" may differ from some or all of the *in vitro* reactions studied, although it possesses similar characteristics.

G. Literature

ARNON, D. I.: Ferredoxin and photosynthesis. Science **149**, 1460—1470 (1965).

— ALLEN, M. B., WHATLEY, F. R.: Photosynthesis by isolated chloroplasts. Nature (Lond.) **174,** 394—396 (1954).

— HORTON, A. A.: Site of action of plastoquinone in the electron transport chain of photosynthesis. Acta chem. scand. **17**, S-135—S-139 (1963).

— WHATLEY, F. R., ALLEN, M. B.: Assimilatory power in photosynthesis. Science **127**, 1026—1034 (1958).

AVRON, M.: A coupling factor in photophosphorylation. In: La photosynthese, pp. 543—555. Paris: Edition Centre National de la Recherche Scientifique 1963.

— Mechanism of photoinduced electron transport in isolated chloroplasts. Current topics in bioenergetics (D. R. SANADI, Ed.) **2**, 1—23 (1967).

— BEN-HAYYIM, G.: Interaction between two photochemical systems in photoreactions of isolated chloroplasts. Proc. International Congress of Photosynthesis, Freudenstadt 1968, pp. 1185—1196.

— CHANCE, B.: The relation of light-induced oxidation reduction changes in cytochrome f of isolated chloroplasts to photophosphorylation. Currents in Photosynthesis, pp. 455—463 (J. B. THOMAS, J. C. GOEDHEER, Eds.). Rotterdam: Ad. Donker 1966.

— — Relation of phosphorylation to electron transport in isolated chloroplasts. Brookhaven Symposia in Biol. **19**, 149—160 (1966).

— JANGENDORF, A. T.: Evidence concerning the mechanism of adenosine triphosphate formation by spinach chloroplasts. J. biol. Chem. **234**, 967—972 (1959).

— KROGMANN, D. W., JAGENDORF, A. T.: The relation of photosynthetic phosphorylation to the Hill reaction. Biochim. biophys. Acta (Amst.) **30**, 144—153 (1958).

— NEUMANN, J.: Photophosphorylation in chloroplasts. Ann. Rev. Plant Physiol. **19**, 137—166 (1968).

— SHAVIT, N.: Inhibition and uncouplers of photophosphorylation. Biochim. biophys. Acta (Amst.) **109**, 317—331 (1965).

BENNUN, A., AVRON, M.: The relation of the light-dependent and light-triggered adenosine triphosphatases to photophosphorylation. Biochim. biophys. Acta (Amst.) **109**, 117—127 (1965).

Boardman, N. K., Anderson, J. M.: Fractionation of the photochemical systems of photosynthesis II. Cytochrome and carotenoid contents of a particle isolated from spinach chloroplasts. Biochim. biophys. Acta (Amst.) **143**, 187—203 (1967).

Carmeli, C., Avron, M.: A light-triggered adenosine triphophate-phosphate exchange reaction in chloroplasts. Europ. J. Biochem. **2**, 318—326 (1967).

Chance, B., San Pietro, A., Avron, M., Hildreth, W. W.: The role of spinach ferredoxin (PPNR) in photosynthetic electron transfer. In: Non-heme iron proteins, pp. 225—236 (A. San Pietro, Ed.). Yellow Springs (Ohio): Antioch Press 1965.

Cramer, W. A., Butler, W. L.: Light-induced absorbance changes of two cytochrome *b* components in the electron-transport system of spinach chloroplasts. Biochim. biophys. Acta (Amst.) **143**, 332—339 (1967).

Duysens, L. N. M., Amesz, J.: Function and identification of two photochemical systems in photosynthesis. Biochim. biophys. Acta (Amst.) **64**, 243—260 (1962).

Emerson, R., Lewis, C. M.: The dependence of quantum yield of *Chlorella* photosynthesis on wavelength of light. Amer. J. Bot. **30**, 165—178 (1943).

Frenkel, A. W.: Light induced phosphorylation by cell-free preparations of photosynthetic bacteria. J. Amer. chem. Soc. **76**, 5568—5569 (1954).

Good, N. E., Izawa, S., Hind, G.: Uncoupling and energy transfer inhibition in photophosphorylation. Current topics in bioenergetics (D. R. Sanadi, Ed.) **1**, 75—112 (1966).

Gorman, D. S., Levine, R. P.: Cytochrome f and plastocyanin: their sequence in the photosynthetic electron transport chain of *Chlamydomonas rehinhardii*. Proc. nat. Acad. Sci. (Wash.) **54**, 1665—1669 (1965).

Govindjee, R., Govindjee, Hoch, G.: Emerson enhancement effect in chloroplast reactions. Plant Physiol. **39**, 10—14 (1964).

Gromet-Elhanan, Z., Avron, M.: Effect of inhibitors and uncouplers on the separate light and dark reactions in photophosphorylation. Plant Physiol. **40**, 1053—1059 (1965).

Hill, R., Scarisbrick, R.: The haematin compounds of leaves. New Phytologist **50**, 98—101 (1951).

Hind, G., Jagendorf, A. T.: Separation of light and dark stages in photophosphorylation. Proc. nat. Acad. Sci. (Wash.) **49**, 715—722 (1963).

— Olson, J. M.: Light-Induced changes in cytochrome b_6 in spinach chloroplasts. Brookhaven Symposia **19**, 188—193 (1966).

Hoch, G., Martin, I.: Two light reactions in TPN reduction by spinach chloroplasts. Arch. Biochem. **102**, 430—438 (1963).

— — Photo-potentiation of adenosine triphosphate hydrolysis. Biochem. biophys. Res. Commun. **12**, 223—228 (1963).

Izawa, S., Winget, G. D., Good, N. E.: Phlorizin, a specific inhibition of photophosphorylation and phosphorylation-coupled electron transport in chloroplasts. Biochem. biophys. Res. Commun. **22**, 223—226 (1966).

Jagendorf, A. T., Uribe, E.: ATP formation caused by acid-base transition of spinach chloroplasts. Proc. nat. Acad. Sci. (Wash.) **55**, 170—177 (1966).

— — Photophosphorylation and the chemiosmotic hypothesis. In: Energy conversion by the photosynthetic apparatus, Brookhaven Symposia in Biol. **19**, 215—245 (1967).

Karlish, S. J. D., Avron, M.: Relevance of proton uptake induced by light to the mechanism of energy coupling to photophosphorylation. Nature (Lond.) **216**, 1107—1109 (1967).

— — Analysis of light-induced proton uptake in isolated chloroplasts. Biochim. biophys. Acta (Amst.) **153**, 878—888 (1968).

Katoh, S., San Pietro, A.: The role of plastocyanin in NADP photoreduction by chloroplasts. In: Biochemistry of copper, pp. 407—422 (J. Peisach, P. Aisen, W. E. Blumberg, Eds.). New York: Academic Press Inc. 1966.

Kok, B.: Photosynthesis: the path of energy. In: Plant biochemistry, 2nd edition, pp. 904—960 (J. Bonner, J. E. Varner, Eds.). New York: Academic Press 1965.

— Dakto, F. A.: Reducing power generated in the second photoact of photosynthesis. Plant Physiol. **40**, 1171—1177 (1965).

— Rurainski, A. J., Owens, O.: The reducing power generated in photoact I of photosynthesis. Biochim. biophys. Acta (Amst.) **109**, 347—356 (1965).

LESTER, R. L., CRANE, F. L.: The natural occurrence of coenzyme Q and related compounds. J. biol. Chem. **234**, 2169—2175 (1959).

LEVINE, R. P.: Analysis of photosynthesis using mutant strains of algae and higher plants. Ann. Rev. Plant Physiol. **20,** 523—540 (1969).

LOSADA, M., ARNON, D. I.: Selective inhibitors of photosynthesis. In: Metabolic inhibitors, pp. 559—593 (R. M. HOCHSTED, J. M. QUASTEL, Eds.). New York: Academic Press 1963.

MACLACHLAN, G. A., PORTER, H. K.: Replacement of oxidation by light as the energy source for glucose metabolism in tobacco leaf. Proc. roy Soc. B **150**, 460—473 (1959).

MARRE, E., FORTI, G., BIANCETTI, R., PARISI, B.: In: La photosynthèse, pp. 557—570. Paris: Edition Centre National Recherche Scientifique 1963.

MCCARTY, R. F., GUILLORY, R. J., RACKER, E.: Dio-9, an inhibitor of coupled electron transport and phosphorylation in chloroplasts. J. biol. Chem. **240**, PC-4822—PC-4823 (1965).

— RACKER, E.: A coupling factor in phosphorylation and hydrogen ion transport. Brookhaven Symposia in Biol. **19**, 202—214 (1966).

NEUMANN, J., JAGENDORF, A. T.: Light induced pH changes related to phosphorylation by chloroplasts. Arch Biochem. **107**, 109—119 (1964).

PACKER, L., CROFTS, A. R.: The energized movement of ions and water by chloroplasts. Current topics in bioenergetics (D. R. SANADI, Ed.) **2**, 24—64 (1967).

PETRACK, B., CARSON, A., SHEPPY, F., FARRON, F.: Studies on the hydrolysis of adenosine triphosphate by spinach chloroplasts. J. biol. Chem. **240**, 906—914 (1965).

RUBENS, S.: Photosynthesis and phosphorylation. J. Amer. chem. Soc. **65**. 279—282 (1943).

SAN PIETRO, A., BLACK, C. C.: Enzymology of energy conversion in photosynthesis. Ann. Rev. Plant Physiol. **16**, 155—174 (1965).

SAUER, K., BIGGINS, J.: Action spectra and quantum yields for nicotinamide adenine dinucleotide phosphate reduction by chloroplasts. Biochim. biophys. Acta (Amst.) **102**, 55—72 (1965).

— KELLY, R. B.: The Hill reaction of chloroplasts. Action spectra and quantum requirements. Biochemistry **4**, 2791—2798 (1965).

SLATER, E. C.: An evaluation of the Mitchell theory of chemiosmotic coupling in oxidative and photosynthetic phosphorylation. Europ. J. Biochem. **1**, 317—326 (1967).

TREBST, A.: The role of benzoquinones in the electron transport system. Proc. roy. Soc. B **157**, 355—366 (1963).

VAMBUTAS, V. K., RACKER, E.: Partial resolution of the enzymes catalyzing photophosphorylation I. Stimulation of photophosphorylation by a preparation of a latent, Ca^{++} dependent adenosine triphosphatase from chloroplasts. J. biol. Chem. **240**, 2660—2667 (1965).

WIESSNER, W.: Relative quantum yields for anaerobic photoassimilation of glucose. Nature (Lond.) **212**, 403—404 (1966).

YOCUM, C. F., SAN PIETRO, A.: Ferredoxin reducing substance from spinach. Biochem. biophys. Res. Commun. **36**, 614—617 (1969).

ZWEIG, G., AVRON, M.: On the oxidation-reduction potential of the photoproduced reductant of isolated chloroplasts. Biochem. biophys. Res. Commun. **19**, 397—400 (1965).

Carbohydrate Metabolism by Chloroplasts

MARTIN GIBBS

The organic nutrition of the chloroplast can be divided into two parts: (a) the autotrophic phase whereby carbon dioxide is converted to a phosphorylated hexose molecule (fructose 6-P) and (b) the heterotrophic phase whereby the carbon of this molecule is elaborated into ultimately usable chemical forms or into structural and other required components of the plant cell. The purpose of this chapter is to present detailed information on the autotrophic phase of chloroplast metabolism including intermediates of the photosynthetic carbon reduction cycle and materials immediately derived from the cycle. In the succeeding chapter of GOODWIN, biosynthesis in the chloroplast of so-called secondary products such as lipids, proteins and pigments will be emphasized.

A. Isolation of Chloroplasts

The bulk of our knowledge concerning chloroplastic CO_2 assimilation has been obtained using spinach *(Spinacia oleracea)* or pea *(Pisum sativum)* chloroplasts isolated by grinding leaf blades in an aqueous buffered solution of suitable osmotic pressure. Chloroplasts prepared in an aqueous medium are exposed to leaching of enzymes and ferredoxin among other water soluble components. A second method was developed —the non-aqueous technique, in order to prevent such losses.

1. Aqueous Methods

Chloroplasts are isolated in buffered isotonic solutions. Chloroplasts prepared in mixtures containing sugars including sucrose, glucose or fructose (WALKER, 1964; SPENCER and UNT, 1965), or sugar alcohols such as sorbitol (WALKER, 1965; JENSEN and BASSHAM, 1966) or mannitol (KALBERER *et al.*, 1967) show high rates of CO_2 fixation and O_2 evolution for periods up to 1 h. Chloroplasts prepared in NaCl (ALLEN *et al.*, 1955; CALO and GIBBS, 1959) show far lower photosynthetic capacity for CO_2 fixation and associated O_2 evolution. The advantage of sugar and sugar alcohols over sodium chloride in maintaining the activity of isolated chloroplasts for CO_2 assimilation is shown in Table 1.

Buffering capacity in the extracting medium has usually been maintained by tris (ALLEN *et al.*, 1955), pyrophosphate (COCKBURN *et al.*, 1968) or the N-substituted taurine, 2-(N-morpholine) ethanesulfonic acid (MES) (JENSEN and BASSHAM, 1965). According to COCKBURN *et al.* (1968), chloroplasts stored at ice temperature in a sorbitol pyrophosphate medium retain their photosynthetic capacity up to 4 h.

The appearance of the chloroplast is greatly modified by the particular isolation procedure employed (WALKER, 1965). When monitored by phase-contrast microscopy, spinach and pea chloroplasts appear to be of two kinds. Those prepared in sugar are shiny, highly refractive, haloed and retain their outer double envelopes.

These are considered ideal for extra-cellular CO_2 fixation. On the other hand, chloroplasts dark and granulated in appearance and whose outer membrane is missing usually function poorly. There is ample evidence correlating the integrity of the bounding membrane and photosynthetic capacity. Nonetheless, this apparent relationship must be regarded with caution. Thus, JENSEN and BASSHAM (1965) prepared chloroplast suspensions containing approximately 75% of the highly refractive type which fixed CO_2 at a rate of 200 μmoles/mg chlorophyll·h. A large proportion of chloroplasts with bounding membranes as monitored by phase-contrast microscopy is not in itself sufficient to bring about high rates of photosynthesis since SPENCER and UNT (1965) recorded a rate of 2.6 μmoles/mg chlorophyll·h with a highly refractive preparation.

Table 1. *Comparison between sugars and related compounds with sodium chloride for CO_2 assimilation by isolated spinach chloroplasts*[a]. *The rates of photosynthesis in the mannitol treatment were 45.5, 42.9 and 34.6 μmoles CO_2 fixed/mg chlorophyll·h, respectively. The values in the table are relative to the mannitol rate*

Compound 0.35 M	Compound in col. 1 used in both preparative and assay solutions	Compound in col. 1 used in assay solution and 0.35 M sorbitol in preparative solution	Compound in col. 1 used in preparative solution and 0.35 M sorbitol in assay solution
Sodium chloride	26	55	84
Mannitol	100	100	100
Inositol	95	88	85
Sorbitol	92	96	89
Sucrose	92	93	92
Glucose	81	90	96

[a] From KALBERER *et al.* (1967).

Recently a great deal of attention has been given to the two classes of chloroplasts in the tropical dicotyledons and grasses such as *Atriplex*, sugarcane and maize (HODGE, McLEAN and MERCER, 1965; LAETSCH and PRICE, 1969). One class, which is predominantly located in the palisade cells of the leaf, have abundant grana but do not store large amounts of starch. The other class, located within the vascular bundle sheath cells are larger in size, richer in starch grains but in most species do not have grana. On the basis of difference in their average densities due to starch content, BALDRY, COOMBS and GROSS (1969) have reported on the separation of the two types of chloroplasts from sugar cane into a fraction (6:1) rich in the grana-type and a fraction containing a large proportion of the non-grana type. As tested by Hill reaction with 2,6-dichlorophenolindophenol as oxidant, their preparations reduced 40 μmoles dye/mg chlorophyll·h. This compares to a rate of 100 in unfractionated chloroplast preparations. Unfractionated chloroplast suspensions of bermudagrass *(Cynodon dactylon)* can catalyze ATP formation coupled to ferricyanide reduction at a rate of 100 μmoles phosphate esterified/mg chlorophyll·h (CHEN *et al.*, 1969). The bermudagrass preparations could also catalyze cyclic photophosphorylation in the presence of phenazine methosulfate. The rates of TPN reduction for fragmented maize and sugarcane chloroplast preparations were found by HEW and GIBBS (1969) to be 140 and 40 μmoles/mg chlorophyll·h, respectively. The ratio for TPN reduction: ATP formation: O_2 evolution was 2:2:1 and the reactions were saturated at approximately 1000 foot-candles.

Data are still fragmentary but the general profile of the light-driven reactions in the dimorphic chloroplasts appear similar to that recorded for spinach and pea preparations. A functional difference in the two types of chloroplasts in sugarcane and maize with respect to photosynthetic carbon metabolism have been envisaged and are discussed in detail elsewhere (see p. 175).

2. Non-Aqueous Methods

These methods were developed in order to prevent the diffusion of hydrophilic components from the chloroplast during isolation. After freeze-drying the leaves, chloroplasts are removed in the dry state with a non-polar solvent such as carbon tetrachloride together with petroleum ether (THALACKER and BEHRENS, 1959) or hexane (STOCKING, 1959). The organelles are separated on the basis of a density gradient. This procedure is well suited for localization studies but suffers from the fact that the particles prepared in non-aqueous solvents have lost the capacity to perform the light-catalyzed steps of photosynthesis. As examples, it has been employed to determine the distribution of enzymes (Chap. C), inorganic ions (STOCKING and ONGUN, 1962), amino acids (AACH and HEBER, 1967), and pyridine nucleotides (HEBER and SANTARIUS, 1965).

B. Compounds Formed during Photosynthetic CO_2 Assimilation and Their Intramolecular Labeling Patterns

1. Spinach and Pea

The products of $^{14}CO_2$ fixation during 2 min of photosynthesis in spinach leaf discs are shown in the radioautograph in Fig. 1. Since the pool sizes of the intermediates of the photosynthetic carbon reduction cycle are small, isotope moves quickly into the "sinks" such as sucrose, glycerate, glycerate 3-P and the amino acids — aspartate, alanine, glycine and serine. In contrast, the distribution of radioactivity in isolated spinach chloroplasts (Fig. 2), shows relatively large amounts of ^{14}C in glycerate 3-P and dihydroxyacetone-P with lesser amounts in ribulose and fructose diphosphates, fructose, ribose and glucose monophosphates as well as glycolate. An insoluble polyglucan is also formed. In some circumstances, sucrose is synthesized (EVERSON *et al.*, 1967, MIYACHI and HOGETSU, 1970). The significance of accumulation of glycolate in the chloroplast preparation but its absence in the intact tissue is treated in detail elsewhere (see p. 203). Major differences in the two labeling patterns are evident. In the chloroplast, a relatively high amount of isotope is located in the intermediates of the carbon reduction cycle while the amino and organic acids usually associated with the citric acid cycle are unlabeled.

Chloroplasts of spinach prepared in NaCl (ALLEN *et al.*, 1955; GIBBS *et al.*, 1967) and of pea in sorbitol (WALKER, 1965) have distribution patterns remarkedly similar to that illustrated in Fig. 2.

Clearly the radioautography of Fig. 1 and 2 are characteristic of the pathway proposed by CALVIN and his associates (BASSHAM *et al.*, 1954) and designated as the photosynthetic carbon reduction cycle (Fig. 3).

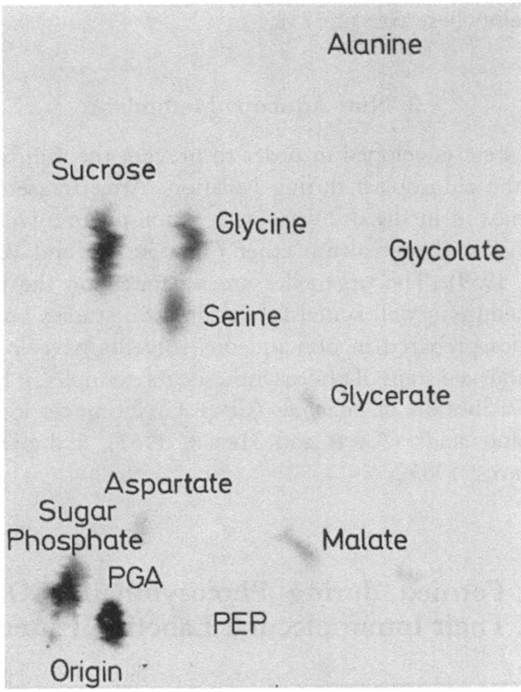

Fig. 1. Radioautography of $^{14}CO_2$ photosynthetic products in spinach discs. The discs were allowed to assimilate $^{14}CO_2$ for 2 min. The products were analyzed by two dimensional paper chromatography and radioautography. *PGA* glycerate 3-P; *PEP* P-enolpyruvate

The appearance of isotope in a limited number of intermediates of the photosynthetic carbon reduction cycle (Fig. 2) is only suggestive evidence that a completely functional reductive CO_2 cycle is present in the isolated organelle. For this reason, it is instructive to determine whether during assimilation of $^{14}CO_2$, isotope spreads throughout the molecules in a manner similar to that found in the intact photosynthetic cell. In a preliminary study, GIBBS and CYNKIN (1958) reported a similarity in photosynthetic $^{14}CO_2$ assimilation between the isolated chloroplast and the leaf of spinach. They observed an asymmetrical and eventual uniform distribution of radiocarbon in the glucosyl residue of chloroplast polysaccharide formed from $^{14}CO_2$ (Table 2). KANDLER and GIBBS (1957) observed this pattern earlier with intact algal and higher plant tissues. Using intact spinach chloroplasts HAVIR and GIBBS (1963) comfirmed and extended the study of GIBBS and CYNKIN (1958). On the other hand, two groups (TREBST and FIEDLER, 1961, 1962; HAVIR and GIBBS, 1963) could observe only a relatively limited spread of tracer in glucose 6-P, fructose 1,6-diP, glycerate 3-P and dihydroxyacetone-P formed by a reconstituted spinach chloroplast preparation fortified with each of the following substrates: ribose 5-P and $^{14}CO_2$, ribulose 1,5-diP and $^{14}CO_2$ or 1-^{14}C-glycerate 3-P. The reconstituted chloroplast preparation unlike the intact chloroplast requires a primer such as ribose

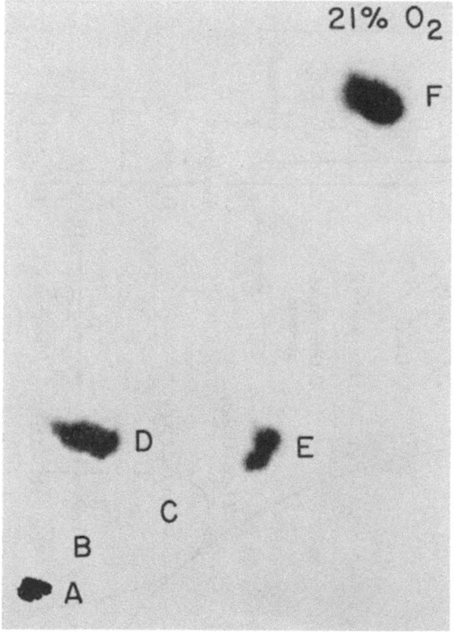

Fig. 2. Radioautography of $^{14}CO_2$ photosynthetic products in spinach chloroplasts isolated by the method of JENSEN and BASSHAM (1966). Photosynthesis was allowed to proceed in air for 15 min in $^{14}CO_2$. *A* insolubles, *B* sugar diphosphates, *C* hexose monophosphates, *D* glycerate 3-P, *E* dihydroxyacetone-P, *F* glycolate

5-P for CO_2 fixation. Furthermore, on a molar basis, in contrast to the whole chloroplast (WALKER, 1967), the amount of CO_2 assimilated by the reconstituted preparation has never been demonstrated to exceed the primer added.

On the basis of the intramolecular distribution data it is concluded that the whole chloroplast possesses a complete photosynthetic carbon reduction cycle for converting CO_2 into carbohydrate. In the reconstituted system, the cycle is operating at a rather limited rate, if at all. The inability of the reconstituted systems to bring about the tracer distribution observed with the intact preparations remains to be elucidated. Most likely, the regeneration of the CO_2-accepting molecule is the rate-limiting factor (KANDLER *et al.*, 1965).

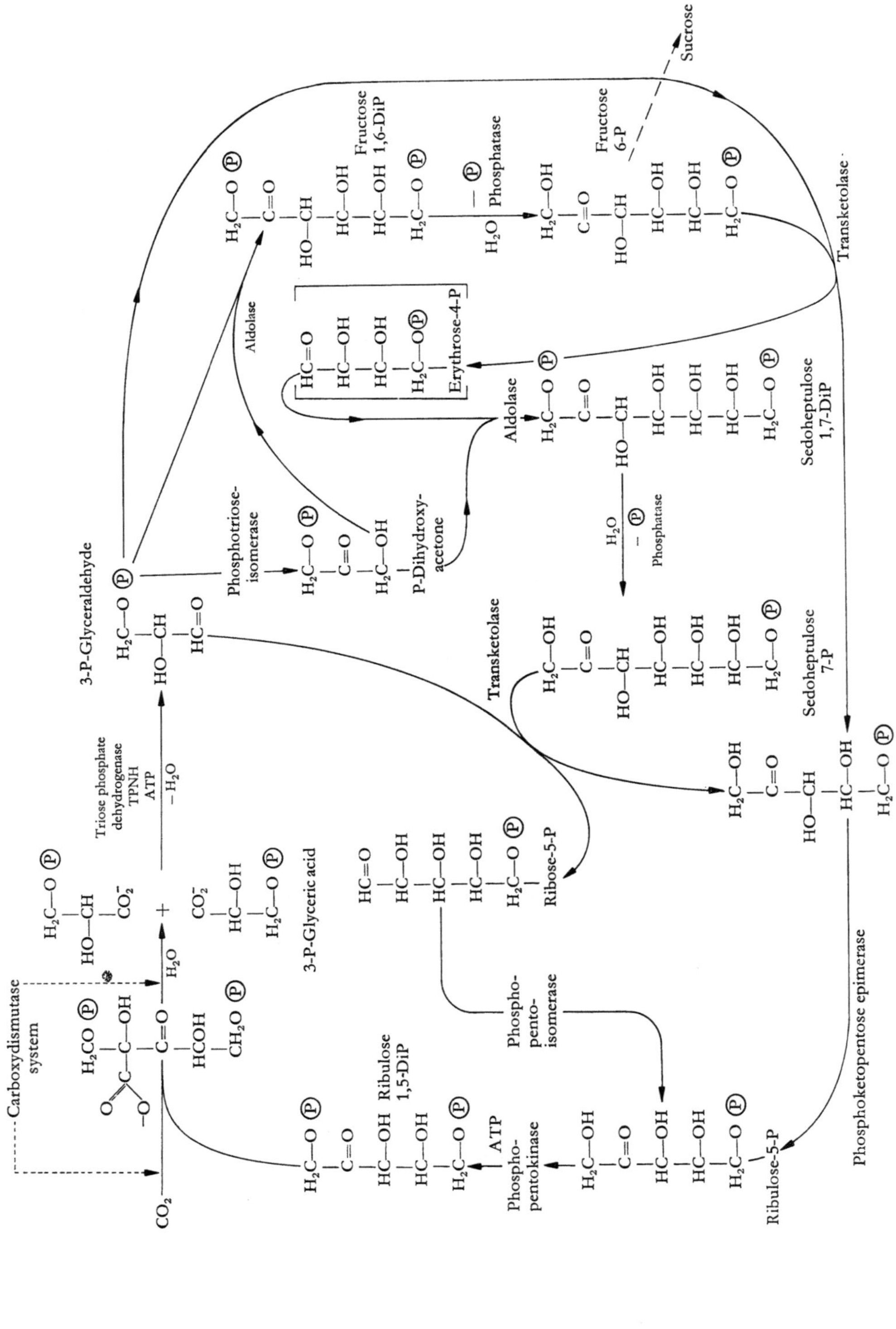

Fig. 3. The photosynthetic carbon reduction cycle. From Calvin and Bassham (1962)

Table 2. *Distribution of ^{14}C in polysaccharide glucose and glycerate 3-P during photosynthetic $^{14}CO_2$ fixation by whole spinach chloroplasts. The 3 and 30 min incubations are those of* GIBBS *and* CYNKIN *(1958); the 4 min and 10 min incubations (*HAVIR *and* GIBBS, *1963)*

Compound degraded	^{14}C content after reaction time of 3 min	30 min	4 min	10 min
Carbons of polysaccharide glucose				
C-1	23[a]	64	72	78
C-2	7	56	62	61
C-3	51	88	86	94
C-4	100	100	100	100
C-5	28	48	40	60
C-6	23	44	45	67
Carbons of glycerate 3-P				
COOH			100	100
CHOH			55	63
CH_2OP			62	63

[a] The carbon content is based on C-4 or COOH = 100.

2. Maize and Related Plants

In contrast to spinach and pea, when $^{14}CO_2$ is fed to a leaf of sugarcane or maize, the first stable compounds to become radioactive are malic and aspartic acids (KORTSCHAK *et al.*, 1965; HATCH and SLACK, 1966). Glycerate 3-P contains radiocarbon after these acids and before hexose phosphates (Table 3). A survey of the early products of photosynthesis in the Australian laboratory has demonstrated this labeling pattern in several plants including sorghum, carpet grass and sedge (HATCH *et al.*, 1967), as well as species of *Amaranthus*, *Cyperus* and *Atriplex* (JOHNSON and HATCH, 1968). Subsequently KANDLER and SENSER (1968) have reported similar data for sorghum. In addition to kinetic data, the Australian group has presented enzymic data and this aspect is dealt with in Section C.2. To account for CO_2 assimilation in these plants

Table 3. *Distribution of isotope in percentage of photosynthetic products in sugarcane. (Adapted from* KORTSCHAK *et al., 1965)*

	Photosynthetic $^{14}CO_2$ fixation for time (sec) 0.6	6.8	10	30
Glycerate 3-P	5.5	20	17	21
Glycerate	0	2	2	1
Aspartate	21	15	18	12
Malate	62	50	42	28
Sucrose	0	1	1	4
Fructose 1,6-diP	0	1	1	2

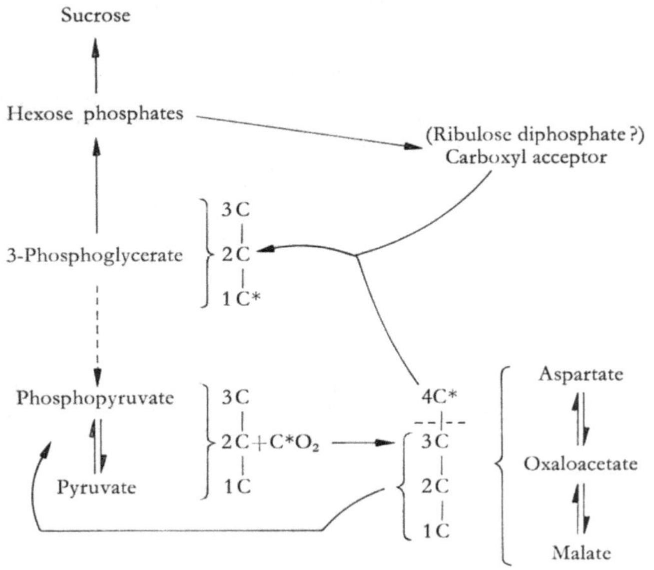

Fig. 4. Pathway proposed by HATCH and SLACK (1966) to account for photosynthetic fixation of carbon dioxide in sugar cane leaves

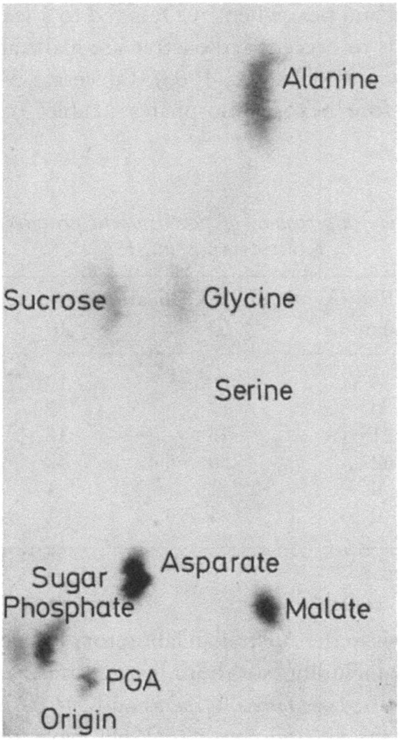

Fig. 5. Radioautography of $^{14}CO_2$ photosynthetic products in maize discs. The period of photosynthesis was 2 min

they have proposed a pathway involving the C_4-dicarboxylic acids (oxaloacetate and malate) in the primary carboxylation reaction (Fig. 4).

In the C_4-dicarboxylic pathway, radioactivity is first incorporated in the C_4-(β-carboxyl) of oxaloacetate or malate and is subsequently transferred to sugar via glycerate 3-P. Furthermore the two cyclic processes summarized in Fig. 4 are thought linked by a transcarboxylation reaction involving a 2-carbon compound such as glycolaldehyde-P or with ribulose 1,5-diP to yield glycerate 3-P. KORTSCHAK (1968)

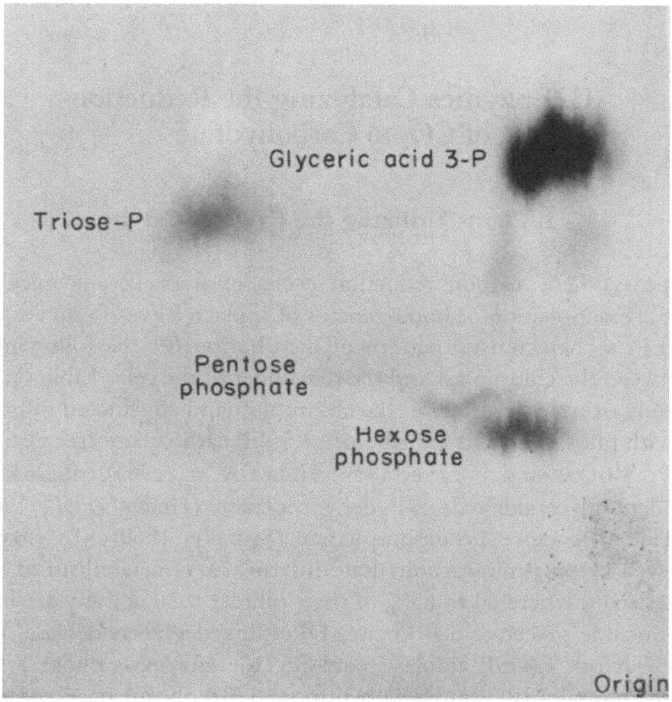

Fig. 6. Radioautography of $^{14}CO_2$ photosynthetic products in maize chloroplasts. The period of photosynthesis was 30 min. Pentose-P (ribose, ribulose); hexose-P (glucose, fructose)

and GALMICHE (1968) have presented experimental evidence casting doubt on the idea of a transcarboxylation reaction. The kinetics of incorporation of ^{14}C from $^{14}CO_2$ into the various carbon atoms of malate, glycerate 3-P and sucrose in sugarcane have been studied by HATCH and SLACK (1966). The labeling patterns in glycerate 3-P and in the hexoses of the sucrose are consistent with a reversal of glycolysis. FARINEAU (1969) has also provided evidence using ^{32}P that the phosphorylated intermediates associated with the Calvin cycle are involved in photosynthesis in the maize plant. The distribution of radiocarbon within the glycerate 3-P and hexose is also consistent to that found in molecules formed in photosynthetic tissue where the reductive carbon cycle of Calvin is quite secure. It is clear that tissues using one pathway or the other cannot be distinguished by intramolecular labeling techniques.

In contrast to the rather large volume of information pertaining to photosynthetic carbon metabolism in spinach and pea, little is available for chloroplasts utilizing the C_4-dicarboxylic acid pathway. In Fig. 5 and 6, the distribution of $^{14}CO_2$ photosynthetic products in isolated maize chloroplasts are contrasted with those in whole leaves. For short periods of photosynthesis, HEW and GIBBS (1969) and GIBBS *et al.* (1970) found the products formed by a maize chloroplast suspension to resemble those characteristic of the cycle depicted in Fig. 3. The rate of dark fixation by the maize chloroplasts was 10% of that in the light.

C. Enzymes Catalyzing the Reduction of CO_2 to Carbohydrate

1. Plants Utilizing the Cycle of Calvin

The photosynthetic carbon reduction cycle involves 12 enzymically catalyzed steps (Fig. 3). Fractionation of homogenates of spinach leaves in non-aqueous media has resulted in a characteristic pattern of distribution for the following groups of enzymes between the chloroplast and the remainder of the cell (Table 4): (a) enzymes located entirely or nearly entirely in the chloroplast and considered to be dominantly associated with photosynthesis are ribulose 1,5-diP carboxylase (see also FULLER and GIBBS, 1959; SMILLIE and FULLER, 1959; HEBER *et al.*, 1963), ribulose 5-P kinase, TPN-dependent glyceraldehyde 3-P dehydrogenase (HEBER *et al.*, 1963) and the alkaline (pH 8.5) fructose 1,6-diphosphatase (SMILLIE, 1960); (b) enzymes usually associated with cytoplasmic carbohydrate intermediary metabolism as well as with photosynthesis and where 60 to 65% of their cellular total activity are located in the chloroplast include glycerate 3-P kinase, DPN-linked glyceraldehyde 3-P dehydrogenase and fructose 1,6-diP aldolase; and (c), the enzymes, ribose 5-P isomerase, transketolase, transaldolase, and sedoheptulose 1,7-diphosphatase considered to be members of the oxidative pentose phosphate cycle as well as photosynthesis and whose chloroplast content is roughly 30 to 35% of the total cellular activity.

Structural integrity has been considered to be related to the ability of the chloroplast to assimilate CO_2 (p. 169). Isolated chloroplasts possessing outer membranes (sorbitol chloroplasts) are thought to retain more of the enzymes catalyzing reduction of CO_2 to carbohydrate than those prepared in NaCl (salt chloroplasts) whose limiting membranes are ruptured. According to experiments of LATZKO and GIBBS (1968), the low rate of CO_2 fixation observed in spinach chloroplasts isolated in 0.35 M NaCl (GIBBS and CALO, 1959) in place of sorbitol (JENSEN and BASSHAM, 1966) is not the result of the leaching out of enzymes of the photosynthetic carbon reduction cycle (Table 5). The major difference between the two preparations was the higher protein to chlorophyll ratio in the sorbitol prepared chloroplasts. On a chlorophyll basis, the specific activities of the enzymes extracted from the sorbitol chloroplast are higher while on a protein basis, enzymes extracted from chloroplasts prepared in

NaCl have higher specific activities. It would appear that proteins other than those directly catalyzing the conversion of CO_2 to carbohydrate are removed during isolation in NaCl.

In an enzyme profile of glycolysis (Wu and Racker, 1959) the least active enzyme had a capacity at least several-fold higher than the overall process. Of the photosynthetic enzymes listed in Fig. 3, Peterkofsky and Racker (1961) found low amounts of ribulose 1,5-diP carboxylase, and of the diphosphatases of sedoheptulose 1,7- and fructose 1,6-diP in extracts of spinach (see also column 1, Table 4 for similar

Table 4. *Specific activity, recovery and distribution of photosynthetic carbon reduction cycle enzymes in young spinach leaves*[a]. *The specific activities (sp. a.) are given as μmoles of substrate consumed per mg protein · h*

	Fresh leaves sp.a.	Freeze dried sp.a.	Recovery after freeze drying %	Total activity located in chloroplast %
Ribulose 5-P kinase	13	7.2	55	102
Ribulose 1,5-diP carboxylase	0.89	0.88	99	100
Glycerate 3-P kinase	92	87	95	61
Glyceraldehyde 3-P dehydrogenase (DPN)	16.4	6.8	41	60
Glyceraldehyde 3-P dehydrogenase (TPN)	28	13.2	47	109
Triose-P isomerase	109	81	74	—
Fructose 1,6-diP aldolase	10.9	9.7	89	65
Fructose 1,6-diPase, pH 8.5	2.4	1.07	45	81
Fructose 1,6-diPase, pH 5.5	0.78	0.75	99	12
Sedoheptulose 1,7-diPase	0.18	0.07	39	28
Transketolase	2.5	0.73	29	35
Ribose 5-P isomerase	25.6	23	90	35
Xylulose 5-P epimerase	36	0	0	—
Transaldolase	0.38	0.1	26	—

[a] From Latzko and Gibbs (1968).

data) and the photoautotrophically grown *Chlorella* and *Euglena* when compared to the overall photosynthetic rate. Latzko and Gibbs (1969) extended their study by comparing the rate of photosynthetic CO_2 fixation with the enzyme activities of the photosynthetic carbon reduction cycle in *Euglena* during chloroplast development, *Chromatium*, in spinach leaves and in *Chlorella pyrenoidosa* grown photoautotrophically in the presence and absence of glucose or heterotrophically with glucose as the sole source of carbon. Only in *Chromatium* were all enzyme activities sufficient to support the *in vivo* rate of CO_2 fixation. In agreement with Peterkofsky and Racker (1961), they found the O_2 evolving cells to be low in ribulose 1,5-diP carboxylase, fructose 1,6-diphosphatase (pH 8.5) and sedoheptulose 1,7-diphosphatase. However, information other than enzyme profile studies lead to the conclusion that at least ribulose

1,5-diP carboxylase and fructose 1,6-diphosphatase are involved in photosynthetic CO_2 fixation. Thus a striking enhancement in the level of carboxylase is observed in greening tissue (HUFFAKER *et al.*, 1966; CHEN *et al.*, 1967) and a suppression of this enzyme is found during heterotrophic growth of *Chlorella* (LATZKO and GIBBS, 1969) and *Rhodospirillum rubrum* (ANDERSON and FULLER, 1967). SMILLIE (1960) has demonstrated a close correspondence between the alkaline fructose 1,6-diphosphatase and chloroplast development.

Table 5. *CO_2 fixation and specific activity of enzymes in spinach chloroplasts prepared by different methods*[a]. *The enzyme specific activities are presented as μmoles of substrate consumed/mg protein·h or chlorophyll·h*

	Salt- 15		Sorbitol-chloroplasts 46	
Ratio: Prot./Chl.	Prot.	Chl.	Prot.	Chl.
CO_2 fixation	0.47	7	2.68	123
Ribulose 5-P kinase	15.4	230	10.5	480
Ribulose 1,5-diP carboxylase	0.57	8.5	0.61	28
Glycerate 3-P kinase	95	1420	61	2800
Glyceraldehyde 3-P dehydrogenase (DPN)	9	135	6.5	300
Glyceraldehyde 3-P dehydrogenase (TPN)	14	210	11	505
Triose-P isomerase	284	4250	210	9650
Fructose 1,6-diP aldolase	5.3	79	4.1	189
Fructose 1,6-diPase, pH 8.5	3.1	46	2.2	101
Sedoheptulose 1,7-diPase	—	—	0.09	4.1
Transketolase	8.1	122	5.8	267
Ribose 5-P isomerase	18.4	276	12.3	565
Xylulose 5-P epimerase	37	550	35	1610
Transaldolase	0.1	1.5	0.1	4.6
P-Glucomutase	1.0	15	0.7	32
Hexose-P isomerase	0.7	10.5	0.8	37
Myokinase	3.8	57	2.2	101
Inorganic pyrophosphatase	10.4	157	8.2	380
Glycerate dehydrogenase	23	350	15	690
Glycerate 3-P dehydrogenase	1.4	21	1.9	87
Glyoxylate reductase	4.3	65	8.7	400
Phosphorylase	0.3	4.3	—	—

[a] From LATZKO and GIBBS (1968).

The data in Table 5 indicate the presence of enzymes in the chloroplast responsible for converting CO_2 to fructose 6-P. BIRD *et al.* (1965) have reported that the enzymes necessary for the conversion of fructose 6-P to UDP-glucose and on to sucrose are localized within the chloroplast. Starting with UDP-glucose, sucrose synthesis can be catalyzed by sucrose synthetase (Eq. 1),

$$\text{UDP-glucose} + \text{fructose} \rightarrow \text{sucrose} + \text{UDP} \quad (1)$$

or by
sucrose phosphate synthetase (2) and sucrose phosphate phosphatase (3)

$$\text{UDP-glucose} + \text{fructose 6-P} \rightleftharpoons \text{sucrose 6-P} + \text{UDP} \quad (2)$$

$$\text{Sucrose 6-P} + H_2O \rightarrow \text{sucrose} + \text{orthophosphate} . \quad (3)$$

Starch synthesis and influence of sucrose on photosynthesis are treated elsewhere (see p. 206).

Another enzyme which may play an important role in photosynthesis is carbonic anhydrase. EVERSON and SLACK (1968) have investigated the intracellular location of this enzyme in leaf extracts of spinach, pea, maize and *Amaranthus*. The distribution of carbonic anhydrase was compared with that of ribulose 1,5-diP carboxylase, a chloroplast enzyme, and a cytoplasmic enzyme, acid phosphatase. Using the non-aqueous technique, the carbonic anhydrase of both pea and spinach leaves fractionated in a manner characteristic of a chloroplast enzyme. In contrast, the smaller amount of carbonic anhydrase in maize and *Amaranthus* behaved as a cytoplasmic enzyme. Diamox (5-acetamido-1,3,4-thiadiazole-2-sulfanomide), an inhibitor of carbonic anhydrase, when used at a concentration which totally inhibits carbonic anhydrase activity in spinach chloroplast extracts was found to inhibit photosynthetic $^{14}CO_2$ fixation by intact chloroplasts about 50% (EVERSON, 1969). Since inhibition was partial, an additional mechanism for the supply of bicarbonate was presumably present in their preparations. These workers have speculated that carbonic anhydrase may function in association with ribulose 1,5-diP carboxylase if CO_2 rather than bicarbonate is the substrate for the carboxylase reaction.

2. Plants Utilizing the C_4-Dicarboxylic Acid Pathway

HATCH and SLACK (1966, 1969) hold the view that the principal enzymes involved in the synthesis and interconversion of the C_4-dicarboxylic acids are:

P-enolpyruvate synthetase (pyruvate, Pi dikinase)

$$\text{Pyruvate} + \text{ATP} + \text{orthophosphate} \rightleftharpoons \text{P-enolpyruvate} + \text{AMP} + \text{pyrophosphate} \quad (4)$$

P-enolpyruvate carboxylase

$$\text{P-enolpyruvate} + CO_2 \rightarrow \text{oxaloacetate} + \text{orthophosphate} \quad (5)$$

Adenylate kinase

$$\text{ATP} + \text{AMP} \rightleftharpoons 2\ \text{ADP} \quad (6)$$

Inorganic pyrophosphatase

$$\text{Pyrophosphate} + H_2O \rightarrow 2\ \text{orthophosphate} \quad (7)$$

and a hitherto uncharacterized enzyme catalyzing the transfer of carbon atom 4 of oxaloacetate to an unspecified acceptor, yielding glycerate 3-P and leaving pyruvate as the other product.

Other enzyme catalyzed reactions which are considered peripheral to the C_4-dicarboxylic acid pathway include:

malate dehydrogenase

$$\text{Oxaloacetate} + \text{DPNH} \rightarrow \text{malate} + \text{DPN} \quad (8)$$

malic enzyme

$$\text{Pyruvate} + CO_2 + \text{TPNH} \rightleftharpoons \text{malate} + \text{TPN} \quad (9)$$

and asparate aminotransferase

$$\text{Oxaloacetate} + \text{glutamate} \rightleftharpoons \text{aspartate} + \alpha\text{-ketoglutarate} . \quad (10)$$

The rates of photosynthetic CO_2 fixation for whole leaves of the tropical grasses, sugar cane and maize and for wheat, oats, and spinach together with activities of enzymes characteristic of the two sets of plants are shown in Table 6.

Table 6. *Rates of photosynthetic carbon dioxide fixation of detached whole leaves and enzyme activities in aqueous leaf extracts*

Enzyme	Sugarcane [a]	Maize [a]	Wheat [a]	Oats [a]	Spinach [b]
	µmoles/mg chlorophyll·h				
Ribulose 1,5-diP carboxylase	18	37	282	270	150
Ribulose 5-P kinase	486	666	1680	1500	600
Glyceraldehyde 3-P dehydrogenase (TPN)	306	270	288	582	180
Fructose 1,6-diphosphatase	46	48	73	69	39
P-Enolpyruvate carboxylase	1110	1050	17	20	—
P-Enolpyruvate synthetase	132	180—240	0	0	—
Adenylate kinase	—	3180	24	—	— [c]
Pyrophosphatase	—	2760	180	180	268 [d]
Malic enzyme	29	27	6	6	—
Malate dehydrogenase	49	39	486	474	—
Asparate aminotransferase	336	216	126	114	—
CO_2 fixation	147	180	102	96	32

[a] From Slack and Hatch (1967), and Hatch, Slack and Bull (1969).
[b] From Peterkofsky and Racker (1961).
[c] Not determined but Santarius and Heber (1965) record a value of 80 in spinach and in *Beta vulgaris* of which about half is located in the chloroplasts.
[d] See Table 5.

The activity of ribulose 1,5-diP carboxylase is much lower than the rate of photosynthesis in extracts from sugar cane and maize. Most likely, extraction of this enzyme is not easily achieved in these plants. On the other hand, the difference in ribulose 5-P kinase activity and photosynthesis is rather small and is indicative of a role for ribulose 1,5-diP carboxylase in both kinds of plants. Malic enzyme, adenylate kinase and pyrophosphatase activities were found to be higher by at least 40-, 35- and 5-fold, respectively, in species with the C_4-dicarboxylic acid pathway. P-enolpyruvate synthetase activity was found in leaf extracts of maize, sugar cane and sorghum but was not detected in extracts of the stem or root of maize or in leaf extracts of silver beet, wheat, oat, sunflower or bean (Hatch and Slack, 1968). Since the different species examined enzymically contained similar amounts of glyceraldehyde 3-P dehydrogenase and fructose 1,6-diphosphatase, Hatch and Slack have concluded that the pathway for the synthesis of hexoses from glycerate 3-P in the tropical grasses is the classical glycolytic one (Embden-Meyerhof pathway). P-enolpyruvate carboxylase, malic enzyme, ribulose 1,5-diP carboxylase and glyceraldehyde 3-P dehydrogenase in contrast to the marker enzyme, invertase, were found to be associated with the chloroplast (Table 7). Recently HATCH and SLACK (1969) were able to separate by non-aqueous density fractionation the chloroplasts of the mesophyll from the larger parenchyma sheath chloroplast. P-enolpyruvate synthetase, TPN-malate dehydrogenase, and P-enolpyruvate carboxylase were found to be associated with the mesophyll chloroplasts while fructose 1,6-diP aldolase, alkaline fructose 1,6-diphosphatase, ribulose 1,5-diP carboxylase and several unspecified enzymes of the photosynthetic reductive carbon cycle were localized completely or predominantly in the bundle-sheath chloroplasts. On the other hand, glycerate 3-P kinase and glyceraldehyde 3-P dehydrogenase were present in both chloroplasts. These authors concluded that the enzymes of the C_4-dicarboxylic acid pathway are divided between

Table 7. *Distribution of some enzymes and chlorophyll in maize leaf fractions prepared in non-aqueous media. (From* HATCH *and* SLACK, *1968)*

Fraction density	Chlorophyll	Ribulose 1,5-diP carboxylase	Glyceraldehyde 3-P dehydrogenase (TPN)	P-enolpyruvate carboxylase	Malic enzyme	Invertase
< 1.37	58	60	55	48	60	26
> 1.37—1.40	27	31	31	34	28	32
1.40	15	9	14	18	12	42

the mesophyll and parenchyma sheath chloroplasts. It is thought that one (mesophyll) contributes by fixing externally or internally-available CO_2 while the other (bundle-sheath) is responsible for the conversion of glycerate 3-P to hexoses and subsequently to starch.

3. Carboxylating Reactions

A discussion of various enzymes which could be envisaged as primary carboxylating enzymes in photosynthetic CO_2 fixation is appropriate at this point. It is clearly important in consideration of photosynthetic carbon dioxide fixation to know whether or not the reactant is CO_2 or HCO_3^- or (H_2CO_3). Four enzymes:

ribulose 1,5-diP carboxylase,

$$\text{Ribulose 1,5-diP} + \text{“}CO_2\text{”} + H_2O \rightarrow 2\ \text{glycerate 3-P} + 2\,H^+ \tag{11}$$

P-enolpyruvate carboxylase,

$$\text{P-enolpyruvate} + \text{“}CO_2\text{”} \rightarrow \text{oxaloacetate} + \text{orthophosphate} \tag{12}$$

P-enolpyruvate carboxykinase,

$$\text{P-enolpyruvate} + \text{“}CO_2\text{”} + \text{ADP} \rightleftharpoons \text{oxaloacetate} + \text{ATP} \qquad (13)$$

and P-enolpyruvate carboxytransphosphorylase,

$$\text{P-enolpyruvate} + \text{“}CO_2\text{”} + \text{orthophosphate} \rightleftharpoons \text{oxaloacetate} + \text{pyrophosphate} \qquad (14)$$

will be considered here. Reactions (13) and (14) resemble Reaction (12) except that, instead of cleavage of the phosphate group of P-enolpyruvate into orthophosphate, the high energy phosphoryl group is preserved in ATP or pyrophosphate. A thorough discussion of carboxylation reactions in plants involving pyruvate has been prepared by WALKER (1962).

Table 8. *Km values for "CO_2" for various carboxylation reactions. The bicarbonate level for half-maximal rate of photosynthesis in spinach chloroplasts is 0.6 mM (* JENSEN *and* BASSHAM, *1966)*

Enzyme	Substrate	*Km*	Source	Ref.
Ribulose 1,5-diP carboxylase		mM		
	HCO_3^-	11	Spinach	[1]
	HCO_3^-	5.6	Spinach	[2]
	HCO_3^-	22	Spinach	[3]
	HCO_3^-	9	Spinach	[4]
	CO_2	0.45	Spinach	[5]
P-Enolpyruvate carboxylase	HCO_3^-	0.31	Liver	[6]
	HCO_3^-	1.2	Thiobacillus	[7]
	HCO_3^-	0.4	Plant	[8]
P-Enolpyruvate carboxykinase	HCO_3^-	20	Liver	[9]
	HCO_3^-	5	Yeast	[10]
	HCO_3^-	10	*R. rubrum*	[11]
P-Enolpyruvate carboxytrans-phosphorylase	HCO_3^-	4	*Propioni bacterium*	[12]

1. WEISSBACH *et al.* (1956)
2. SUGIYAMA *et al.* (1968)
3. PAULSEN and LANE (1966)
4. AKOYUNOGLOU and CALVIN (1963)
5. COOPER *et al.* (1969)
6. MARUYAMA *et al.* (1966)
7. SUZUKI and WERKMAN (1958)
8. EVERSON and SLACK (1968)
9. CHANG *et al.* (1966)
10. CANNATA and STOPPANI (1963)
11. COOPER and BENEDICT (1968)
12. LOCHMÜLLER *et al.* (1966)

An understanding of the mechanistic details of these carboxylations requires the identification of the active species of "CO_2". It has been common practice to invoke *Km* values as a guideline in determining whether one or another particular carboxylation reaction functions in whole plant photosynthesis. Until the mechanism of each reaction is elucidated, this procedure must be viewed with caution. The Michaelis constants for "CO_2" for the four reactions are listed in Table 8.

The active species of "CO_2" involved in the enzymic carboxylation of P-enolpyruvate catalyzed by P-enolpyruvate carboxylase is HCO_3^- or (H_2CO_3) and not CO_2 (MARUYAMA *et al.*, 1966). Propionyl CoA carboxylase (KAZIRO *et al.*, 1962) and pyruvate carboxylase (COOPER *et al.*, 1968), both biotin dependent enzymes also utilize HCO_3^- as the active species. In contrast, CO_2 rather than HCO_3^- is the active species of the reactions catalyzed by P-enolcarboxykinase and P-enolpyruvate carboxytransphosphorylase (COOPER *et al.*, 1968) and by ribulose 1,5-diP carboxylase (COOPER *et al.*, 1969). Recognition that CO_2 rather than HCO_3^- is co-reactant of ribulose 1,5-diP resulted in the low *Km* value recorded in Table 8 for Reaction (11). A re-evaluation of the active species of "CO_2" may possibly result in a *Km* for P-enolpyruvate carboxykinase and P-enolpyruvate carboxytransphosphorylase approaching that of the other two carboxylating enzymes. An inspection of the values recorded in Table 8 indicates that the affinities of ribulose 1,5-diP carboxylase and P-enolpyruvate carboxylase for their respective substrates are approximately similar. Furthermore, the free energy changes of the two carboxylating enzymes are of similar magnitude. Thus, the free energy change of the ribulose 1,5-diP carboxylase catalyzed reaction is estimated at − 9,000 cal (BASSHAM and CALVIN, 1957) while for the carboxylation of P-enolpyruvate carboxylase

$$\text{P-enolpyruvate}^{3-} + ADP^{3-} + H^+ + HCO_3^- \rightarrow \text{oxaloacetate}^{2-} + ATP^{4-} + H_2O \quad \Delta F = -9200 \text{ cal} \tag{15}$$

$$ATP^{4-} + H_2O \rightarrow ADP^{-3} + HPO_4^{2-} + H^+ \quad \Delta F = -8000 \text{ cal} \tag{16}$$

$$\text{P-enolpyruvate}^{3-} + HCO_3^- \rightarrow \text{oxaloacetate}^{2-} + HPO_4^{2-} \quad \Delta F = -17{,}200 \text{ cal} \tag{17}$$

the value is roughly − 17,000 cal. It should be noted, however, that the ribulose 5-P kinase reaction has a free energy change of − 3,000 cal, favoring ribulose 1,5-diP while the phosphorylation of pyruvate to P-enolpyruvate strongly favors pyruvate formation. On the other hand, HATCH and SLACK (1968) have estimated the equilibrium constant of Reaction (4) to be 2×10^{-4} and the traditional pyruvate kinase

$$\text{Pyruvate}^- + ATP^{4-} \rightleftharpoons \text{P-enolpyruvate}^{3-} + ADP^{3-} + H^+ \tag{18}$$

has an equilibrium constant of 10^{-8} (KRIMSKY, 1959). The free energy change of the overall reaction

$$\text{Oxaloacetate} + H_2O \rightarrow \text{pyruvate} + HCO_3^- \tag{19}$$

has been calculated to be − 5300 (EVANS *et al.*, 1943).

In addition, GOLDSWORTHY (1968) showed that the apparent Michaelis constant for photosynthesis with maize, sugar cane and tobacco are of the same magnitude, the values being approximately equal to the natural CO_2 concentration (0.03%) in the atmosphere. Whatever, the mechanism by which plants photoassimilate CO_2, a higher photosynthetic capacity ascribed to plants using the C_4-dicarboxylic acid pathway, is apparently not due to a carboxylating enzyme or combination of enzymes more efficient in scavenging for "CO_2."

D. Stoichiometry of Photosynthesis Measured in the Chloroplast

ALLEN *et al.* (1955) provided direct evidence that the overall reaction of photosynthesis by the isolated spinach chloroplast may be represented by the traditional photosynthetic equation:

$$CO_2 + 2\,H_2O \rightarrow (CH_2O) + H_2O + O_2\,. \tag{20}$$

In their study, CO_2 uptake was measured with radiocarbon while O_2 evolution was measured manometrically and identified with luminescent bacteria. The chloroplasts were isolated in 0.35 M NaCl and showed rates in the order of 1 to 2 μmoles CO_2/mg chlorophyll·h. Subsequently, WALKER *et al.* (1968) were able to measure simultaneously photosynthetic carbon assimilation and its associated O_2 evolution in a single reaction mixture. They also determined CO_2 fixation using radiocarbon but O_2 evolution was measured polarographically. The photosynthetic quotient was unity during illumination (Fig. 7). Following illumination, an O_2 uptake was observed in the dark which was not accompanied by any net release of newly fixed $^{14}CO_2$. Ascorbate enhanced the rate of post-illumination O_2 uptake. The rate of post-illumination O_2 uptake was about 3 times the dark uptake and about 25% of the preceeding evolution.

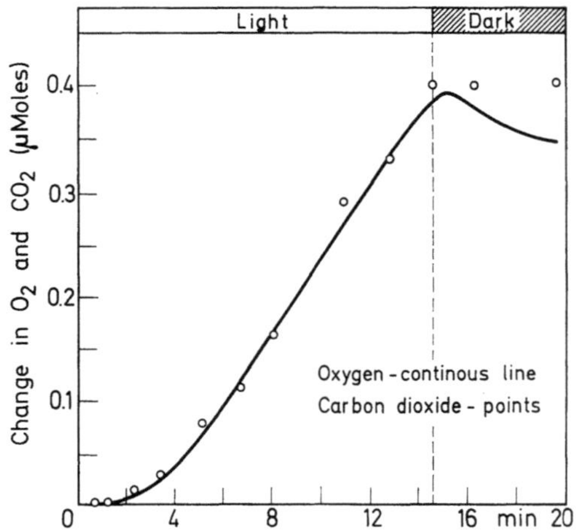

Fig. 7. Simultaneous measurement of the oxygen evolved and the $^{14}CO_2$ assimilated by isolated spinach chloroplasts. Samples (open circles) were assayed for radioactivity. Oxygen (continuous line) was measured simultaneously in the same vessel. From WALKER *et al.* (1968)

The rate was observed to return within 3 to 5 min to that of the dark uptake. Since the photosynthetic quotient is unity during illumination, it is probable that post-illumination O_2 uptake starts when photosynthesis ceases. In an earlier study, WALKER and HILL (1967) had observed a molar ratio of 2 CO_2:1 O_2 for chloroplasts photosynthesizing in a mixture of reduced glutathione and ascorbate. When the rate of dost-illumination O_2 uptake was added to the rate of net evolution, this gave an O_2:CO_2 stoichiometry of unity. Peroxidation of ascorbate enhanced by light is well known and ascorbic acid oxidase might also be a contributory factor. Most important was their finding that the induction periods in O_2 evolution illustrated in Fig. 7 closely resemble those observed in CO_2 fixation in similar reaction mixtures (see p. 189).

It is appropriate here to consider the relationship between CO_2 fixation, O_2 evolution and photophosphorylation. Illuminated chloroplasts generate ATP by two types of photophosphorylation, cyclic and non-cyclic. Cyclic phosphorylation yields only ATP and produces no net change in the oxidation-reduction status of any component.

$$n(ADP) + n(\text{orthophosphate}) \xrightarrow{\text{light}} n(ATP)\,. \quad (21)$$

By contrast, in non-cyclic photophosphorylation ATP formation is coupled to a net reduction of pyridine nucleotide, associated with an evolution of O_2.

$$TPN + H_2O + ADP + \text{orthophosphate} \xrightarrow{\text{light}} TPNH + ATP + 1/2\,O_2\,. \quad (22)$$

For every 3 moles of CO_2 converted to glyceraldehyde 3-P (triose-P), the scheme depicted in Fig. 3 requires 9 moles of ATP and 6 moles of TPNH.

$$3\,CO_2 + 9\,ATP + 6\,TPNH \rightarrow \text{triose-P}\,(C_3H_5\text{-}O_3\text{-}PO_3H_2) + 9\,ADP + 6\,TPN\,. \quad (23)$$

Presumably 2 moles each of ATP and TPNH come from the Reaction (22) and the third ATP is photogenerated by Reaction (21) (RAMIREZ *et al.*, 1968).

Referring to Fig. 3, there are two reactions in which ATP is consumed. The sequence whereby glycerate 3-P is reduced to glyceraldehyde 3-P in which glycerate 1,3-diP is an intermediate requires phosphate but is not phosphate consuming [Reaction (24)].

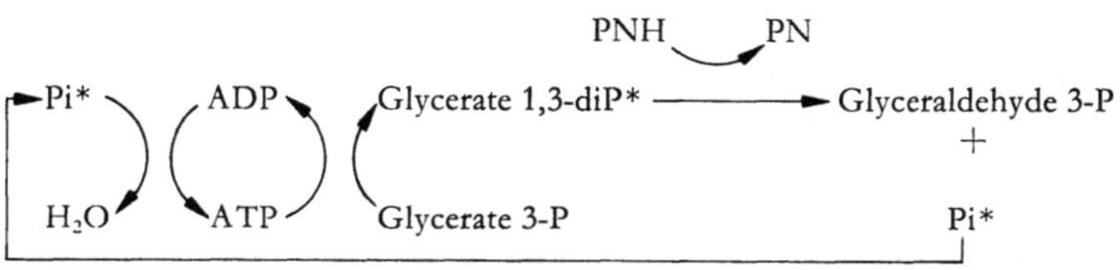

(24)

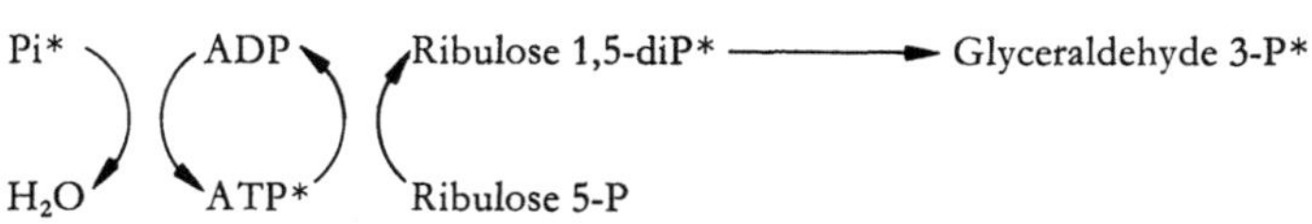

(25)

Reaction (25) in which ribulose 5-P is converted to ribulose 1,5-diP represents the single step leading to the incorporation of phosphate into the photosynthetic carbon reduction cycle. In the complete cycle, sedoheptulose 1,7-diphosphatase and fructose 1,6-diphosphatase catalyze the release of orthophosphate from their corresponding substrates. Therefore, for every 9 moles of ATP consumed only 1 mole of phosphate is esterified into triose-P according to the equation

$$3\,CO_2 + 3\,H_2O + H_3PO_4 \rightarrow C_3H_5O_3 - PO_3H_2 + H_2O + 3\,O_2\,. \quad (26)$$

WALKER and his associates were the first to measure the extent to which phosphorylation and CO_2 are linked in the chloroplast during photosynthesis. BALDRY *et al.* (1966) observed that the characteristic early lag in CO_2 fixation (see p. 189) was paralleled by a similar lag in the uptake of phosphate. Subsequently the overall rate of organic phosphate formation was found to be roughly one-sixth of the rate of carbon assimilation into acid stable products. When fructose 1,6-diP was added to the chloroplast suspension, the ratio of carbon to phosphate was close to 11:1. In later publications, COCKBURN *et al.* [1967 (1, 2, 3)] used chloroplast suspensions where O_2 evolution was dependent upon exogenous orthophosphate and observed a stoichiometry of approximately 3 moles of O_2 evolved for each mole of phosphate added. Their results satisfy Eq. (26). They cannot be interpreted to indicate the pathway of phosphate esterification into ATP but they do imply that in their spinach chloroplast preparations, triose-P is probably synthesized according to the scheme in Fig. 3.

To conclude, the intact leaf can reduce CO_2 to a compound at the level of carbohydrate at rates as high as 180 to 200 µmoles/mg chlorophyll·h (RABINOWITCH, 1956). This would require a photophosphorylation rate of about 600 µmoles/mg chlorophyll·h and a rate of pyridine nucleotide reduction equivalent to 400 µmoles/mg chlorophyll·h. Rates of this magnitude have been demonstrated and it is now quite clear that the spinach chloroplast is self-sufficient in supplying assimilatory power which is more than ample for the *in vivo* system. Like whole cell photosynthesis, the photosynthetic quotient in the spinach chloroplast is unity and the uptake of inorganic phosphate appears to be in agreement with the present view of carbon reduction into carbohydrates in this plant. These results and the close similarity between the photosynthetic carbon products of intact cell and of isolated chloroplast make it most likely that the chloroplast is the site of the complete photosynthetic process in green plants.

The basic reactions of the C_4-dicarboxylic acid cycle, as HATCH and SLACK view it (Fig. 4) consists of four Reactions (4, 5, 8), and (26) in conjuction with the photosynthetic carbon reduction cycle of CALVIN (Fig. 3)

$$\text{Oxaloacetate or malate} + C_2\text{—O—P} \rightarrow \text{glycerate 3-P} + \text{pyruvate}\,. \qquad (26)$$

Reaction (26) is the proposed carboxylation of a phosphorylated 2 carbon piece (C_2—O—P). Reaction (27) is the sum of Reactions (4, 5, 8), and (26), and shows that one mole of

$$C_2\text{—O—P} + CO_2 + \text{ATP} \rightarrow \text{glycerate 3-P} + \text{AMP} + \text{pyrophosphate}\,. \qquad (27)$$

However, by some mechanism, the AMP must be returned to a usable form either by involving the concerted effort of a second molecule of ATP and adenylate kinase or perhaps by a direct incorporation of pyrophosphate into ATP by photophosphorylation.

An alternative pathway involves P-enolpyruvate carboxykinase [Reaction (13)] (UTTER and KURAHASHI, 1954) or P-enolpyruvate carboxytransphosphorylase [Reaction (14)] (LOCHMÜLLER *et al.*, 1966) thereby coupling the fixation of CO_2 and retaining the energy of the phosphoryl residue. P-enolpyruvate carboxykinase was shown by MAZELIS and VENNESLAND (1957) to be widely distributed in plant tissue including both monocotyledonous and dicotyledonous plants and active preparations were obtained from fruits, roots, buds, cotyledons, seeds and leaves. For example, they measured the enzyme in extracts of maize seedlings, radish leaf, cabbage leaf, turnip and wheat germ. The enzyme was absent in spinach leaf extracts. HATCH and SLACK (1957) were not able to detect P-enolpyruvate carboxykinase in leaf extracts of sugar cane, maize, sorghum, wheat, oats and silver beet.

In contrast to the Calvin cycle, at least four moles of ATP are required per mole of CO_2 assimilated by the overall C_4-dicarboxylic acid pathway. On the other hand, both pathways consume an equivalent amount of reduced pyridine nucleotide. On the basis of cyclic and non-cyclic photophosphorylation kinetic experiments with chloroplasts of bermudagrass *(Cynadon dactylon)*, a plant which assimilates newly incorporated $^{14}CO_2$ into malate and aspartate, CHEN *et al.* (1969) concluded that the photosynthetic capacity of the intact leaves could be accounted for by their *in vitro* preparations.

Whatever the mechanism of phosphate participation in the C_4-dicarboxylic acid pathway, the only step leading to the incorporation of isotopically-labeled inorganic phosphate is the formation of the phosphorylated 2-carbon acceptor either involving Reaction (25) or its equivalent. In this respect, the two pathways are similar.

E. Kinetics of CO_2 Fixation

1. Induction Phenomenon

A property of freshly isolated spinach or pea chloroplasts is an induction phase or "lag period" in CO_2 fixation (BAMBERGER and GIBBS, 1965; BUCKE *et al.*, 1966; BALDRY *et al.*, 1966). Chloroplasts whether isolated in tris-NaCl (GIBBS *et al.*, 1967), orthophosphate-sugar mixtures [COCKBURN *et al.*, 1967 (2)] or according to the technique of JENSEN and BASSHAM (1966) exhibit photosynthetic induction phenomena which are broadly similar. When small quantities of certain intermediates of the photosynthetic carbon cycle are incubated with chloroplast suspensions, or if the parent tissue is illuminated prior to removal of the chloroplasts, the lag period is shortened or, at times, eliminated (Fig. 8). When supplied with catalytic quantities of ribose 5-P or fructose 1,6-diP the maximum rate is only slightly less than with substrate amounts of phosphorylated sugar. Fixation is found to continue at an unchanged rate when sufficient CO_2 has been assimilated to account for the stoichiometric conversion of the added substrate to ribulose 1,5-diP. Such an effect is in accord with the occurrence of a completely functioning photosynthetic carbon reduction cycle.

Not all intermediates of the photosynthetic carbon reduction cycle of CALVIN are effective in increasing total fixation or in eliminating the "lag period" in spinach and pea chloroplasts (Table 9). In substrate amounts (1 to 5 mM) the most effective are ribose 5-P, fructose 1,6-diP, glycerate 3-P, dihydroxyacetone-P and glyceraldehyde 3-P. Glycerate 3-P in contrast to the others gives variable results since on occasion, inhibition rather than stimulation is observed (SCHACTER, 1969). Results with fructose 6-P and glucose 6-P are variable but generally these compounds either stimulate to a lesser extent than the preceeding groups or have no effect. According to BUCKE *et al.* (1966), ribose and fructose but not their isomers glucose and xylose increase fixation. A kinase specific for fructose has been reported by MEDINA and SOLS (1956). In general, mixtures do not enhance the uptake of CO_2 greater than brought about by a single compound alone. D-Sedoheptulose 7-P, D-sedoheptulose 1,7-diP, and glyceraldehyde inhibited CO_2 fixation. SCHULMAN and GIBBS (1968) have reported that the D-glyceraldehyde 3-P dehydrogenase of pea seed and leaves are competitively inhibited by D-sedoheptulose 7-P and D-sedoheptulose 1,7-diP. The site of glyceraldehyde inhibition is unknown. From the data presented in Table 9 and from the literature (BAMBERGER and GIBBS, 1965; WALKER, 1967; BUCKE *et al.*, 1966) it is clear that the general pattern of effectiveness of these phosphorylated intermediates for chloroplasts isolated in saline, sucrose or sorbitol is the same. It would appear, therefore, this phenomenon is a property of chloroplasts functioning at high or comparatively low rates of photosynthesis. The data are consistent with the view that the chloroplast envelope is selectively permeable to several intermediates of the photosynthetic carbon reduction cycle and that this selectivity is not dependent on the compound used to maintain the osmotic pressure in the chloroplast suspension.

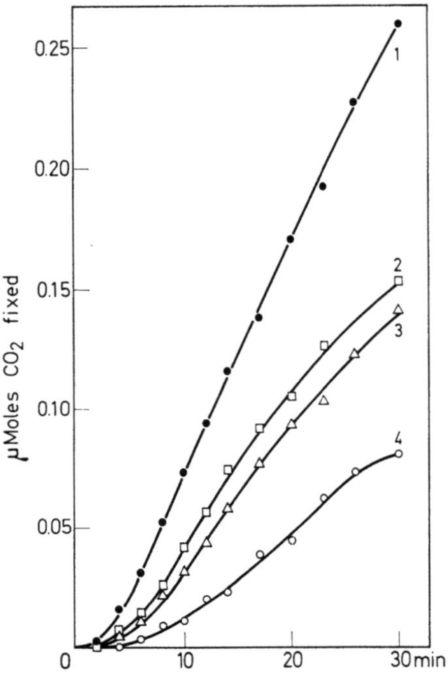

Fig. 8. Progress curves of $^{14}CO_2$ fixation by isolated pea chloroplast in the presence of catalytic (curves *2* and *3*) and substrate amounts (curve *1*) and absence (curve *4*) of ribose 5-P. From Walker (1967)

The results illustrated in Fig. 8 (see also numerous examples in Walker, 1965, and in Ellyard, 1968) suggest that isolated spinach and pea chloroplasts initially lack a sufficient amount of a component of the photosynthetic carbon reduction cycle needed to bring about the maximum rate of CO_2 fixation but that this stuff is synthesized during illumination autocatalytically. Thus, in time, the maximum rate is essentially equal in chloroplasts where the induction period is lengthened by washing or in preparations to which effective intermediates are added. Furthermore, when steady-state photosynthesis in a chloroplast suspension, provided with ribose 5-P or fructose 1,6-diP, is interrupted by a dark period of 5 to 10 min, fixation is resumed without a lag at the same rate when illumination is resumed. This is also characteristic of induction phenomena in intact organisms. This component required to erase the lag phase is retained within the chloroplast to permit resumption of fixation at the maximum rate. However, some of the component apparently can diffuse out of the intact chloroplast since the medium in which the chloroplasts were illuminated shortened the lag period when added back to fresh chloroplasts (BALDRY *et al.*, 1966).

It has been considered that the induction phase may be governed by dark rather than light-catalyzed reactions. ATP may be ruled out by the fact that glycerate 3-P which in contrast to glyceraldehyde 3-P, ribose 5-P or fructose 1,6-diP is not capable of yielding ATP through substrate phosphorylation in these preparations is as effective in enhancing CO_2 fixation justified the conclusion that ATP is not limiting in the lag phase. In addition, BALDRY *et al.* (1966) found that the lag period is appreciably lengthened by a decrease in temperature but is less affected by a decrease in light intensity. These results strengthen the view that the length of the lag phase is primarily dependent on a material synthesized in the dark reactions of photosynthesis. In agreement with this conclusion are the data of HEBER and his associates who have measured the levels of ATP and reduced pyridine nucleotide during the induction phase. On illumination, the increase in the level of ATP is rapid and reaches a maximum in the chloroplast after about 30 sec in light (SANTARIUS and HEBER, 1965). This coincides with an increase in reduced triphosphopyridine nucleotide (HEBER and SANTARIUS, 1968) and a decrease in the pool size of glycerate 3-P (URBACH *et al.*, 1965). Therefore, after the level of glycerate 3-P has been reduced upon illumination, ATP and reductant reach a steady state and ribulose 1,5-diP is regenerated from triose-P and the uptake of CO_2 can also proceed at a steady rate.

Table 9. *Effect of additives on photosynthetic CO_2 fixation*

Addittion	BUCKE, WALKER and BALDRY[a]	BAMBERGER and GIBBS[b]
	µmoles CO_2 fixed/mg chl.·h	
None	2.5	1.3
Glycerate 3-P	9.6	2.6
Ribose 5-P	44.4	3.0
Fructose 1,6-diP	43.8	3.0
Dihydroxyacetone-P	13.2	—
Glucose 6-P	9.6	1.3
Fructose 6-P	6.6	1.3
Glucose 1-P	4.8	—
Sedoheptulose 7-P	—	0.3
Sedoheptulose 1,7-diP	—	0.3
Ribose	7.8	—
Fructose	16.2	—

[a] BUCKE, WALKER and BALDRY (1966), pea chloroplasts in 0.33 M sorbitol.
[b] BAMBERGER and GIBBS (1965), spinach chloroplasts in 0.35 M NaCl.

The studies discussed in this section make it quite clear that the lag phase is the result of a shortage of intermediates of the photosynthetic carbon reduction cycle in the chloroplast and that this shortage can be eliminated either by addition of an effective intermediate (Table 9) or by synthesis from newly assimilated CO_2. LATZKO and Gibbs (1969) have measured enzymically changes in the levels of ribulose 1,5-diP, pentose monophosphates (mixture of ribose 5-P, ribulose 5-P and xylulose 5-P), triose-P (essentially dihydroxyacetone-P) and ATP during the initial phase of CO_2 fixation in spinach chloroplasts isolated in NaCl (Fig. 9). Their results indicate that the lag is associated with the level of pentose monophosphate and in their preparations when this level approached 0.1 μmole/3 mg chlorophyll, the rate of CO_2 fixation reaches the highest rate.

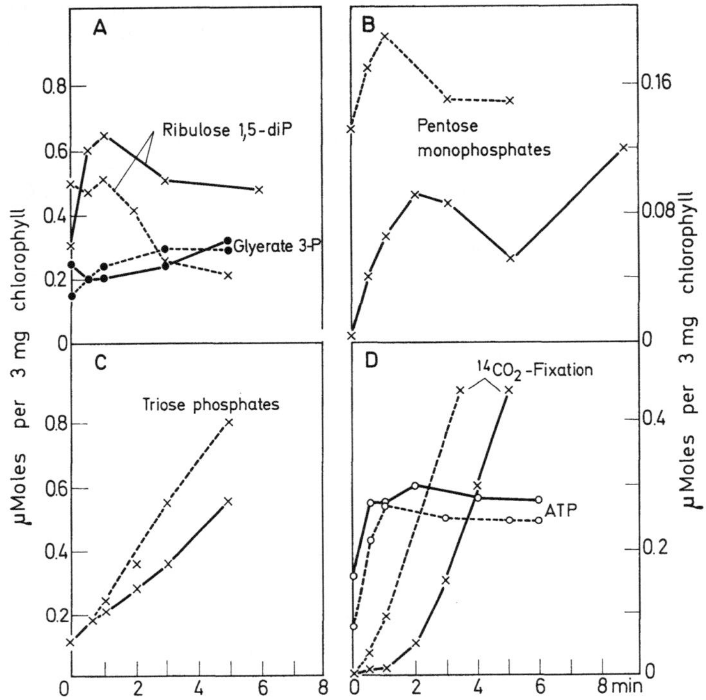

Fig. 9. Changes in the level of photosynthetic intermediates during the initial phase of CO_2 fixation in isolated spinach chloroplasts. ×——— ×, preparations showing an initial lag in CO_2 fixation; × - - - - - - ×, no initial lag. From Latzko and Gibbs (1969)

A delayed start of photosynthesis after a period of darkness is well known in intact tissue photosynthesis. The extent of delay depends on the duration of the dark period and the conditions to which the tissue has been subjected during this period. It is probable that the lag periods described in this section for the isolated chloroplast is related to the "short" induction phenomena enumerated by Rabinowitch (1956) for whole cell photosynthesis.

2. CO_2 Dependent and Light-Dark Transients

Transient changes in the photosynthetic carbon reduction cycle which occur as a result of altering the partial pressure of CO_2 are primarily those involving the intermediates of the carboxylation step whereas the light-dark switch stops the formation of ATP and reduced pyridine nucleotide and consequently the reductive part of the cycle. WILSON and CALVIN (1955) used the technique of CO_2 transients to identify the reciprocal relationship between glycerate 3-P and ribulose 1,5-diP in intact *Chlorella*. In isolated spinach chloroplasts, a rapid increase in bicarbonate concentration results in a striking increase in dihydroxyacetone-P and is paralleled by a sharp

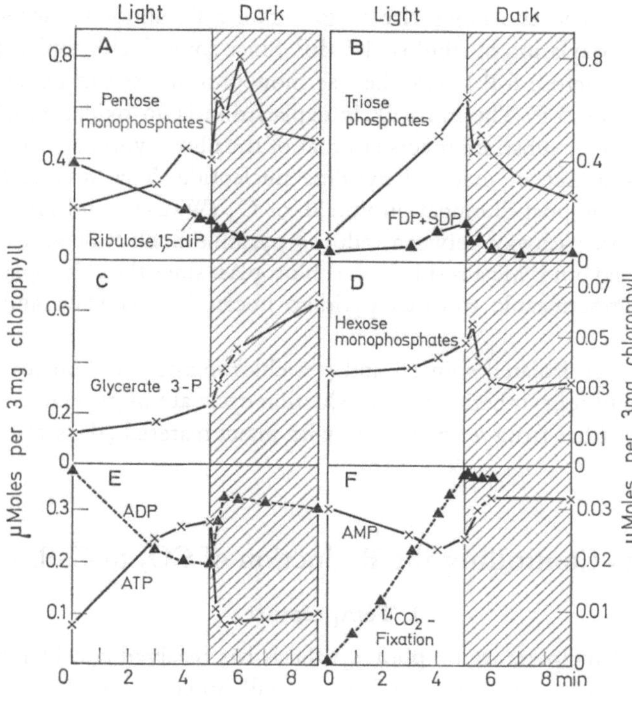

Fig. 10. Light-dark transient changes in the level of photosynthetic intermediates in spinach chloroplasts. Abbreviations in B are FDP (fructose 1,6-diP) and SDP (sedoheptulose 1,7-diP). From LATZKO and GIBBS (1969)

decrease in the concentration of pentose monophosphate (LATZKO and GIBBS, 1969). In contrast, the change in level of glycerate 3-P is limited to a brief overshoot and the ribulose 1,5-diP level undergoes a small decrease. Presumably the formation of reduced pyridine nucleotide and ATP are not rate-limiting. Complicating a study in transients is the fact that selected intermediates of the cycle can diffuse out of the chloroplast while others are well retained with the chloroplast (see p. 201). Thus pentose monophosphate, a compound that is permeable to the chloroplast envelope, accumulates in a limiting concentration but as noted above, rapidly disappears on addition of CO_2. The pentose monophosphate transient is apparently due to its diffusion out of and then re-entry into the chloroplast for subsequent metabolism.

As a result of a light-dark transition (Fig. 10), the levels of glycerate 3-P and pentose monophosphate increase sharply while those of dihydroxyacetone-P and fructose 1,6-diP decrease rapidly (Heber *et al.*, 1967; Bassham, 1968; Latzko and Gibbs, 1969). In the dark, the bulk of the ATP is consumed rapidly. Increase of ADP is reciprocal to the oscillation of ATP whereas the level of AMP apparently as a result of adenylate kinase activity follows with a short delay. It is clear that the sharp drop in the level of ATP is not the result of its metabolism in the ribulose 5-P kinase reaction since CO_2 metabolism ceases immediately after turning off the light. A light dependent ATPase may be responsible for the rapid loss of ATP (Petrack and Lipmann, 1961). The enormous rise in glycerate 3-P cannot be accounted for in terms of the carboxylation of ribulose 1,5-diP. Since spinach chloroplast preparations do not contain fructose 6-P kinase, hexose monophosphates can be eliminated as a source of glycerate 3-P. The most likely explanation is an oxidation of dihydroxyacetone-P. A flow of substrate reverse to that of the photosynthetic carbon reduction cycle implies a source of oxidized pyridine nucleotide. This reverse flow may be associated with the post-illumination uptake of O_2 (Walker and Hill, 1967). The electron transport pathway between reduced pyridine nucleotide and O_2 cannot be presently assigned but it must reside in the chloroplast since the chloroplast membrane is apparently impermeable towards pyridine nucleotides (Heber and Santarius, 1965).

In summary, in isolated spinach and pea chloroplasts, the kinetics of the intermediates of the photosynthetic carbon reduction cycle are in general agreement with those reported earlier from *in vivo* studies with intact material (Bassham and Calvin, 1957).

F. Factors Controlling the Reduction of CO_2 to Carbohydrate

1. Temperature

The effect of temperature on photosynthesis has received considerable attention. In weak light where the velocity of the overall process is limited by the photochemical reactions, the rate of photosynthesis should be insensitive to the temperature changes or a $Q_{10} = 1$. Under conditions of high light intensities and saturating levels of CO_2, the rate is temperature dependent. The temperature dependence of photosynthesis is most readily characterized by the Q_{10} (temperature coefficient) value. According to studies carried out with intact organisms, the temperature dependence of photosynthesis does not always follow the Arrhenius equation; an apparent increase in activation energy usually is found below 10° and a decrease about 20° (Table 10). In contrast, the rate of respiration tends to increase with a Q_{10} of about 2. Baldry, Bucke and Walker (1966) observed a pattern similar to intact organisms when measuring photosynthetic CO_2 fixation in isolated pea chloroplast in saturating light and high CO_2. The chloroplast system has the decided advantage that photosynthesis can be measured, uncomplicated by respiration and other physiological phenomena of whole organisms or detached organs. In the pea chloroplast system, over the range 30° to 5° the lag increased by a factor of three while the velocity decreased by a factor of 10. The results of Baldry *et al.* (1966) support the view of Warburg (1919) that the rate of photosynthesis in excess illumination and an ample supply of CO_2 is a linear function of temperature. The effects of temperature on photosynthesis are, therefore, a property of the carbon fixation cycle. To account for the variation in Q_{10} values, SELWYN (1966) has proposed a model for photosynthetic fixation which takes into account the cyclic and autocatalytic features of the carbon reduction cycle and includes a positive feedback metabolic regulation involving the regeneration of intermediates and the reactions removing intermediates.

Table 10. *Temperature dependence of photosynthesis in whole organisms and in isolated chloroplasts*

Reference	Plant	Temp. range	Q_{10}
EMERSON (1929)	*Chlorella*	4—8	7.6
		8—12	4.8
		12—16	2.5
		16—26	1.54
EMERSON and GREEN (1934)	*Gigartina*	4—8	13.5
		8—12	4.2
		12—16	1.7
VAN DER PAAUW (1934)	*Chlamydomonas*	10—20	2.5
		15—25	2.1
		20—30	1.7
BALDRY *et al.* (1966)	Pea chloroplast	5—7	9.0
		7—10	5.3
		10—15	2.9
		15—20	1.91
		20—25	1.5
		25—30	1.28

2. pH

The incorporation rate of CO_2 fixation by chloroplasts isolated in 0.35 M saline proceeds at maximal velocity in the vicinity of pH 7.4 (ELBERTZHAGEN and KANDLER, 1962; GIBBS and CALO, 1959). Varying the pH of the preparative solution between 7.2 and 8.0 and of the assay solution between 7.6 and 8.0 has little effect on CO_2 assimilation (KALBERER *et al.*, 1967). The fixation rate falls off slowly at neutrality but declines sharply on the alkaline side.

3. Enzyme Activation

In his discussion on "long induction" periods, RABINOWITCH (1956) speculated upon the deactivation, in the dark, of some enzymes needed for photosynthesis and their gradual reactivation in light. Clearly, following illumination differences in the rate of enzyme activation could account for transients in the levels of intermediates until a final steady level is attained. Oscillations in the concentration of intermediates of the cycle can be affected by a sudden change in the supply of CO_2 but correlation with enzyme activation has not been established.

Buchanan *et al.* (1967) have demonstrated that ferredoxin reduced photochemically or by means of H_2 and hydrogenase is required in the cleavage of fructose 1,6-diP by fructose 1,6-diphosphatase in accordance with Reaction (28).

$$\text{Fructose 1,6-diP} + H_2O \xrightarrow[\text{ferredoxin}]{\text{reduced}} \text{fructose 6-P} + \text{orthophosphate}\,. \tag{28}$$

The time course of activation is given in Fig. 11. Dithiothreitol may act as a substitute for reduced ferredoxin (Buchanan *et al.*, 1968).

An activation up to 6-fold of ribulose 5-P kinase during illumination of intact chloroplasts has also been reported (Gibbs *et al.*, 1968). Photosynthesis was required since activation by light was blocked by arsenite, iodoacetamide or dichlorophenyl-

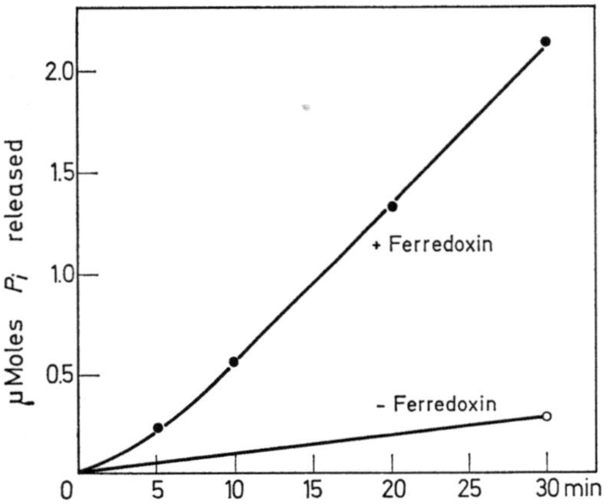

Fig. 11. Time course of ferredoxin-activated inorganic phosphate (Pi) release from fructose 1,6-diP. From Buchanan *et al.* (1967)

dimethylurea (DCMU). Similar to fructose 1,6-diphosphatase, full activation was attained by adding dithiothreitol to the reaction mixture incubated in the dark. Dithiothreitol could not be replaced by reduced glutathione.

Light is required for the maintenance of P-enolpyruvate synthetase but not for ribulose 1,5-diP carboxylase and adenylate kinase within the leaf of *Amaranthus palmeri* (Slack, 1969). When the leaf is placed into darkness, half-loss for activity is 15 min. Without 2-mercaptoethanol, P-enolpyruvate synthetase is rapidly and irreversibly inactivated, indicating that the enzyme is a sulfhydryl protein (Hatch and Slack, 1968). Whether dithiothreitol could substitute for light was not reported.

To account for the observation that on cessation of illumination, photosynthetic CO_2 fixation in spinach chloroplast virtually stops in the dark, even though there remain adequate amounts of ribulose 1,5-diP and CO_2 for the reaction to continue, Jensen and Bassham (1968) have proposed a loss of activity of ribulose 1,5-diP carboxylase. They observed, however, that addition of ATP to the chloroplast suspension did stimulate the formation of ribulose 1,5-diP. Contrary to the experience of others (Walker, 1965; Ellyard, 1968), when the phase was interrupted by a period of darkness, the fixation rate was diminished as compared with that of the first light period. The rate-limiting step, they contend, is the carboxylation reaction. In studying light-dark regulation in the spinach chloroplast, Latzko and Gibbs (1969) also observed an immediate stoppage of CO_2 uptake in the dark even though the ribulose 1,5-diP remained relatively high. Dilley and Vernon (1965) and others have reported proton transfer within the chloroplast during photosynthesis. Perhaps, regulation does not reside with the activity of the ribulose 1,5-diP carboxylase but rather in the concentration of CO_2 in the vicinity of the enzyme.

The increase in activity observed on transferring darkened leaves or chloroplasts to light may represent either the reactivation of inactivated enzyme or newly synthesized protein. If light caused the production of *de novo* synthesis, a lag period would be expected after exposure to light due to time required for the cell to prepare an inducer. With respect to P-enolpyruvate synthetase (Slack, 1968) and to ribulose 5-P kinase (Latzko and Gibbs, 1968), the maximum increase occurred during the first 2 min. More likely the increase is due to activation rather than enzyme synthesis.

According to Buchanan *et al.* (1968), reduced ferredoxin is required for the assembly of an active fructose 1,6-diphosphatase from its components, possibly by converting -S-S- to -SH groups. The fact that dithiothreitol can substitute for the light reactions suggests the possibility of an increasing concentration of a reductant in the chloroplast during illumination. This reductant may be reduced pyridine nucleotide. Hatch and Turner (1960) have purified from pea and detected in extracts of many seedlings an enzyme (protein disulfide reductase) which catalyzes the reduction of disulfide bonds of glyceraldehyde 3-P dehydrogenase in the presence of reduced triphosphopyridine nucleotide. Equilibrium of the reaction favored reduction and the enzyme was inhibited by arsenite and Cd^{2+}, specific inhibitors of reactions involving the dithiol linkage.

Another type of enzyme activation has been described by Fuller and Hudock (1967), and Ziegler and associates (Ziegler *et al.*, 1968; Müller *et al.*, 1969). They have demonstrated a possible interconversion of DPN and TPN glyceraldehyde 3-P dehydrogenases activities by the metabolic products of environment changes. This type of activation of the "dark enzyme" resulting in the "light-enzyme" is not due to an unspecific reduction of the disulfide grouping in the reaction center.

4. Oxygen

The inhibition of photosynthesis by oxygen was discovered by Warburg (1920) in algae (Warburg effect) and has been demonstrated in higher plants (McAlister and Myers, 1940) and in isolated chloroplasts (Ellyard and Gibbs, 1967). The studies of Wilson and Calvin (1955) with intact *Scenedesmus* and by Bassham and Kirk (1962) as well as Coombs and Whittingham (1966) with intact *Chlorella* established a correlation between photosynthesis in excess O_2 or in low concentrations of CO_2 and high yields of glycolate. Another well documented property of the Warburg effect in intact organisms is that the magnitude of inhibition is inversely related to the bicarbonate concentration (Tamiya and Huzisige, 1949). A search of the literature indicates an interdependence among the photosynthetic carbon reduction cycle, photorespiration and the Warburg effect with glycolate as the substance common to each.

The Warburg effect was characterized in the isolated spinach chloroplast by ELLYARD and GIBBS (1969). The magnitude of O_2 inhibition was found to be inversely related to the bicarbonate concentration (Fig. 12). Similar to intact

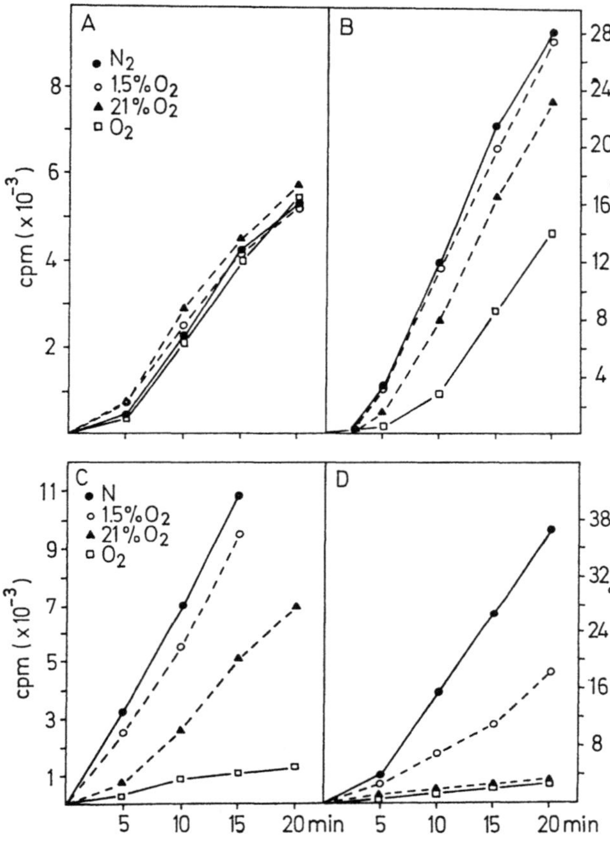

Fig. 12. Influence of oxygen and bicarbonate on $^{14}CO_2$ fixation by chloroplasts. In each case gas phases were 100% N_2 (●), 1.5% O_2 + 98.5% N_2 (o); 21% O_2 + 79% N_2 (Δ) and 100% O_2 (□). Bicarbonate concentrations were: *A* 13 mM, *B* 10 mM, *C* 4 mM, *D* 1 mM. From ELLYARD and GIBBS (1969)

Chlorella (TAMIYA and HUZISIGE, 1949), the inhibition of photosynthesis by O_2 in the chloroplast was reversible. The Warburg effect has been envisioned by some workers (TURNER *et al.*, 1958) to occur at the enzymic level in the photosynthetic carbon reduction cycle. Although glyceraldehyde 3-P dehydrogenase and ribulose 5-P kinase but not aldolase extracted from photosynthesizing chloroplast are inactivated slowly, nonetheless, their specific activities are sufficient to account for the observed rates of CO_2 uptake (GIBBS *et al.*, 1968). Sulfhydryl reagents such as reduced lipoate, 2,3-dimercaptopropanol or glutathione do not overcome the inhibition by O_2. On the other hand, a variety of sugar phosphate including fructose 1,6-diP and ribose 5-P eliminate completely the inhibition in 100% O_2.

Several investigators (TURNER and BRITTAIN, 1962; HEBER and FRENCH, 1968) have interpreted the Warburg effect in terms of a Mehler type reaction whereby an intermediate or the ultimate electron acceptor, TPN, of the photochemical act is produced in diminished quantities under O_2. ELLYARD and GIBBS (1969) found the rate of photoreduction of TPN down to 0.5 micromolar TPN to be essentially the same under an atmosphere of N_2 or O_2. Nonetheless, under conditions which are favorable for the Warburg effect, such as high light intensity and rate limiting concentrations of CO_2, it is clearly possible that the level of TPN might be diminished to the point where reduced ferredoxin or ferredoxin-TPN reductase (flavoprotein) may accumulate and then react with O_2 rather than with TPN (COOMBS and WHITTINGHAM, 1966).

From the studies with intact cells, there is little doubt that glycolate plays a pivotal role in any attempt to account for the Warburg effect. The chloroplast responded similarly to increasing partial pressures of O_2 and to decreasing levels of bicarbonate by producing increased amounts of glycolate.

ELLYARD and GIBBS (1969) concluded that the major aspects of the Warburg effect are associated with the chloroplast and furthermore that glycolate is formed in the chloroplast and is closely associated with the photosynthetic carbon reduction cycle. The depression of photosynthesis by O_2 has been explained in terms of a shift of a major portion of the total carbon assimilated into glycolate and thereby, impairing the functioning of the photosynthetic carbon cycle (BASSHAM and KIRK, 1962). Introduction of carbon into the cycle either in terms of a high concentration of CO_2 or an intermediate of the cycle eliminates the effect of O_2 by off-setting a drainage resulting from the formation of glycolate.

BRADBEER and RACKER (1961) singled out dihydroxyethylthiamine pyrophosphate (glycolaldehyde-transketolase addition complex) as the precursor of glycolate. The fact that glycolate is produced solely during the photochemical act suggested a link between the photosynthetic carbon reduction cycle and the electron transport system. The inhibition of photosynthetic CO_2 assimilated by DCMU is relieved by ascorbate (BAMBERGER and GIBBS, 1965). Under N_2 and an ample supply of bicarbonate, glycolate in contrast to all other products failed to accumulate (PLAUT and GIBBS 1969). When conditions were more favorable for glycolate formation such as a high partial pressure of oxygen or low concentrations of bicarbonate, the addition of ascorbate to the DCMU-inhibited chloroplast suspension restored glycolate synthesis. These authors interpreted their data to indicate two separate mechanisms catalyzing the oxidation of the transketolase addition complex (Fig. 13).

Spinach chloroplast fragments were found to catalyze the light dependent formation of glycolate from fructose 6-P in the presence of transketolase, ferredoxin and TPN (SHAIN and GIBBS, 1969). The reconstituted preparation is DCMU sensitive and is saturated at 1500 foot candles. The data recorded in Table 11 illustrate that the yield of glycolate enhanced as the O_2 tension is increased, is diminished by the presence of catalase. Furthermore, the 2-carbon piece could be converted to glycolate by supplying electrons through phenylenediamine into photosystem 2 and is coupled to TPN reduction with a stoichiometry of one mole of glycolate synthesized per mole of TPN reduced. Maize particles could substitute for spinach fragments. This suggests that the difference with respect to glycolate formation between photo-

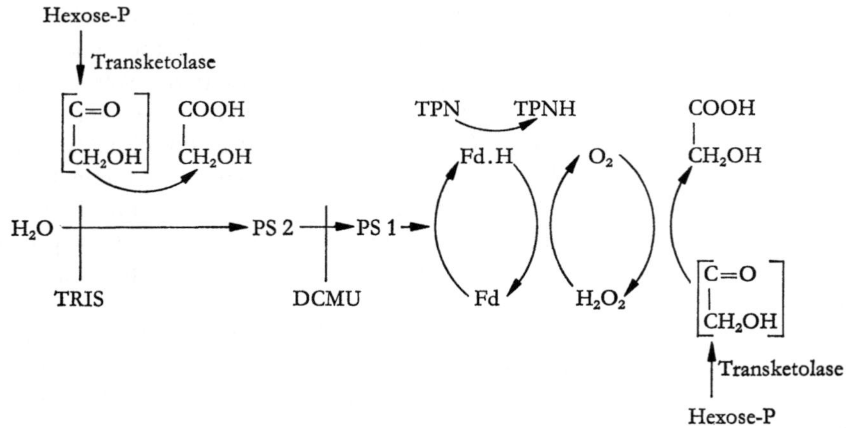

Fig. 13. Proposed mechanism for oxidation of transketolase addition complex to glycolate. Fd and FdH are oxidized and reduced ferredoxin. [C = O | CH_2OH] is the transketolase addition complex. Tris signifies chloroplasts washed with tris buffer whereby electron transport between water and PS 2 is inhibited but electron donors such as reduced phenylenediamine can donate electrons to PS 2 (Yamashita and Butler, 1968). The addition complex can be oxidized by donating electrons into PS 2 or by H_2O_2 generated in PS 1

Table 11. *Glycolate formation with a fragmented chloroplast preparation under nitrogen and oxygen as effected by the presence of catalase. (From* Shain *and* Gibbs, *1969)*

Gas	Experimental conditions	Glycolate formation μmoles/mg chlorophyll · h
Nitrogen	Dark	0.22
	Dark + catalase	0.14
	Light	3.04
	Light + catalase	2.75
Oxygen	Dark	0.24
	Dark + catalase	0.01
	Light	6.17
	Light + catalase	2.60

respiring (spinach) and non-photorespiring (maize) plants may be in the formation of the transketolase addition complex and may be dependent on the content of transketolase within their chloroplasts (also see p. 203).

5. Transport of Metabolites

a) Intermediates of the Photosynthetic Carbon Reduction Cycle

SACHS (1862) considered the chloroplast as an organelle whose primary responsibility was the reduction of CO_2 to starch. He speculated that all of the carbon mobilized by the leaf during photosynthesis passed through the polysaccharide in the chloroplast. Starch was the end product of carbon metabolism in photosynthesis and this polyglucan was the starting point for the synthesis of all other organic compounds in the plant. It is now clear that an intracellular transport of selected

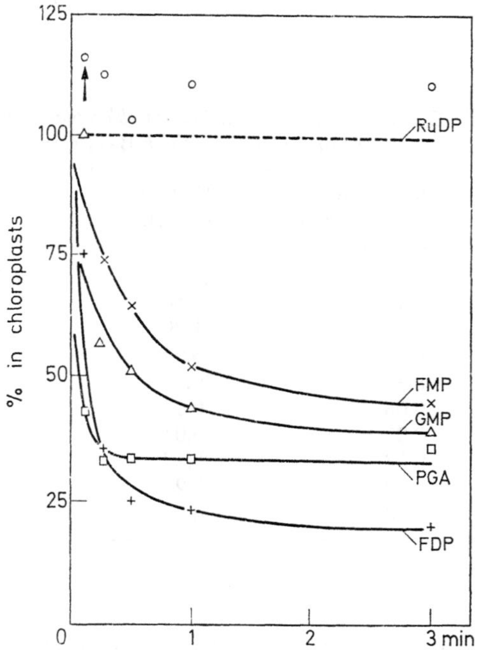

Fig. 14. Distribution of intermediates of the photosynthetic carbon reduction cycle between chloroplast and cytoplasm of leaf cells from *Elodea densa*. From HEBER (1967)

intermediates of the photosynthetic carbon reduction cycle and other substances such as glycolate exists.

HEBER (1967) has investigated compartmentation of photosynthetic intermediates by extracting chloroplasts from the remainder of the cellular components by the non-aqueous method after the leaf was fed photosynthetically $^{14}CO_2$ (Fig. 14). From Fig. 14, it is evident that ribulose 1,5-diP (RuDP) is present in the chloroplast even after 3 min of photosynthesis. Sedoheptulose 1,7-diP (not shown) acts in a similar fashion. In contrast, in the same period of illumination roughly half of the fructose 6-P (FMP), glucose 6-P (GMP), fructose 1,6-diP (FDP) and glycerate 3-P (PGA) appear in the cytoplasm. A serious drawback of this approach is the difficulty in evaluating whether transfer of radiocarbon is direct or occurs via transport metabolites and subsequent cytoplasmic metabolism. Indeed, in other experiments, the conclusion was reached that glycerate 3-P (URBACH *et al.*, 1965), dihydroxyacetone-P and possibly fructose 1,6-diP but not the hexose monophosphates function as transport metabolites in photosynthesis (HEBER *et al.*, 1967).

BASSHAM *et al.* (1968) investigated diffusion of photosynthetic intermediates by directly measuring the distribution of labeled compounds between the chloroplast photosynthesizing in $^{14}CO_2$ and the medium in which they were suspended. Over 80% of the carbon appeared in the medium. Of this, the greater portion was in glycerate 3-P and dihydroxyacetone-P with smaller amounts also in fructose 1,6-diP, sedoheptulose 1,7-diP and glycolate. In contrast, ribulose-diP, fructose 6-P, glucose 6-P and sedoheptulose 7-P were retained within the chloroplasts (Table 12). On inspection of the table, the reason for the difference in level of dihydroxyacetone-P between isolated chloroplast and intact plant (Figs. 1 and 2) becomes clear. At least one assumption must be made in arriving at definitive conclusions about metabolite distribution in chloroplast suspensions. Enzymic catalyzed interconversions in the suspending media cannot be discounted. The preparations of JENSEN and BASSHAM (1966) contain a mixed population of chloroplasts. About 30% of the chloroplasts have lost their envelopes and as pointed out by BUCKE *et al.* (1966) enzymes could escape from these damaged particles. Thus, the concerted effort of fructose 1,6-diP aldolase and glyceraldehyde 3-P isomerase favors the conversion of fructose 1,6-diP to dihydroxyacetone-P. Nevertheless, it would seem that selected intermediate compounds of the photosynthetic carbon reduction cycle migrate rapidly from chloroplasts to the suspending medium.

Table 12. *Appearance of ^{14}C labeled compoumds in isolated chloroplasts and in suspending medium after 3 min of photosynthesis in $^{14}CO_2$. (From* BASSHAM *et al., 1968)*

Compound	Chloroplast	Medium	Chlorplast: Medium
	µmoles ^{14}C (mg chlorophyll · h)$^{-1}$		Ratio
Dihydroxyacetone-P	0.56	23.8	42.5
Fructose 1,6-diP	0.19	6.66	34.0
Pentose mono-P	0.21	1.79	8.5
Glycolate	0.71	3.94	5.5
Glycerate 3-P	8.3	39.6	4.8
Sedoheptulose 1,7-diP	0.06	0.12	2.0
Ribulose 1,5-diP	0.07	0.03	0.4
Fructose 6-P	0.38	0.17	0.4
Glucose 6-P	1.60	0.21	0.13
Sedoheptulose 7-P	4.30	0.27	0.06

In agreement with the data of BASSHAM *et al.* (1968), are the findings mentioned earlier (BAMBERGER and GIBBS, 1965; BUCKE *et al.*, 1966) on the types of photosynthetic intermediate compounds which are capable of eliminating the lag in CO_2 fixation (p. 189).

All of the results are consistent with the view that the chloroplast membrane is permeable or partly permeable to several phosphorylated intermediates of the photosynthetic carbon reduction cycle. BASSHAM (1968) has pointed out that those intermediates lying between the ribulose 1,5-diP carboxylase and fructose 1,6-diphosphatase and including pentose monophosphates diffuse rapidly from the chloroplast to the suspending medium and presumably cytoplasm (Fig. 3). The significance of this selectivity is unclear.

b) Glycolate

Many plants, but not all, exhibit photorespiration which is attributed to the oxidation of glycolate (ZELITCH, 1958). Glycolate is a product of chloroplast metabolism (p. 197).

Table 13. *Specific activity of peroxisomal enzymes related to glycolate metabolism. (From* TOLBERT *et al., 1969)*

Enzyme	Specific activitiy µmoles/mg protein · min
Glycolate oxidase	18—60
Glutamate-glyoxylate aminotransferase	0.3—0.96
DPNH: glyoxylate reductase	18—60
Malate dehydrogenase (DPN)	$1.2—1.8 \times 10^3$
Catalase	$3—3.6 \times 10^{-3}$

Examination of the intracellular localization of glycolate oxidase, the plant enzyme initiating the further metabolism of the β-hydroxy acid, resulted in the conclusion that glycolate oxidase is not located in the chloroplast (THOMPSON and WHITTINGHAM, 1968) but in a microbody designated as peroxisome (TOLBERT *et al.*, 1968). The peroxisome, 0.5 to 1.0 µ in diameter containing a dense granular stroma is surrounded by a single membrane. On the other hand, the chloroplasts of higher plants, bounded by an envelope consisting of two membranes, are discs or flat ellipsoids, 3 to 10 µ across.

The function of the peroxisome in photorespiration and in the glycolate pathway (glycolate to glycerate 3-P) has been studied by TOLBERT and his associates (TOLBERT *et al.*, 1968; KISAKI and TOLBERT, 1969). The specific activities of some of the peroxisomal enzymes pertinent to this discussion are cited in Table 13. A hypothesis is shown in Fig. 15 to account for the interrelationships among the leaf cellular fractions concerned with glycolate metabolism. According to TOLBERT and coworkers, in the peroxisome glycolate is oxidized to glyoxylate and H_2O_2, the latter being destroyed by the presence of 7,000-fold more catalase activity than glycolate oxidase activity. The peroxisome, they found, does not oxidize glycolate or glyoxylate to CO_2. ZELITCH (1964), on the other hand, has envisaged the H_2O_2 as a source for peroxidation of the glyoxylate to yield formation and CO_2. As pictured in the Fig. 15, glyoxylate is converted to glycine which diffuses from the peroxisome to the mitochondria. In the mitochondria, 2 glycine molecules are converted to 1 serine and

1 CO_2, this reaction being the source of CO_2 observed in photorespiration. Subsequently, serine is converted to glycerate (STAFFORD *et al.*, 1954) and eventually re-enters chloroplastic metabolism as glycerate 3-P (HATCH and SLACK, 1969).

Glycolate oxidase is a peroxisomal enzyme and since glycolate oxidase appears to be obligatory for the phenomenon termed photorespiration, then it follows that a correlation should exist between the presence of peroxisomes (or a marker enzyme such as glycolate oxidase) and light-enhanced respiration. Leaf homogenates of

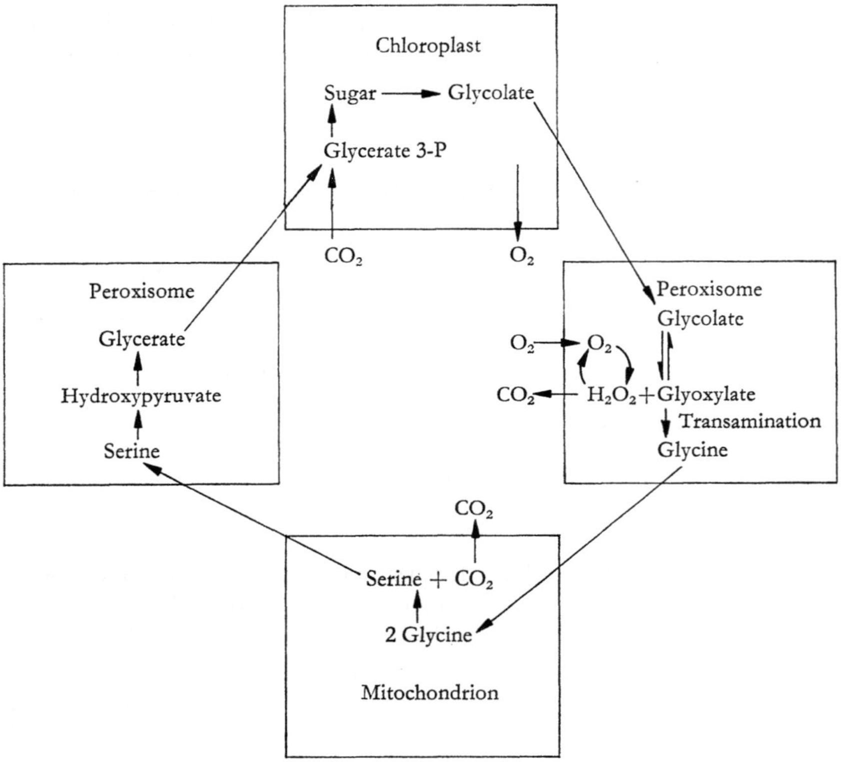

Fig. 15. Glycolate pathway and its distribution in the cell. From KISAKI and TOLBERT (1961). See also ZELITCH (1964) and HATCH and SLACK (1969)

plants with photorespiration (spinach, sunflower, tobacco, pea, wheat, bean and Swiss chard) were found to have glycolate oxidase activities ranging from 30 to 90 μmoles/g fresh weight·h or a specific activity of 1.2 to 3.0 μmoles/mg protein·h. In sharp contrast, homogenates of maize and sugar cane leaves (without photorespiration) have 2 to 5% as much glycolate oxidase (TOLBERT *et al.*, 1968). This amount of activity in maize and sugar cane is still substantial and indicates that the absence of photorespiration cannot be accounted solely in terms of the glycolate pathway.

Like photosynthesis, photorespiration can be envisaged in terms of a "light" and a "dark" set of reactions. The "light" phase located in the chloroplast comprises the reactions involved in the conversion of CO_2 to glycolate while the result of the "dark" phase is the subsequent metabolism of glycolate to CO_2 involving the concerted effort of the peroxisome and perhaps of the mitochondrion.

6. Orthophosphate and Pyrophosphate

At low concentrations, orthophosphate stimulates the rate of photosynthetic CO_2 assimilation in spinach chloroplasts, however, a diminution occurs as the concentration is increased (GIBBS and CALO, 1959; LOSADA *et al.*, 1960). JENSEN and BASSHAM (1966) also reported a stimulation by pyrophosphate. If the concentration of orthophosphate in the reaction mixture is raised above about 0.05 millimolar, the induction period is prolonged and the maximum rate is decreased [COCKBURN *et al.*, 1967 (1)]. Further increases in orthophosphate lengthen the lag period and a corresponding decrease in rate [COCKBURN *et al.*, 1967 (2, 3)]. A concentration of 2 millimolar was usually sufficient to quench photosynthesis with CO_2 as the sole added substrate. On the other hand, pyrophosphate failed to produce these effects at a 100 times the concentration and indeed, alleviated orthophosphate inhibition. Orthophosphate inhibition could be reversed (COCKBURN *et al.*, 1968) by all the compounds that have been shown to stimulate CO_2 fixation (p. 189). A difference between the compounds in kinetics of reversal was found which probably reflected differences in in their ability to enter the chloroplast. The actual mechanism of reversal is not clear.

Ribose 5-P isomerase (ANDERSON *et al.*, 1967), ribulose 1,5-diP carboxylase (WEISSBACH *et al.*, 1956) and ADP-glucose pyrophosphorylase (SANWAL *et al.*, 1968) are sensitive to orthophosphate but only inhibition of the latter is relieved by intermediates of the carbon cycle. It is noteworthy that ribose 5-P or fructose 1,6-diP could relieve the inhibition of CO_2 fixation and associated O_2 evolution of spinach chloroplasts caused by inhibitors of photophosphorylation including desaspidin, nigericin, arsenate and trifluorocarbonylcyanide phenylhydrazone as well as iodoacetamide and arsenite, inhibitors of carbon metabolism (SCHACTER *et al.*, 1968). Also JENSEN and BASSHAM (1968) have shown that fructose 1,6-diP helps to sustain a high rate of chloroplast activity when the carbon supply was adequate. They propose that the stimulatory effect of added fructose 1,6-diP is exerted on the carboxylation reaction rather than in the regeneration of ribulose 1,5-diP.

In the total absence of orthophosphate, O_2 evolution ceased but could be restarted by the addition of orthophosphate [COCKBURN *et al.*, 1967 (3)]. Pyrophosphate, ATP or ADP but not AMP could substitute for orthophosphate. The response to ATP was slower, implying slow penetration or most likely external hydrolysis followed by penetration of orthophosphate. The amount of O_2 evolved was in accordance with Eq. (26) and inorganic pyrophosphate produced twice as much O_2 as equimolar orthophosphate. This might imply hydrolysis but the results of JENSEN and BASSHAM (1968) and COCKBURN *et al.* [1967 (6)] suggest a more direct participation of pyrophosphate in photosynthesis.

Oxygen evolution in the isolated chloroplast in the presence of substrate concentrations of glycerate 3-P (WALKER and HILL, 1967) is not initially dependent on the presence of CO_2, presumably because glycerate 3-P is converted to glycerate 1,3-diP which then serves as hydrogen acceptor. In mixtures containing both CO_2 and glycerate 3-P, O_2 evolution but not carbon uptake is linear (Fig. 16); the latter showing a lag, more pronounced by the presence of orthophosphate [Cockburn *et al.*, 1968 (2)]. Furthermore, with dihydroxyacetone-P, the rate of CO_2 evolution agreed more closely with that of O_2. With ribose 5-P, both O_2 and CO_2 responses were slower but parallel. Also there was greater discrepancy in the photosynthetic quotient with glycerate 3-P than with dihydroxyacetone-P or with ribose 5-P.

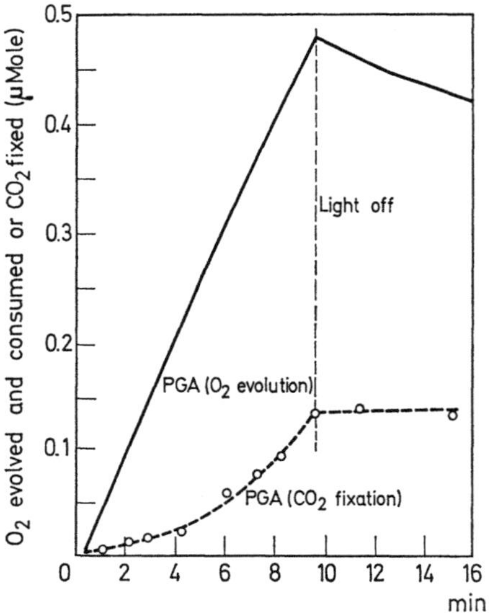

Fig. 16. Simultaneous measurements of O_2 evolution (———) and $^{14}CO_2$ fixation (- - - - -) in the presence of glycerate 3-P and an "inhibitory" amount of orthophosphate. From Cockburn *et al.* [1968 (2)]

7. Intermediates of the Photosynthetic Carbon Reduction Cycle

The close connection between starch and photosynthesis was noted by Sachs (1862). Starch resulting from photosynthesis is stored in the chloroplast and, hence, would be expected to be closely connected with the traffic of carbon within that organelle.

Starting with glucose 6-P, three reactions are pertinent to starch synthesis. They are catalyzed by phosphoglucomutase, ADP-glucose pyrophosphorylase and starch synthetase [Reactions (29, 30) and (31)]:

$$\text{Glucose 6-P} \rightleftharpoons \text{glucose 1-P}\,, \tag{29}$$

$$\text{ATP} + \alpha\text{-D-glucose 1-P} \rightleftharpoons \text{ADP-D-glucose} + \text{pyrophosphate}\,, \tag{30}$$

$$\text{ADP-D-glucose} + \text{starch primer} \rightarrow \text{ADP} + \text{glucosyl-primer}\,. \tag{31}$$

The role of phosphorylase [Reaction (32)] has been questioned in starch synthesis since the equilibrium constant for the phosphorylase reaction is roughly unity in contrast to 250 for the nucleoside sugar reaction.

$$\text{Starch} + \alpha\text{-D-glucose 1-P} \rightleftharpoons \text{glucosyl-starch} + \text{phosphate}\,. \tag{32}$$

The regulation of starch synthesis is apparently not accomplished by controlling the activity of starch synthetase but rather of the ADP-glucose pyrophosphorylase. Analogous to bacterial ADP-glucose pyrophosphorylase (SHEN and PREISS, 1964), the spinach chloroplast enzyme is stimulated by a number of intermediates of the photosynthetic carbon reduction cycle and related compounds (GHOSH and PREISS, 1965). Activation is highest with glycerate 3-P (Table 14).

Leaf ADP-glucose pyrophosphorylases (tobacco, tomato, barley, sorghum, sugar beet, rice and spinach) generally have three common properties (SANWAL *et al.*, 1968). These enzyme are activated 5- to 17-fold by glycerate 3-P, inhibited by inorganic phosphate, and the inhibition caused by inorganic phosphate is reversed by glycerate 3-P. Clearly these findings form the basis for a hypothesis of the regulation of starch formation. During photosynthesis the level of inorganic phosphate would be kept low while glycerate 3-P increases slightly. This situation would therefore contribute to an environment necessary for active starch synthesis. In the dark the level of inorganic phosphate rises rapidly and thereby inhibits this biosynthetic pathway.

Table 14. *Activation of spinach leaf ADP-glucose pyrophosphorylase by various metabolites. (From* PRIESS, *1967)*

Activator 1.5 mM	ADP-glucose formed μmoles/mg protein · h	Relative increase
None	22.6	1
Glycerate 3-P	2.0	9.3
Glycerate 2-P	187	8.3
Acetyl CoA	102	4.5
Glycerate 2,3-diP	93	4.1
Fructose 6-P	81	3.6
P-Enolpyruvate	79	3.5
Deoxyribose 5-P	72	3.2
Ribose 5-P	63	2.8
α-Glycerol-P	54	2.4
P-Hydroxypyruvate	52	2.3
Fructose 1,6-diP	50	2.2
Glyceraldehyde 3-P	—	1
Dihydroxyacetone-P	—	1
Glucose 6-P	—	1

An end product of photosynthesis, sucrose, is able to influence the rate of photosynthesis. HARTT (1963) recorded a decrease in photosynthesis with increasing sucrose content. This observation may be explained by HAWKER's (1967) finding that sucrose phosphatase is inhibited by sucrose.

G. Literature

AACH, H. G., HEBER, U.: Kompartimentierung von Aminosäuren in der Blattzelle. Z. Pflanzenphysiol. **57**, 317—328 (1967).

AKOYUNOGLOU, G., CALVIN, M.: Mechanism of the carboxydismutase reaction. II. Carboxylation of the entyme. Biochem. Z. **338**, 20—30 (1963).

ALLEN, M. B., ARNON, D. I., CAPINDALE, J. B., WHATLEY, F. R., DURHAM, L. J.: Photosynthesis by isolated chloroplasts. III. Evidence for complete photosynthesis. J. Amer. chem. Soc. **77**, 4149—4155 (1955).

ANDERSON, L., FULLER, R. C.: Photosynthesis in *Rhodospirillum rubrum*. III. Metabolic control of reductive pentose phosphate and tricarboxylic acid cycle enzymes. Plant Physiol. **42**, 497—502 (1967).

— WORTHEN, L. E., FULLER, R. C.: Ribose 5-P isomerase from *Rhodospirillum rubrum*. Plant Physiol. **42**, 48 (1967).

BALDRY, C. W., BUCKE, C., WALKER, D. A.: Some effects of temperature on carbon dioxide fixation by isolated chloroplasts. Biochim. biophys. Acta (Amst.) **126**, 207—213 (1966).

— Incorporation of inorganic phosphate into sugar phosphates during carbon dioxide fixation by illuminated chloroplasts. Nature (Lond.) **210**, 793—796 (1966).

— COOMBS, J., GROSS, D.: Isolation and separation of chloroplasts from sugar cane. Z. Pflanzenphysiol. **60**, 78—81 (1969).

— WALKER, D. A., BUCKE, C.: Calvin-cycle intermediates in relation to induction phenomena in photosynthetic carbon dioxide fixation by isolated chloroplasts. Biochem. J. **101**, 642—646 (1966).

BAMBERGER, E. S., GIBBS, M.: Effects of phosphorylated compounds and inhibitors on CO_2 fixation by intact spinach chloroplasts. Plant Physiol. **40**, 919—926 (1965).

BANDURSKI, R. S., GREINER, C. M.: The enzymatic synthesis of oxalacetate from phosphorylenol-pyruvate and carbon dioxide. J. biol. Chem. **204**, 781—786 (1953).

BASSHAM, J. A., BENSON, A. A., KAY, L. D., HARRIS, A. Z., WILSON, A. T., CALVIN, M.: The path of carbon in photosynthesis XXI. The cyclic regeneration of the carbon dioxide acceptor. J. Amer. chem. Soc. **76**, 1760—1770 (1954).

— JENSEN, R. A.: Photosynthesis of carbon compounds. In: Harvesting the sun, pp. 79—110 (A. SAN PIETRO, F. A. GREER, T. J. ARMY, Eds.). New York: Academic Press 1967.

— KIRK, M.: The effect of oxygen on the reduction of CO_2 to glycolic acid and other products during photosynthesis by *Chlorella*. Biochem. biophys. Res. Commun. **9**, 376—380 (1962).

— — JENSEN, R. G.: Photosynthesis by isolated chloroplasts. I. Diffusion of labeled photosynthetic intermediates between isolated chloroplasts and suspending medium. Biochim. biophys. Acta (Amst.) **153**, 211—218 (1968).

BIRD, I. F., PORTER, H. K., STOCKING, C. R.: Intracellular localization of enzymes associated with sucrose synthesis in leaves. Biochim. biophys. Acta (Amst.) **100**, 365—375 (1965).

BRADBEER, J. W., RACKER, E.: Glycolate formation from fructose 6-P by cell free preparations. Fed. Proc. **20**, 88 (1961).

BROOKS, K., CRIDDLE, R. S.: Enzymes of the carbon cycle of photosynthesis. I. Isolation and properties of spinach chloroplast aldolase. Arch. Biochem. **117**, 650—659 (1966).

BUCHANAN, B. B., KALBERER, P. P., ARNON, D. I.: Ferredoxin-activated fructose diphosphatase in isolated chloroplasts. Biochem. biophys. Res. Commun. **29**, 74—79 (1967).

— Ferredoxin-activated fructose diphosphatase of isolated chloroplasts. Fed. Proc. **27**, 344 (1968).

BUCKE, C., WALKER, D. A., BALDRY, C. W.: Some effects of sugars and sugar phosphates on carbon dioxide fixation by isolated chloroplasts. Biochem. J. **101**, 636—641 (1966).

CALVIN, M., BASSHAM, J. A.: The photosynthesis of carbon compounds. New York: W. A. Benjamin, Inc. 1962.

CANNATA, J., STOPPANI, A. M. O.: Phosphopyruvate carboxylase from Bakers' yeast II. Properties of enzyme. J. biol. Chem. **238**, 1208—1212 (1963).

CHANG, H. C., MARUYAMA, H., MILLER, R., LANE, M. D.: The enzymatic carboxylation of phosphoenolpyruvate II. Purification and properties of liver mitochondrial phosphoenolpyruvate carboxykinase. J. biol. Chem. **241**, 2413—2420 (1966).

CHEN, S., MCMAHON, D., BOGORAD, L.: Early effects of illumination on the activity of some photosynthetic enzymes. Plant Physiol. **42**, 1—5 (1967).

CHEN, T. M., BROWN, R. H., BLACK, C. C., JR.: Photosynthetic activity of chloroplasts isolated from bermudagrass (*Cynodon dactylon* L.) a species with a high photosynthetic capacity. Plant. Physiol. **44**, 649—654 (1969).

COCKBURN, W., BALDRY, C. W., WALKER, D. A.: (1) Oxygen evolution by isolated chloroplasts with carbon dioxide as hydrogen acceptor. A requirement for orthophosphate or pyrophosphate. Biochim. biophys. Acta (Amst.) **131**, 594—596 (1967).

— — — (2) Photosynthetic induction phenomena in spinach chloroplasts in relation to the nature of the isolating medium. Biochim. biophys. Acta (Amst.) **143**, 606—613 (1967).

— — — (3) Some effects of inorganic phosphate on O_2 evolution by isolated chloroplasts. Biochim. biophys. Acta (Amst.) **143**, 614—624 (1967).

— WALKER, D. A., BALDRY, C. W.: (1) The isolation of spinach chloroplasts in pyrophosphate media. Plant Physiol. **43**, 1415—1418 (1968).

— — — (2) Photosynthesis by isolated chloroplasts. Reversal of orthophosphate inhibition by CALVIN cycle intermediates. Biochem. J. **107**, 89—95 (1968).

COOMBS, J., WHITTINGHAM, C. P.: The mechanism of inhibition of photosynthesis by high partial pressures of oxygen in *Chlorella*. Proc. roy. Soc. B **165**, 511—520 (1966).

COOPER, R. A., KORNBERG, H. C.: Net formation of phosphoenolpyruvate from pyruvate by *Escherichia coli*. Biochim. biophys. Acta (Amst.) **104**, 618—620 (1965).

COOPER, T. G., BENEDICT, C. R.: PEP carboxykinase exchange reaction in photosynthetic bacteria. Plant Physiol. **43**, 788—792 (1968).

— FILMER, D., WISHNICK, M., LANE, M. D.: The active species of "CO_2" utilized by ribulose diphosphate carboxylase. J. biol. Chem. **244**, 1081—1083 (1969).

— TCHEN, T. T., WOOD, H. G., BENEDICT, C. R.: The carboxylation of phosphoenolpyruvate and pyruvate. I. The active species of "CO_2" utilized by phosphoenolpyruvate carboxykinase, carboxytransphophorylase, and pyruvate carboxylase. J. biol. Chem. **243**, 3857—3863 (1968).

DILLEY, R. A., VERNON, L. P.: Ion and water transport processes related to the light-dependent shrinkage of spinach chloroplasts. Arch. Biochem. **111**, 365—375 (1965).

DOWNTON, W. J. S., BISALPUTRA, T., TREGUNNA, E. B.: The distribution and ultrastructure of chloroplasts in leaves differing in photosynthetic carbon metabolism. II. *Atriplex rosea* and *Atriplex hastata* (Chenopodiaceae). Canad. J. Botany **47**, 915—919 (1969).

ELBERTZHAGEN, H., KANDLER, O.: Differences in the pH optimum of the photosynthetic fixation of carbon dioxide in isolated whole and broken chloroplasts. Nature (Lond.) **194**, 312—313 (1962).

ELLYARD, P. W.: The Warburg Effect, investigations using isolated spinach chloroplasts. Ph. D. Thesis. Cornell University, Ithaca, New York, 1968.

— GIBBS, M.: The effect of oxygen on photosynthesis in chloroplasts. Plant Physiol. **42**, S-33 (1967).

— — Inhibition of photosynthesis by oxygen in isolated spinach chloroplasts. Plant Physiol. **44**, 1115—1121 (1969).

EMERSON, R.: Photosynthesis as a function of light intensity and of temperature with different concentrations of chlorophyll. J. gen. Physiol. **12**, 623—631 (1929).

— GREEN, L.: Manometric measurements of photosynthesis in the marine alga *Gigartina*. J. gen. Physiol. **17**, 817—842 (1934).

EVANS, E. A., JR., VENNESLAND, B., SLOTIN, L.: Mechanism of carbon dioxide fixation in cell-free extracts of pigeon liver. J. biol. Chem. **147**, 771—784 (1943).

EVERSON, R. G.: Bicarbonate equilibria and apparent Km(HCO_3) of isolated chloroplasts. Nature (Lond.) **222**, 876 (1969).

— COCKBURN, W., GIBBS, M.: Sucrose as a product of photosynthesis in isolated spinach chloroplasts. Plant Physiol. **42**, 840—844 (1967).

— SLACK, C. R.: Distribution of carbonic anhydrase in relation to the C_4 pathway of photosynthesis. Phytochemistry **7**, 581—584 (1968).

FARINEAU, J.: Metabolism of some phosphorylated compounds and photophosphorylation in maize leaves. Planta **85**, 135—156 (1969).

FULLER, R. C., GIBBS, M.: Intracellular and phylogenetic distribution of ribulose 1,5-diphosphate carboxylase and D-glyceraldehyde-3-phosphate dehydrogenases. Plant. Physiol. **34**, 324—329 (1959).

GALMICH, J. M.: Photosynthetic incorporation of carbon from CO_2 into malic acid and phophoglyceric acid in *Zea mays*. In: Photosynthesis in sugar cane, pp. 36—39. (J. COOMBS, Ed.), London: Tate and Lyle 1968.

GHOSH, H. P., PREISS, J.: The biosynthesis of starch in spinach chloroplasts. J. biol. Chem. **240**, P. C. 960—962 (1965).

GIBBS, M., BAMBERGER, E. S., ELLYARD, P. W., EVERSON, R. G.: Assimilation of carbon dioxide by chloroplast preparations. In: Biochemistry of chloroplasts, Vol. II, p. 3—35 (T. W. GOODWIN, Ed.). London: Academic Press 1967.

— CALO, N.: Factors affecting light induced fixation of carbon dioxide by isolated spinach chloroplasts. Plant Physiol. **34**, 318—323 (1959).

— CYNKIN, M. A.: The conversion of carbon-14 dioxide to starch glucose during photosynthesis by spinach chloroplasts. Nature (Lond.) **182**, 1241—1242 (1958).

— ELLYARD, P. W., LATZKO, E.: Warburg Effect: Control of photosynthesis by oxygen. In: Comparative biochemistry and biophysics of photosynthesis, pp. 387—399. (K. SHIBATA, A. TAKAMIYA, A. T. JAGENDORF, JR., R. C. FULLER, Eds.), Tokyo, Japan: Univ. Tokyo Press 1968.

— LATZKO, E., O'NEAL, D., HEW, C. S.: Photosynthetic carbon fixation by isolated maize chloroplasts. Biochem. biophy. Res. Commun. **40**, 1356—1361 (1970).

— KANDLER, O.: Asymmetric distribution of C^{14} in sugars formed during photosynthesis. Proc. nat. Acad. Sci. (Wash.) **43**, 446—451 (1957).

— PLAUT, Z., LATZKO, E., ELEY, J., SCHACTER, B.: Carbon fixation by isolated chloroplasts. In: Photosynthesis in sugar cane, pp. 49—60 (J. COOMBS, Ed.). London: Tate and Lyle Ltd. 1968.

GOLDSWORTHY, A.: Comparison of the kinetics of photosynthetic carbon dioxide fixation in maize, sugar cane and its relation to photorespiration. Nature (Lond.) **217**, 62 (1968).

HARTT, C. E.: Translocation as a factor in photosynthesis. Naturwissenschaften **50**, 666—667 (1963).

HATCH, M. D., SLACK, C. R.: A new carboxylation reaction and the pathway of sugar formation. Biochem. J. **101**, 103—111 (1966).

— — A new enzyme for the intervonversion of pyruvate and phosphopyruvate and its role in the C_4 dicarboxylic acid pathway of photosynthesis. Biochem. J. **106**, 141—146 (1968).

— — NADP-specific malate dehydrogenase and glycerate kinase in leaves and evidence for their location in chloroplasts. Biochem. biophys. Res. Commun. **43**, 589—593 (1969).

— — BULL, T. A.: Light induced changes in the content of some enzymes of the C_4-dicarboxylic acid pathway of photosynthesis and its effect on other characteristics of photosynthesis. Photochemistry **8**, 697—706 (1969).

— — JOHNSON, H. S.: Further studies on a new pathway of photosynthetic carbon dioxide fixation in sugar-cane and its occurrence in other plant species. Biochem. J. **102**, 417—422 (1967).

— — TURNER, J. F.: A protein disulfide reductase from pea seeds. Biochem. J. **76**, 556—562 (1960).

HAVIR, E. A., GIBBS, M.: Studies on the reductive pentose phosphate cycle in intact and reconstituted chloroplast systems. J. biol. Chem. **238**, 3183—3187 (1963).

HAWKER, J. S.: Inhibition of sucrose phosphatase by sucrose. Biochem. J. **102**, 401—406 (1967).

HEBER, U. W.: Photosynthetic activity of isolated pea chloroplasts. In: Biochemistry of chloroplasts, Vol. II, p. 71—78 (T. W. GOODWIN, Ed.). London: Academic Press 1967.

— FRENCH, C. S.: Effects of oxygen on the electron transport chain of photosynthesis. Planta **79**, 99—112 (1968).

— PON, N. G., HEBER, M.: Localization of carboxydismutase and triose phosphate dehydrogenases in chloroplasts. Plant Physiol. **38**, 355—360 (1963).

HEBER, U. W., SANTARIUS, K. A.: Compartmentation and reduction of pyridine nucleotides in relation to photosynthesis. Biochim. biophys. Acta (Amst.) **109**, 390—408 (1965).

— — HUDSON, M. A. HALLIER, U. W.: I. Intrazellularer Transport von Zwischenprodukten der Photosynthese im Photosynthese-Gleichgewicht und im Dunkel-Licht-Dunkel-Wechsel. Z. Naturforsch. **22b**, 1189—1199 (1967).

HEW, C. S., GIBBS, M.: A study of corn and sugar cane chloroplasts. Proc. XIth Int. Bot. Congress, 1969, p. 90.

HODGE, A. J., MCLEAN, J. D., MERCER, F. V.: Ultrastructure of the lamellae and grana in the chloroplast of *Zea mays* L. Biophys. Biochem. Cytol. **1**, 605—614 (1955).

HOOD, W. CARR, N. G.: Association of NAD and NADP linked glyceraldehyde-3-phosphate dehydrogenase in the blue-green alga *Anabaena variabilis*. Planta **86**, 250—258 (1969).

HUFFAKER, R. C., OBENDORF, R. L., KELLER, C. J., KLEINKOPF, G. E.: Effects of light intensity on photosynthetic carboxylative phase enzymes and chlorophyll synthesis in greening leaves of *Hordeum vulgare* L. Plant Physiol. **41**, 913—918 (1966).

JENSEN, R. G., BASSHAM, J. A.: Photosynthesis by isolated chloroplasts. Proc. nat. Acad. Sci. (Wash.) **56**, 1095—1101 (1965).

— — Photosynthesis by isolated chloroplasts. II. Effects of addition of cofactors and intermediate compounds. Biochim. biophys. Acta (Amst.) **153**, 219—226 (1968).

— — Photosynthesis by isolated chloroplasts. III. Light activation of the carboxylation reaction. Biochim. biophys. Acta (Amst.) **153**, 227—234 (1968).

JOHNSON, H. S., HATCH, M. D.: Distribution of the C_4-dicarboxylic acid pathway of photosynthesis and its occurrence in dicotyledonous plants. Phytochemistry **7**, 375—380 (1968).

KALBERER, P. P., BUCHANAN, B. B., ARNON, D. I.: Rates of photosynthesis by isolated chloroplasts. Proc. nat. Acad. Sci. (Wash.) **57**, 1542—1549 (1967).

KANDLER, O., ELBERTZHAGEN, H., HABERER-LIESENKÖTTER: Rate-limiting factors in the photosynthesis of isolated chloroplasts. In: Biochemistry of chloroplasts, Vol. II, p. 39—51 (T. W. GOODWIN, Ed.). London: Academic Press 1967.

— SENSER, M.: Differences in the pattern of products of photosynthesis after short term photosynthesis, in $^{14}CO_2$, of *Oryza* and *Sorghum*. In: Photosynthesis in sugar cane, pp. 30—32 (J. COOMBS, Ed.). London: Tate and Lyle 1968.

KAZIRO, Y., HASS, L. F., BOYER, P. D., OCHOA, S.: Mechanism of propionyl carboxylase reaction II. Isotopic exchange and tracer experiments. J. biol. Chem. **237**, 1460—1468 (1963).

KISAKI, T., TOLBERT, N. E.: Glycolate and glyoxylate metabolism by isolated peroxisomes or chloroplasts. Plant Physiol. **44**, 242—250 (1969).

KORTSCHAK, H. P.: Photosynthesis in sugar cane and related species. In: Photosynthesis in sugar cane, pp. 18—24 (J. COOMBS, Ed.). London: Tate and Lyle 1968.

— HARTT, C. E., BURR, G. O.: Carbon dioxide fixation in sugar cane leaves. Plant Physiol. **40**, 209—213 (1965).

KRIMSKY, I.: Phosphorylation of pyruvate by the pyruvate kinase reaction and reversal of glycolysis in a reconstructed system. J. biol. Chem. **234**, 232—236 (1959).

LAETCH, W. M., PRICE, I.: Development of the dimorphic chloroplasts of sugar cane. Amer. J. Bot. **56**, 77—87 (1969).

LATZKO, E., GIBBS, M.: Distribution and activity of enzymes of the reductive pentose phosphate cycle in spinach leaves and in chloroplasts. Z. Pflanzenphysiol. **59**, 184—194 (1968).

— — Effect of O_2, arsenite and sulfhydryl compounds and light on the activity of ribulose 5-phosphate kinase. In: Int. Congress of Photosynthesis Research, Freudenstadt 1968, pp. 1624—1630.

— — Enzyme activities of the carbon reduction cycle in some photosynthetic organisms. Plant Physiol. **44**, 295—300 (1969).

— — Level of photosynthetic intermediates in isolated spinach chloroplasts. Plant Physiol. **44**, 396—402 (1969).

LOCHMÜLLER, H., WOOD, H. G., DAVIS, J. J.: Phosphoenolpyruvate carboxytransphosphorylase. II. Crystallization and properties. J. biol. Chem. **241**, 5678—5691 (1966).

LOSADA, M., TREBST, A. V., ARNON, D. I.: Photosynthesis by isolated chloroplasts XI. CO_2 assimilation in a reconstituted chloroplast system. J. biol. Chem. **235**, 832—839 (1960).

MARUYAMA, H., EASTERDAY, R. L., CHANG, H., LANE, M. D.: The enzymatic carboxylation of phosphoenolpyruvate. I. Purification and properties of phosphoenolpyruvate carboxylase. J. biol. Chem. **241**, 2405—2412 (1966).

MAZELIS, M., VENNESLAND, B.: Carbon dioxide fixation into oxalacetate in higher plants. Plant Physiol. **32**, 591—600 (1957).

MCALISTER, E. D., MYERS, J.: The time course of photosynthesis and fluorescence observed simultaneously. Smithsonian Inst. Publs. Misc. Collections **99**, No. 6 (1940).

MEDINA, A., SOLS, A.: A specific fructokinase in peas. Biochim. biophys. Acta (Amst.) **19**, 378—379 (1956).

MIYACHI, S., HOGETSU, D.: Light-enhanced carbon dioxide fixation in isolated chloroplasts. Plant Cell Physiol. **11**, 927—936 (1970).

MÜLLER, B., ZIEGLER, I., ZIEGLER, H.: Lichtinduzierte reversible Aktivitätssteigerung der NADP-abhängigen Glycerinaldehyd-3-phosphat-Dehydrogenase in Chloroplasten. Europ. J. Biochem. **9**, 101—106 (1969).

PAULSEN, J. M., LANE, M. D.: Spinach ribulose diphosphate carboxylsae. I. Purification and properties of the enzyme. Biochemistry **5**, 2350—2357 (1966).

PETERKOFSKY, A., RACKER, E.: The reductive pentose phosphate cycle III. Enzyme activities in cell free extracts of photosynthetic organisms. Plant Physiol. **36**, 409—414 (1961).

PETRACK, B., LIPMANN, F.: Photophosphorylation and photohydrolysis in cell-free preparations of blue-green alga. In: Light and life, pp. 621—630 (W. D. MCELROY, B. GLASS, Eds.). Baltimore, Maryland: Johns Hopkins Press 1961.

PLAUT, Z., GIBBS, M.: Glycolate formation in spinach chloroplasts. Plant Physiol. **45**, 470—474 (1970).

PREISS, J.: Regulation of the biosynthesis of starch in spinach leaf chloroplasts. In: Biochemistry of chloroplasts, Vol. II, pp. 131—154 (T. W. GOODWIN, Ed.). London: Academic Press 1967.

RABINOWITCH, E. I.: Photosynthesis and related processes, Vol. II, part A. New York: Interscience Publishers, Inc. 1956.

RAMIREZ, J. M., DEL CAMPO, F. F., ARNON, D. I.: Photosynthetic phosphorylation as energy source for protein synthesis and carbon dioxide assimilation by chloroplasts. Proc. nat. Acad. Sci. (Wash.) **59**, 606—612 (1968).

SACHS, J.: Über den Einfluß des Lichtes auf die Bildung des Amylums in den Chlorophyllkörnern. Botan. Z. **20**, 365—373 (1862).

SANTARIUS, K. A., HEBER, U.: Changes in intracellular levels of adenosine 5′-triphosphate, adenosine 5′-diphosphate, adenosine 5′-monophosphate and inorganic phosphate and regulatory function of adenylate system in leaf cells during photosynthesis. Biochim. biophys. Acta (Amst.) **102**, 39—54 (1968).

SANWAL, G. G., GREENBERG, E., HARDIE, J., CAMERON, E. C., PREISS, J.: Regulation of starch biosynthesis in plant leaves: activation and inhibition of ADP-glucose pyrophosphorylase. Plant Physiol. **43**, 417—427 (1968).

SCHACTER, B.: Studies on the control of photosynthesis in isolated spinach chloroplasts. Ph. D. Thesis, Brandeis University 1970.

— ELEY, J. H., JR., GIBBS, M.: Effect of sugar phosphates and photosynthesis inhibitors on CO_2 fixation and O_2 evolution by chloroplasts. Plant Physiol. **43**, S-30 (1968).

SCHULMAN, M. D., GIBBS, M.: D-Glyceraldehyde 3-phosphate dehydrogenases of higher plants. Plant. Physiol. **43**, 1805—1812 (1968).

SELWYN, M. J.: Temperature and photosynthesis. II. A mechanism for the effects of temperature on carbon dioxide fixation. Biochem. biophys. Acta (Amst.) **126**, 214—224 (1966).

SHAIN, Y., GIBBS, M.: Formation of glycolic acid by spinach chloroplast fragments. Proc. XIth Int. Bot. Congress 1969.

SHEN, L., PREISS, J.: The activation and inhibition of bacterial adenosine-diphosphoglucose pyrophosphorylase. Biochem. biophys. Res. Commun. **17**, 424—429 (1964).

SLACK, C. R.: The photoactivation of a phosphopyruvate synthase in leaves of *Amaranthus palmeri*. Biochem. biophys. Res. Commun. **30**, 483—488 (1968).

— HATCH, M. D.: Comparative studies on the activity of carboxylases and other enzymes in relation to the new pathway of photosynthetic carbon dioxide fixation in tropical grasses. Biochem. J. **103**, 660—665 (1967).

SLACK, C. R., HATCH, M. D.: Localization of enzymes of the C_4-dicarboxylic acid pathway of photosynthesis in maize leaves. Proc. XIth Int. Bot. Congress 1969.

SMILLIE, R.: Alkaline C-1 fructose, 1,6-diphosphatase. Evidence for its participation in photosynthesis. Nature (Lond.) **187**, 1024—1025 (1962).

— FULLER, R. C.: Ribulose 1,5-diphosphate carboxylase activity in relation to photosynthesis by intact leaves and isolated chloroplasts. Plant Physiol. **34**, 651—656 (1959).

SPENCER, D., UNT, H.: The effect of molybdate on the activity of tomato acid phosphatases. Aust. J. biol. Sci. **18**, 197—210 (1965).

STAFFORD, H. A., MAGALDI, A., VENNESLAND, B.: The enzymatic reduction of hydroxypyruvic acid to D-glyceric acid in higher plants. J. biol. Chem. **207**, 621—630 (1954).

STOCKING, C. R.: Chloroplast isolation in nonaqueous media. Plant Physiol. **34**, 56—61 (1959).

— UNGUN, A.: The intracellular distribution of some metallic elements in leaves. Amer. J. Bot. **49**, 284—289 (1962).

SUGIYAMA, T., NAKAYAMA, N., AKAZAWA, T.: Structure and function of chloroplast proteins. V. Homotropic effect of bicarbonate in the RuDP carboxylase reaction and the mechanism of activation by magnesium ions. Arch. Biochem. **126**, 737—745 (1968).

TAMIYA, M., HUZISIGE, H.: Effect of oxygen on the dark reaction of photosynthesis. Stud. Tokugawa Inst. **6**, 83—104 (1949).

THALACKER, R., BEHRENS, M.: Über den Reinheitsgrad der in einem nichtwäßrigen spezifischen Gewichtsgradienten gewonnenen Chloropasten. Z. Naturforsch. **14b**, 443—446 (1959).

THOMPSON, C. M., WHITTINGHAM, C. P.: Glycolate metabolism in photosynthesizing tissue. Biochim. biophys. Acta (Amst.) **153**, 260—269 (1968).

TOLBERT, N. E., OESER, A., YAMAZAKI, R. K., HAGEMAN, R. H., KISAKI, T.: A survey of plants for leaf peroxisomes. Plant Physiol. **44**, 135—147 (1969).

TREBST, A., FIEDLER, F.: ^{14}C-Verteilung in der Glucose bei der Photosynthese mit Chloroplasten. Z. Naturforsch. **16b**, 284—285 (1961).

— Über die Ursache der asymmetrischen ^{14}C-Verteilung in der Hexose bei der Photosynthese mit Chloroplasten. Z. Naturforsch. **17b**, 553—558 (1962).

TURNER, J. S., BRITTAIN, E. G.: Oxygen as a factor in photosynthesis. Biol. Rev. **37**, 130—170 (1962).

— TURNER, J. F., SHORTMAN, K. D., KING, J. E.: The inhibition of photosynthesis by oxygen. II. The effect of oxygen on glyceraldehyde phosphate dehydrogenase from chloroplasts. Aust. J. biol. Sci. **11**, 336—342 (1958).

URBACH, W., HUDSON, M. A., ULLRICH, W., SANTARIUS, K. A., HEBER, U.: Verteilung und Wanderung von Phosphoglycerat zwischen den Chloroplasten und dem Cytoplasma während der Photosynthese. Z. Naturforsch. **20b**, 890—898 (1965).

UTTER, M. F., KURAHASHI, K.: Mechanism of action of oxalacetic carboxylase. J. biol. Chem. **207**, 821—841 (1954).

VAN DER PAAUW, F.: Der Einfluß der Temperatur auf Atmung und Kohlensäure-Assimilation einiger Grünalgen. Planta **22**, 396 (1934).

WALKER, D. A.: Pyruvate carboxylation and plant metabolism. Biol. Rev. **37**, 215—256 (1962).

— Improved rates of carbon dioxide fixation by illuminated chloroplasts. Biochem. J. **92**, 22c (1964).

— Correlation between photosynthetic activity and membrane integrity in isolated pea chloroplasts. Plant Physiol. **40**, 1157—1161 (1965).

— Photosynthetic activity of isolated pea chloroplasts. In: Biochemistry of chloroplasts, Vol. II, pp. 53—70 (T. W. GOODWIN, Ed.). London: Academic Press 1967.

— BALDRY, C. W., COCKBURN, W.: Photosynthesis by isolated chloroplasts, simultaneous measurement of carbon assimilation and oxygen evolution. Plant Physiol. **43**, 1419—1422 (1968).

— HILL, R.: The relation of oxygen evolution to carbon assimilation with isolated chloroplasts. Biochim. biophys. Acta (Amst.) **131**, 330—338 (1967).

WARBURG, O.: Über die Geschwindigkeit der photochemischen Kohlensäurezersetzung in lebenden Zellen. Biochem. Z. **100**, 258—272 (1919).

WARBURG, O.: Über die Geschwindigkeit der photochemischen Kohlensäurezersetzung in lebenden Zellen. Biochem. Z. **103**, 188—217 (1920).

WEISSBACH, A., HORECKER, B. L., HURWITZ, J.: The enzymatic formation of phosphoglyceric acid from ribulose diphosphate and carbon dioxide. J. biol. Chem. **218**, 795—810 (1956).

WILSON, A. T., CALVIN, M.: The photosynthetic cycle. CO_2 dependent transients. J. Amer. chem. Soc. **77**, 5948—5957 (1955).

WU, R., RACKER, E.: Regulatory mechanisms in carbohydrate metabolism III. Limiting factors in glycolysis of ascites tumor cells. J. biol. Chem. **234**, 1029—1035 (1959).

YAMASHITA, T., BUTLER, W. L.: Photoreduction and photophosphorylation with tris-washed chloroplasts. Plant Physiol. **43**, 1978—1986 (1968).

ZELITCH, I.: Organic acids and respiration in photosynthetic tissues. Ann. Rev. Plant Physiol. **15**, 121—142 (1964).

— The role of glycolic oxidase in the respiration of leaves. J. biol. Chem. **233**, 1299—1303 (1958).

ZIEGLER, H., ZIEGLER, I., SCHMIDT-CLAUSEN, H. J.: Die lichtinduzierte Aktivitätssteigerung der $NADP^+$-abhängigen Glycerinaldehyd-3-Phosphat-Dehydrogenase. VII. Die Abhängigkeit der Enzymaktivität von Bestrahlungsstärke und -dauer bei *Lemma*. Planta **81**, 181—192 (1968).

Biosynthesis by Chloroplasts

TREVOR W. GOODWIN

A. Proteins

For a cell or a sub-cellular particle such as a chloroplast to synthesize proteins it must able to produce or obtain from outside a) DNA; b) messenger-RNA, transfer-RNA and ribosomal RNA; c) a source of available energy, that is ATP; d) enzymes concerned with amino-acid activation; e) the particles, polysomes, on which the amino-acids are assembled into proteins, and f) the component amino acids themselves. These various requirements will be considered separately before the overall process of protein biosynthesis in chloroplasts is discussed.

1. The Presence of DNA and RNA

The evidence that *Acetabularia* chloroplasts contain DNA is that bacteria-free enucleate *Acetabularia* still possess DNA which is located in the chloroplasts (1×10^{-16} g/chloroplast) (GIBOR and GRANICK, 1964; GIBOR, 1965) and that enucleate organisms can replicate their chloroplasts (SHEPHARD, 1965). The evidence with *Euglena* is equally strong.

The DNA from *Euglena* shows three peaks in CsCl gradients (EDELMAN *et al.*, 1965; RAY and HANAWALT, 1965); one with a buoyant density of 1.686 g/cc is enriched in the extracts from green cells and absent from mutants unable to form proplastids or chloroplasts (SCHIFF and EPSTEIN, 1966). According to GIBOR and GRANICK (1964) one *Euglena* chloroplast contains 4×10^{-15} g DNA. This is apparently sufficient to code for ribosomes and many proteins if the chloroplast is haploid or diploid but not if it is polyploid like the mitochondrion (see LOENING, 1968).

Radioautography has also provided evidence for chloroplast DNA in algae. After incorporating tritiated guanine and adenine into *E. gracilis* v. *bacillaris*, SAGAN *et al.* (1965) found blackening all over the cell; this was removed almost completely by the combined action of RNAase and DNAase. RNAase alone removed all but black clumps of DNA over the nucleus and chloroplast, while DNAase treatment left a light blackening all over the cell. However, in the algae *Dictyota* and *Padina* (^{3}H) thymidine is incorporated into the chloroplast only during certain stages of chloroplast division and mature, non-dividing chloroplasts incorporate no label (STEFFENSEN and SHERIDAN, 1965). The chloroplast DNA of *Chlamydomonas* replicates semi-conservatively and at a different time in the cell-cycle from the nuclear DNA (CHIANG and SUEOKA, 1967). A DNA polymerase which synthesizes DNA from its constituent nucleotides is probably present in *Euglena* chloroplasts, but has not yet been unequivocally demonstrated (SCHIFF and EPSTEIN, 1966). However the DNA synthesized by isolated *Euglena* chloroplasts has the same buoyant density in CsCl as chloroplast DNA which is much less than that of nuclear DNA (SCOTT *et al.*, 1968).

Direct isolation of DNA from chloroplasts of higher plants has not been reported but certainly leafy material contains minor DNA fractions in CsCl gradients which could be of chloroplast origin. Other indirect evidence will be considered later that isolated chloroplasts from tobacco leaves synthesize DNA *in vivo* (Tewari and Wildman, 1967) as do spinach chloroplasts (Spencer and Whitfeld, 1967).

Experiments which further demonstrate the presence of DNA in chloroplasts are those of Wollgiehn and Mothes (1964) and Kirk (1964). The former showed a DNAase-sensitive incorporation of (^{3}H) thymidine into chloroplasts of immature leaves of tobacco while the latter found a DNA-dependent RNA polymerase in broad bean chloroplasts. It is important that no incorporation of (^{3}H) thymidine was found in chloroplasts from mature leaves.

The chloroplast RNA polymerase can be differentiated from the nuclear system in broad beans by the ratio of moles of adenine to guanine incorporated, 2.02 and 1.56, respectively. The chloroplast enzyme from tobacco is not complexed with histone as is the nuclear enzyme, is located in the supernatant fraction of the chloroplast (Wildman, 1967), and is actively synthesized when a plastid develops into a chloroplast [Bogorad, 1967 (1, 2)]. Although there is little chloroplast development in the bean leaf during the first few days of germination (Loening and Ingle, 1967) chloroplast ribosomal RNA is synthesized in the dark in bean leaves, but on illumination of the leaves it is stimulated to a somewhat greater extent than is cytoplasmic ribosomal RNA (Boardman, 1966; Gyldenholm, 1966; Loening and Ingle, 1967). About 50% of the RNA synthesized during illumination of an etiolated seedling is in the newly developed chloroplast and there is a marked synthesis of RNA during light-induced chloroplast formation in *Euglena* (Zeldin and Schiff, 1967, 1968). Gibbs (1967, 1968) has provided electron microscopic evidence that most of the chloroplast RNA is synthesized *in situ* on the chloroplast DNA template.

The experiments of Kirk (1964) also indicated that messenger RNA was synthesized in chloroplasts and similar results have been obtained with isolated chloroplasts from *Acetabularia* (Schweiger and Berger, 1964; Schweiger *et al.*, 1967) and from tobacco (Tewari and Wildman, 1967).

Messenger-RNA from *Euglena*, which will stimulate amino acid incorporation into *Escherichia coli* ribosomes, is present in the 100,000 × g supernatant and sediments as a 12S component (Eisenstadt and Brawerman, 1964).

Transfer-RNA is present in chloroplasts from pea seedlings (Sissakian *et al.*, 1965) and tobacco leaves (Spencer and Wildman, 1964). That from peas (3.25S) will accept amino acids in the presence of amino acid activating enzymes and ATP. Transfer-RNA for N-formylmethionine, the initiating amino acid in protein synthesis in bacteria, has been demonstrated in French bean chloroplasts (Burkard *et al.*, 1967).

Two types of ribosome, of approximately 70S and 80S, have been detected in leaf tissue but it is the 70S type which is concentrated in the chloroplast (Karlsson *et al.*, 1966). A similar situation is observed in *Chlorella* (Wildman, 1967); in *Euglena* there appears to be 70S and 87S ribosomes (Rawson and Stutz, 1968), although earlier work had suggested that the two components sedimented at 60S (chloroplast) and 70S (cytoplasm) (Brawerman and Eisenstadt, 1964).

The 70S particles in the plastids of etiolated bean leaves increase only slightly during greening and Boardman (1967) concludes that the synthesis of these particles is not a light-dependent process; however, the effect depends upon the age of the leaf; light stimulates synthesis of ribosomal-RNA in greening apical sections of barley leaves, but has little effect on synthesis of older etiolated leaves as they green up (SMITH *et al.*, 1970). Furthermore light affected the synthesis of a number of enzymes equally in both young and old leaves. As it is now well established that greening is inhibited by actinomycin D, it is possible that the greening process is dependent on light-activated synthesis of messenger-RNA and WILLIAMS and NOVELLI (1964) showed that ribosomes from etiolated maize shoots have less amino acid incorporating activity than ribosomes from similar plants grown in the light.

The 70S ribosomes are structurally similar to those from bacteria; average dimensions are 268 ± 14 Å $\times$ 214 ± 20 Å with a distinct cleft about 140 Å from one end. This gives some support to the view that chloroplasts have evolved by symbiotic invasion of a fungal cell by a blue-green alga or a photosynthetic bacterium.

Apart from size the chloroplast ribosomes differ from cytoplasmic ribosomes in a number of ways (WILDMAN, 1967; EISENSTADT, 1967; BOARDMAN *et al.*, 1966; BOARDMAN, 1967): (i) the specific activity for protein synthesis is much higher in chloroplast ribosomes; (ii) dialysis against *tris* buffer results in almost complete dissociation of the 70S particles but not of the 80S particles; (iii) in *Euglena* chloramphenicol binds much more slowly to the chloroplast ribosomes than to the cytoplasmic ribosomes; (iv) the amino acid incorporating activity of 70S ribosomes is maximal at 15 mM Mg^{2+} while that of the 80S ribosomes is maximal at 5 mM Mg^{2+}.

A full description of a *Chlamydomonas* mutant which is deficient in chloroplast ribosomes is awaited with interest (see e.g. JOHNSON *et al.*, 1968).

The RNA from higher plant ribosomes is mainly of two species 23S and 16S and the general view is that the 16S predominates in chloroplasts and the 23S is mainly in the cytoplasm (LOENING and INGLE, 1967; STUTZ and NOLL, 1967; POLLARD *et al.*, 1966), and similar results were obtained with *Acetabularia* (BALTUS and QUERTIER, 1966). However, SPENCER and WHITFELD (1966) state that chloroplast RNA from spinach, tobacco, radish, pea and tomato is exclusively of the 16S type. The situation in *Euglena* is somewhat different: 19S and 14S components were found in the chloroplast while only the 19S component was found in the cytoplasm (BRAWERMAN and EISENSTADT, 1964; ZELDIN and SCHIFF, 1967). Furthermore the chloroplast RNA has a high adenine and guanine content and the cytoplasmic RNA a high guanine and cytosine content (BRAWERMAN and EISENSTADT, 1964).

Chloroplast polysomes are more unstable on isolation than cytoplasmic polysomes but in the Pinto bean they are clearly polymers of the 78S ribosomes because these are formed on treating the polysomes with ribonuclease.

Apparently such polysomes hardly exist in plants kept in the dark for 14 h, but after 40 min. illumination they become the major polysome component (CLARK *et al.*, 1964; STUTZ and NOLL, 1967). Isolated tobacco polysomes are active in protein synthesis (CHEN and WILDMAN, 1967). Polysomes in chloroplasts have also been observed by CLARK *et al.* (1964), GUNNING (1965) and BAMJI and JAGENDORF (1966).

2. Energy

ATP is abundantly available via photophosphorylation (see AVRON, p. 149), and SPENCER (1965) has shown that ATP formed during photophosphorylation can replace exogenous ATP in the protein-synthesizing system in chloroplasts. According to Ramirez *et al.* (1968) it is cyclic photophosporylation which is the main source of ATP.

3. Amino Acid-Activating Enzymes

Amino acid-activating enzymes have been demonstrated in isolated chloroplasts from a number of sources (Bové and Raacke, 1958; Marcus, 1959; Henshall and Goodwin, 1964; Sissakian *et al.*, 1965). They are concentrated in the supernatant fraction of the chloroplast. Non-aqueously prepared chloroplasts contain some 15 to 40% of the total leaf activity (Heber, 1962).

4. Amino Acids

A number of amino acids become rapidly labeled from $^{14}CO_2$ in Calvin-type experiments with algae; it was therefore assumed that they are formed in the chloroplasts and that the necessary enzymes are located in the stroma. This view was confirmed by the incorporation of $^{14}CO_2$ into alanine and aspartate by isolated chloroplasts (Tolbert, 1959). These experiments also indicate that an amino acid pool exists in the chloroplasts. All the early Calvin experiments indicated that incorporation into alanine, aspartic acid, and serine is particularly rapid (Smith, Bassham and Kirk, 1961). In *Chlorella pyrenoidosa*, for example, the labelling of alanine is faster than that of sucrose. As the rate of alanine synthesis reaches a maximum after 3 to 5 min exposure to $^{14}CO_2$ when the intermediates of the carbon reduction cycle become saturated then, as compounds such as sucrose are not saturated until much later, it is concluded that alanine is formed directly from components of the cycle. Furthermore, variations in the level of activity in 3-phosphoglyceric acid (PGA) are parallelled by changes in the level of activity in alanine (Smith *et al.*, 1961). This suggests that the reaction takes place via phosphoenolpyruvate [Reaction (1)]. This reaction scheme assumes that amination or transamination occurs in the chloroplasts. Rosenberg, Capindale and

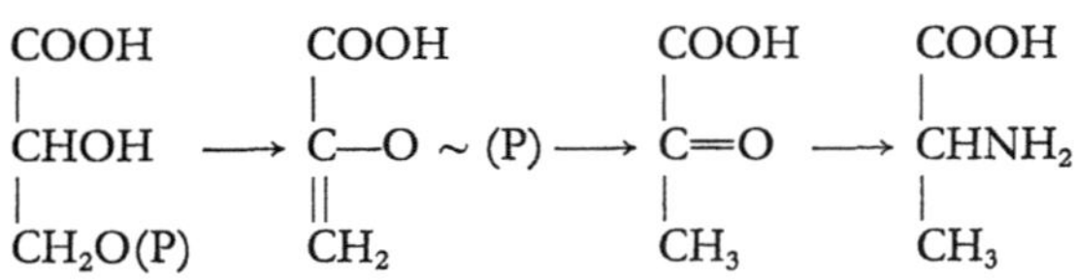

(1)

Whatley (1958) provided direct evidence for such reactions when they showed that aspartate can be formed from oxaloacetate in isolated chloroplasts. They also showed that oxaloacetate is formed by fixation of CO_2 into phosphoenolpyruvate by phosphoenolpyruvate carboxylase which is also present in chloroplasts. It is this reaction [Reaction (2)] which accounts for the rapid labelling of aspartate in Calvin's early experiments and not the fact, as with alanine, that the precursor is a component of the

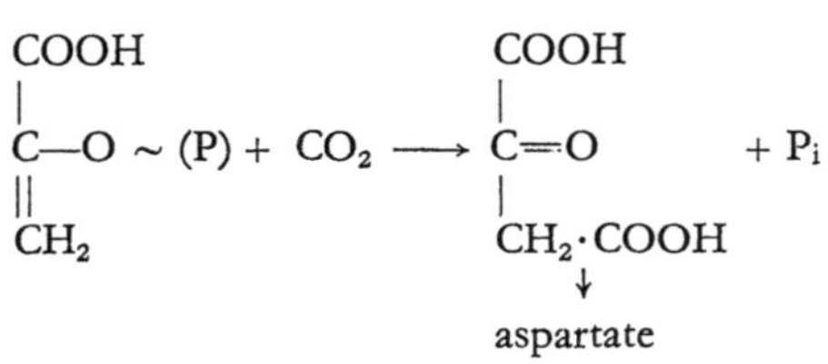

(2)

carbon reduction cycle. Oxaloacetate can also arise from malate in the chloroplast because of the presence there of malate dehydrogenase (PIERPOINT, 1963). This is unlikely to be a primary source of oxaloacetate.

Glycine is also rapidly labeled from $^{14}CO_2$ and the main pathway probably involves glyoxylic acid, the formation of which is favoured at low partial pressure of O_2 (PRITCHARD, WHITTINGHAM and GRIFFIN, 1961); glycolic acid, which can be synthesized by isolated chloroplasts (VANDOR and TOLBERT, 1968) could then undergo oxidation to glyoxylic acid which on transamination forms glycine [Reaction (3)]. However, some organisms (algae) are said to be devoid of glycolic oxidase (HESS *et al.*, 1965), and in the others the reaction may not occur in the chloroplasts but in the closely associated particles the peroxisomes (TOLBERT and YAMAZAKI, 1969). The source of glycolic acid is discussed in detail on p. 203.

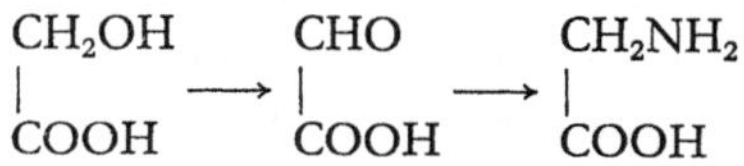

(3)

The parallel pattern of labelling observed with PGA and serine in experiments with relatively high CO_2 levels (1 to 2%) indicates that under these conditions serine is formed from PGA via hydroxypyruvate [Reaction (4)] (SMITH *et al.*, 1961), and 70

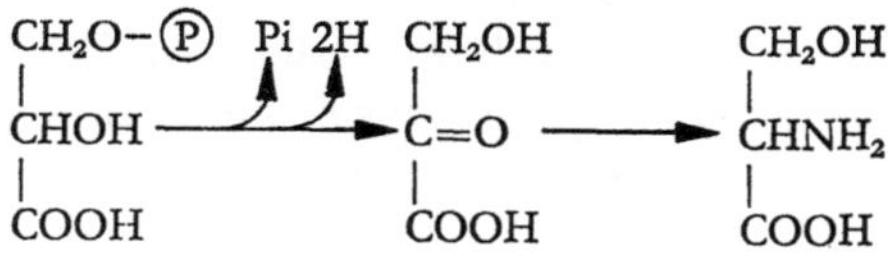

(4)

to 80% of the label from $^{14}CO_2$ is found in the carboxyl group of serine (CHANG and TOLBERT, 1965). Under low partial pressures of oxygen a pathway involving the hydroxymethylation of glycine with hydroxymethyltetrahydrofolic acid (CH_2OH-THFA) may predominate [Reaction (5)]. This reaction has been recently demonstrated in pea chloroplasts (SHAH and COSSINS, 1970).

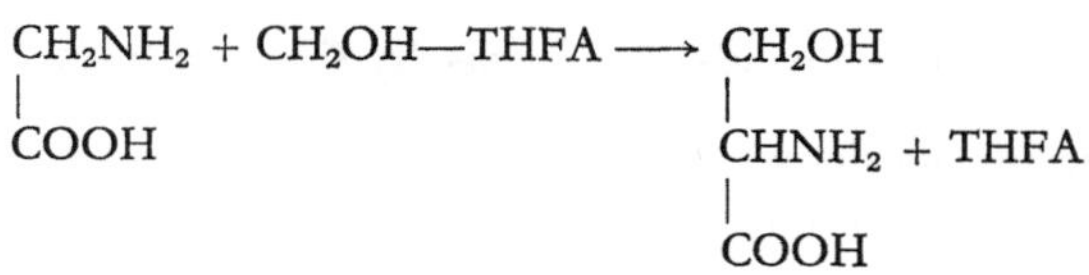

(5)

Glutamate is probably formed in chloroplasts by reductive amination of α-ketoglutarate [Reaction (6)], the NADPH required arising from the light reaction in photosynthesis. The obvious source of α-ketoglutarate is citrate formed by the condensing enzyme (oxaloacetate + acetyl-CoA → citrate + CoASH). This enzyme is

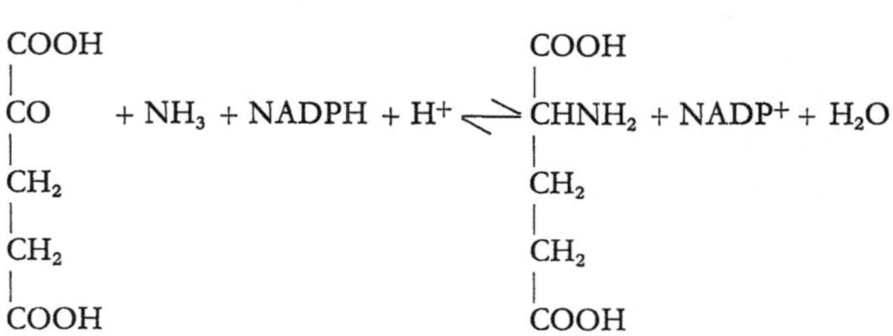

(6)

present in green leaves, and as only 54% of its activity is associated with the tricarboxylic acid cycle in the mitochondria, some of the remainder is possibly located in the chloroplast (Hiatt, 1962). The conversion of citrate into α-ketoglutarate would involve the enzymes aconitase and isocitrate dehydrogenase, in that order. Aconitase occurs in the leaves of a number of plants but according to Bacon *et al.* (1963) it is not associated with cytoplasmic organelles. Isocitrate dehydrogenase has also not yet been demonstrated in chloroplasts.

Other arguments against the citrate route to glutamate in chloroplasts are based on the kinetic observations of Smith *et al.* (1961) that glutamate was labelled much more quickly than citrate. However, Hiller (1964) has described entirely different experiments which he considers demonstrate that citrate could be a precursor of glutamate. He found that in the presence of fluoroacetate, which inhibits the conversion of citrate into oxaloacetate, $^{14}CO_2$ fixation into citrate by *Chlorella* approaches that found in glutamate in uninhibited cultures under otherwise identical conditions. This led to the view that the citrate pool in chloroplasts is so small and turns over so quickly that even in short term experiments a major part of the label passes through into glutamate. The general conclusion must be that as the tricarboxylic acid cycle is not present in chloroplasts most of the α-ketoglutarate used in the chloroplast is synthesized in another part of the cell.

The observations that the aromatic residues of plastoquinone and related compound arise from CO_2 in the chloroplast (Griffiths *et al.*, 1967) (see also p. 257) via shikimate [Threlfall *et al.*, 1966 (3)] and tyrosine (Whistance *et al.*, 1967) strongly suggest that tyrosine, at least of the aromatic amino-acids, is synthesized within the chloroplast.

No demonstration of synthesis of sulphur-containing amino acids in chloroplasts has yet been reported but "active sulphur", 3′phosphoadenosine-5′phosphosulphate, synthesis occurs in *Euglena* chloroplast fragments (Davies *et al.*, 1966) and the enzyme ATP-adenylsulphate-3′-phosphotransferase is present in chloroplasts of higher plants (Mercer and Thomas, 1969).

5. Overall Protein Synthesis by Chloroplasts

Many early experiments demonstrated that $^{14}CO_2$ was rapidly incorporated into chloroplast protein and that enzymes, which were confined to the chloroplast, increased in activity on illumination of tissue grown in the dark. Carboxydismutase, for example, increased 4 to 6 times on illuminating etiolated leaves of barley (Huffaker *et al.*, 1964) and bean (Margulies, 1964); similar observations have been made with ribulose 5-phosphate kinase, ribose 5-phosphate isomerase in maize and RNA polymerase [Bogorad, 1967 (1)]. Isolated pea chloroplasts also synthesize carboxydismutase (Margulies and Parenti, 1968) and isolated bean chloroplasts make the coupling factor (Ca^{2+}-dependent ATPase) (Ranaletti *et al.*, 1969). In *Euglena* similar light-induced increases have been recorded for NADPH-dependent glyceraldehyde 3-phosphate dehydrogenase (Hudock and Fuller, 1965) and fructose diphosphatase [App and Jagendorf, 1963 (1); Smillie, 1963]. An interesting exception is the NADPH-dependent glyceraldehyde 3-phosphate dehydrogenase in higher plants. This appears to be formed by the transformation of the NADH-dependent enzyme already present in etiolated tissues. The conversion is not inhibited by chloramphenicol and the enzyme reverts to the NADH-dependent type if the illuminated plant is returned to darkness (HUDOCK and FULLER, 1965). Ribosomes from *Euglena* chloroplasts will respond to a foreign message in the form of messenger RNA from bacteriophage f_2 and synthesize what appears to be phage protein. In this system the initial amino acid laid down was N-formylmethionine (SCHWARTZ *et al.*, 1967).

BRACHET (1967) and his colleagues have demonstrated net synthesis in enucleate *Acetabularia* of proteins which are usually considered to be under nuclear control. This is explained by the proposed existence of a stable messenger RNA, from the nucleus, which has a half life of 2 to 3 weeks. However, actinomycin D inhibits amino acid incorporation into isolated chloroplasts from 11 days enuclated organisms. This suggests that chloroplast DNA is directing the synthesis of chloroplast protein because actinomycin-D inhibits DNA-dependent RNA polymerase.

If pieces of leaf lamina are exposed to labelled amino acids for 10 min then only 5% of the label appears in the nuclear fraction, and the remainder is equally distributed between chloroplast and cytoplasm. In the chloroplast 3% of the activity was in the structural protein, 80% of the remainder in the 4-18S protein fraction and only 20% of the remainder in the 70S ribosomal fraction. When the exposure time was increased to 30 min the amount in the ribosomal fraction remained the same while that in the 4-18S fraction was 10 times greater (WILDMAN, 1967). BAMJI and JAGENDORF (1966) found that 50% of the protein formed in 20 min in wheat chloroplasts is liberated from the polysomes as soluble protein. BEZINGER *et al.* (1964) claim that lipoproteins are involved as the first steps in the synthesis of simple proteins.

At the molecular level chloroplast polysomes from tobacco are active *in vitro* in synthesizing protein (CHEN and WILDMAN, 1967) and the degraded polysomes, ribosomes, from higher plants are also effective [APP and JAGENDORF, 1963 (2); SISSAKIAN *et al.*, 1965; WILDMAN, 1967] as are ribosomes from *Euglena* (EISENSTADT and BRAWERMAN, 1963, 1964). The chloroplast ribosomes are much more active per mg ribosome than are cytoplasmic extracts in tobacco leaf preparations and it is the 70S ribosomes which are responsible for the major part of the activity (BOARDMAN, 1967).

Considerable evidence now exists that chloramphenicol inhibits protein synthesis by isolated chloroplasts (EISENSTADT and BRAWERMAN, 1963, 1964; SPENCER and WILDMAN, 1964), and by 70S ribosomes but not by 80S ribosomes (SPENCER and WILDMAN, 1964), while cycloheximide (actidione) does not inhibit protein synthesis in isolated chloroplasts (ELLIS, 1969) or 70S ribosomes (STUTZ and NOLL, 1967) but does inhibit synthesis by the 80S ribosomes (WETTSTEIN *et al.*, 1964; SIEGEL and SISLER, 1965). Recent work indicates that although this is basically true the situation is a little more complex. Cycloheximide certainly blocks chlorophyll synthesis in greening *Euglena* and this is considered to be a secondary effect resulting from the inhibition of synthesis of the 'structural protein' in the thylakoid membrane by cytoplasmic ribosomes (KIRK and ALLEN, 1965; KIRK, 1968). Experiments with *Chlamydomonas reinhardi* (HOOBER and SIEKEVITZ, 1968) tend to support this idea. In the presence of chloramphenicol synthesis of chlorophyll and thylakoids is only slightly reduced (10%) although the Photosystem I and II activities were considerably reduced; on the other hand full inhibition of chlorophyll and thylakoid formation was achieved with actidione (5 mg/ml). In the presence of lower concentrations of actidione (1 mg/ml) only partial inhibition of structural protein occurred but the thylakoids so formed had full Photosystem I and II activities. If cells inhibited with chloramphenicol are washed free of this inhibitor and resuspended in actidione, no further thylakoid membranes were synthesized but those already present gained full Photosystem II activity. Thus it seems that in the presence of chloramphenicol structural protein synthesis was not inhibited but that electron-transport protein synthesis on chloroplast ribosomes was inhibited. When chloramphenicol was replaced by cycloheximide further synthesis of structural protein was stopped but the synthesis of the electron transport proteins was allowed to continue.

A similar situation is revealed by work on higher plants. Chloramphenicol inhibits the synthesis of the following chloroplast proteins: carboxydismutase, glyceraldehyde 3-phosphate dehydrogenase (NADP-dependent), ferredoxin, ferredoxin-NADP reductase, cytochrome *f* and a *b*-type cytochrome. Cycloheximide also inhibits the last three, which are usually bound to thylakoid membranes (SMILLIE *et al.*, 1967; GIBBONS *et al.*, 1969). In developing pea buds the appearance of chloroplast ribosomal RNA before chlorophyll (POLLACK and DAVIES, 1970) support the view that the structural proteins involved in chlorophyll formation are formed on chloroplast ribosomes. In excised maize shoots nitrite reductase, a chloroplast enzyme, is inhibited by chloramphenicol while nitrate reductase, an extra-plastidic enzyme, is not inhibited (SCHRADER *et al.*, 1967).

It is reasonably certain from these substantial numbers of observations that chloroplasts have all the genetic machinery for synthesizing protein. It would seem, however, from some of Brachet's experiments, for example, that some of the messenger RNA used by the chloroplast is of nuclear origin. How large a part nuclear messenger RNA plays in chloroplast biochemistry remains to be determined.

B. Lipids

1. General

The major lipid components of green tissues are located in the chloroplast mostly in the lamellae (see p. 40); they represent a complex mixture, some of which are unique. For convenience of reference the main components are summarized in Table 1, from which it will be noted that the esterifying fatty acids are also often unique to chloroplasts.

2. Fatty Acids

a) General

SMIRNOV (1960, 1962) first demonstrated that acetate is incorporated into fatty acids by isolated sunflower chloroplasts and this was soon confirmed and extended by MUDD and MCMANUS (1962) for spinach chloroplasts and STUMPF and JAMES (1963) for lettuce chloroplasts. The co-factor requirements for the system are ATP, CoASH, Mg^{2+}, HCO_3^-, and NADPH. With soluble preparations from chloroplasts, substitution of acetyl-CoA for acetate removes the requirement for ATP and substitution of malonyl-CoA for acetate removes the requirement for both ATP and CO_2. With intact chloroplasts the coenzyme-A derivatives are ineffective, probably because of a permeability barrier imposed by the outer membrane of the chloroplast (BROOKS

Table 1. *The major lipids of chloroplast lamellae of spinach* (ALLEN *et al.*, *1966*)

Lipid[a]	Acyl components (% distribution)[b]								
	14:0	16:0	16:1	16:2	16:3	18:0	18:1	18:2	18:3
Monogalactosyl diglyceride					25			2	72
Digalactosyl diglyceride		3			5		2	2	87
Trigalactosyl diglyceride	1	9		1	15		1	1	70
Phosphatidyl glycerol	1	11	32[c]		2		2	4	47
Sulpholipid		39						6	52
Lecithin		12			4		9	16	58
Phosphatidyl inositol	4	34			3	2	7	15	27

[a] For full structures see pp. 131, 134

[b] The nomenclature: number before colon is number of carbons in the fatty acid; number after colon the number of unsaturated double bonds in the fatty acids

[c] This is the Δ^3-*trans*-isomer

and STUMPF, 1966). There is a marked photostimulation of fatty acid synthesis by isolated chloroplasts (SMIRNOV, 1962) which occurs even in the presence of ATP and when the synthesis is reduced by inhibitors of photophosphorylation (e.g. CMU) (STUMPF *et al.*, 1967). These observations would suggest that photoproduction of NADPH is essential for synthesis and this has been demonstrated by STUMPF *et al.* (1963) for systems in which non-cyclic photophosphorylation is taking place. With broken chloroplasts MUDD and MCMANUS (1964, 1965) showed that the photostimulation effect is concerned only with the production of sufficient NADPH. The situation in intact chloroplasts is more complex, because light still has a marked but transitory stimulatory effect when sufficient NADPH and ATP are available for the amount of lipid synthesis going on (STUMPF *et al.*, 1963). In spite of much careful work the full explanation of this effect is not yet forthcoming.

In fragmented chloroplasts, incorporation of (^{14}C) acetate into long chain acids requires both chloroplast fragments and soluble protein (MUDD and MCMANUS, 1962); if, however, acetyl-CoA is the substrate, a soluble system, which does not sediment at 100,000 *g* for 4 h, can be prepared provided that HCO_3^-, ATP, NADPH, NADH and reduced glutathione are present. This system appeared to be very similar to the *E. coli* system but markedly different from yeast and mammalian fatty acid synthesizing systems, and later work by STUMPF and his colleagues has emphasized this similarity. The fatty acid synthetases from yeast and liver have now been isolated as large multi-enzyme complexes which sediment at high speed (see LYNEN, 1967, for a review). However, the system from chloroplasts of higher plants (BROOKS and STUMPF, 1966), from avocado mesocarp (OVERATH and STUMPF, 1964) and from bacteria (LENNARZ *et al.*, 1962; ALBERTS *et al.*, 1963; PUGH *et al.*, 1964) can be separated into a number of fractions which can be combined to give full synthetic activity. One of the components is a heat-stable low molecular weight protein called acyl carrier protein (ACP). The ACP's from avocado and spinach have now been obtained in a homogeneous state and studied in detail (SIMONI *et al.*, 1967).

The molecular weight of the spinach ACP is 10,300 and that of the avocado ACP 11,400; this is similar to the molecular weight of the bacterial protein and the amino-acid composition of all these proteins is similar. The bacterial ACP contains phosphopantetheine attached as a phosphodiester linkage via the hydroxyl group of serine (I) (SAUER *et al.*, 1964; MAJERUS *et al.*, 1965). The ACP from avocado contains the same prosthetic group (SIMONI *et al.*, 1967) and one assumes that this is also the case with

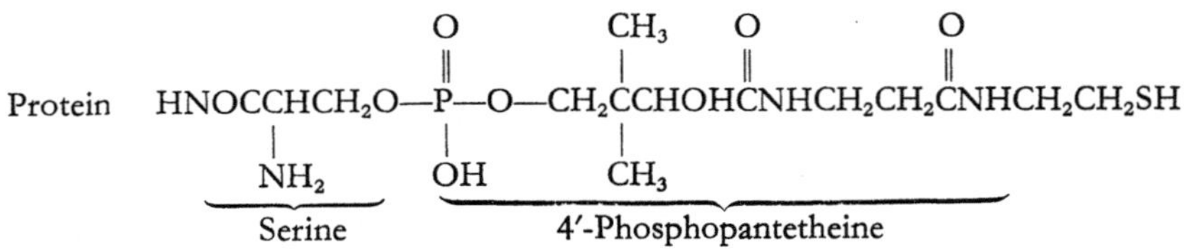

(I)

the spinach protein. All the ACP's have one active SH group, but the avocado ACP has a second SH group which appears to play no direct part in its synthetase activity. Bacterial ACP's are active in the plant synthetase systems, indeed more active than the plant ACP's themselves, so that it can reasonably be assumed that the pathway outlined here [Reactions (7) to (13)], which has been demonstrated in bacteria, also occurs in chloroplasts. The details, however, have still to be examined.

$$CH_3COSCoA + ACP \rightleftharpoons CH_3COS{-}ACP + CoASH \qquad (7)$$

Acetyl ACP transacylase

$$\underset{\displaystyle COOH}{CH_2}COSCoA + ACP \rightleftharpoons \underset{\displaystyle COOH}{CH_2}COS{-}ACP + CoASH \qquad (8)$$

Malonyl ACP transacylase

$$CH_3COS{-}ACP + HS\text{-}Enz. \rightleftharpoons CH_3COS\text{-}Enz + ACP \qquad (9)$$

$$CH_3COS\text{-}Enz + \underset{\displaystyle COOH}{CH_2}COS{-}ACP \rightleftharpoons CH_3COCH_2COS{-}ACP + HS\text{-}Enz + CO_2 \qquad (10)$$

Condensing enzyme

$$CH_3COCH_2COS{-}ACP + NADPH + H^+ \rightleftharpoons CH_3\underset{\displaystyle OH}{CH}CH_2COS{-}ACP + NADP^+ \qquad (11)$$

β-Ketoacylreductase

$$CH_3\underset{\displaystyle OH}{CH}\cdot CH_2COS{-}ACP \rightleftharpoons CH_3CH = CHCOS{-}ACP + H_2O \qquad (12)$$

αβ-Hydroxyacyldehydrase

$$CH_3CH{=}CHCOS{-}ACP + NADP + H^+ \rightleftharpoons CH_3CH_2CH_2COS{-}ACP + NADP^+ \qquad (13)$$

Enoyl ACP reductase

The butyryl-ACP derivative formed by this series of reactions then re-enters the scheme at Reaction (9) with the result that after passing through the subsequent reactions hexanoyl-ACP (C-6) is formed. Continuation of this spiral process leads to longer chain acyl-ACP's which are then liberated as free fatty acids. As indicated above the major products of the spinach system are palmitic acid (C-16; 20%) and stearic acid (C-18; 80%) (SIMONI *et al.*, 1967). The regulatory mechanism which leads to a mixture such as this is unknown, but it is interesting to find that if spinach ACP is replaced by bacterial ACP in this system, then the product consists entirely of stearic acid. On the other hand, the *Euglena* ACP produces fatty acids from C-8 to C-18 (see next section).

Although, as already pointed out, direct demonstration of all the steps from 7 to 13 has not yet been made, it is very likely that the synthesis proceeds in this way. The direct intervention of malonyl-CoA and ACP has, however, been demonstrated by BROOKS and STUMPF (1966) who showed that in the presence of ACP and malonyl-CoA, CO_2 was rapidly fixed into malonyl-CoA. This is due to an exchange reaction (GOLDMAN *et al.*, 1963) which involves the rapid reversal of Reactions (10) and (8). No significant incorporation is observed in the absence of malonyl-CoA and only about 9% of the fixation obtained in the complete system is noted in the absence of ACP.

It is appropriate here to consider briefly the source of the starting material, acetyl-CoA, in chloroplasts. This is still undecided; BASSHAM (1966) suggests that the action of pyruvic acid oxidase is unlikely because of the reducing environment in the chloroplast but the possibility that conversion takes place outside the chloroplasts by migration out of pyruvate and in of acetyl-CoA appears to be ruled out because the chloroplast membrane is impermeable to acetyl-CoA (STUMPF *et al.*, 1967). Another possibility is that acetyl phosphate arises by phosphorolysis of the thiamine pyrophosphate (TPP)-glycolaldehyde complex formed during the action of phosphoketolase in the Calvin cycle (Reaction) [p. 200 (14)]:

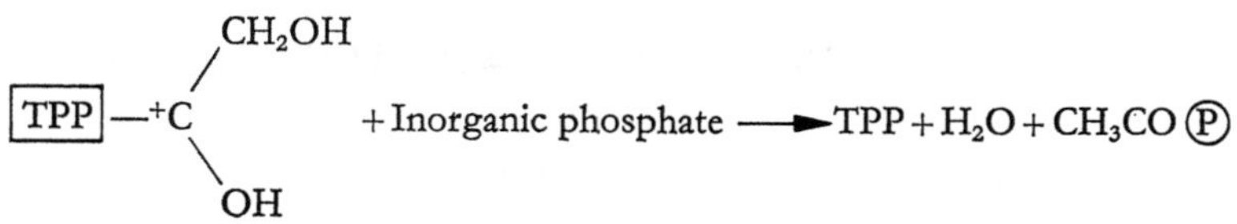

(14)

The acetyl phosphate formed could then undergo transacylation with coenzyme A. However, the existence of acetyl phosphate in plants has not yet been demonstrated. A final possibility is that it arises from hydroxypyruvate or pyruvate produced by the deamination of serine which is rapidly synthesized via the glycolate pathway (p. 219).

So far the general mechanism concerned with the synthesis of fatty acids has been considered. The problem of the formation of the various types of fatty acid, in particular unsaturated fatty acids which are abundant in chloroplasts, must now be considered.

b) Specific Fatty Acids

(i) *Monoenes.* The biosynthetic pathway for synthesis of oleic acid (18:1; *cis*, Δ^9) in *Euglena* chloroplasts is quite distinct from that outside the chloroplast. Extracts from green *Euglena* contain a soluble enzyme which will convert stearyl-ACP into oleate, probably as oleyl-ACP [Reaction (15)]. The enzyme requires O_2 and NADPH and is specific for stearyl-ACP, neither palmityl-ACP nor stearyl-CoA can act as substrate; it is absent from etiolated cells and is therefore almost certainly localized in the chloroplast (Bloch *et al.*, 1967). Outside the chloroplast the desaturation of stearate takes place with the CoA ester and not the ACP derivative as the substrate; the enzyme, which is particulate, will also desaturate palmityl-CoA to palmitoleyl-CoA (Nagai and Bloch, 1965). When *Euglena* is grown in the light on a heterotrophic medium, both routes are functional.

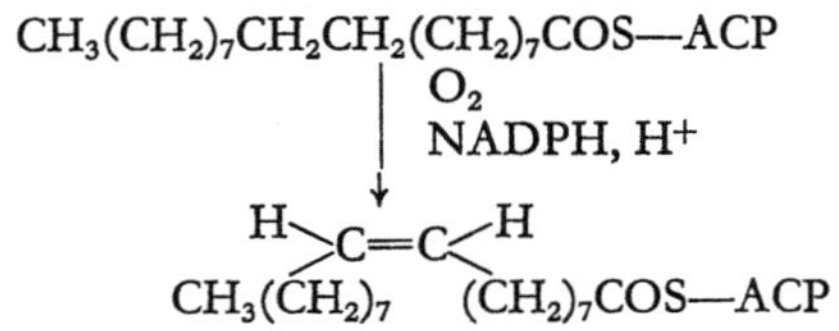

(15)

Other mono-unsaturated fatty acids are present in *Euglena*, particularly in photoauxotrophic cells where Δ^5- and Δ^7- tetradecenoic acids (14:1), Δ^7- and Δ^9-hexadecenoic acids (16:1), and Δ^{11}-octadecenoic acids (18:1) are found. As extracts of *Euglena* will elongate ACP derivatives of the homologous series C-8 to C-16 to produce both unsaturated and saturated acids, Nagai and Bloch (1965) consider that the unsaturated fatty acids of intermediate chain length are synthesized by oxidative desaturation of the C-10 or C-12 ACP derivatives followed by the series of reactions indicated in Fig. 1. The overall pattern of synthesis in *Euglena* can be

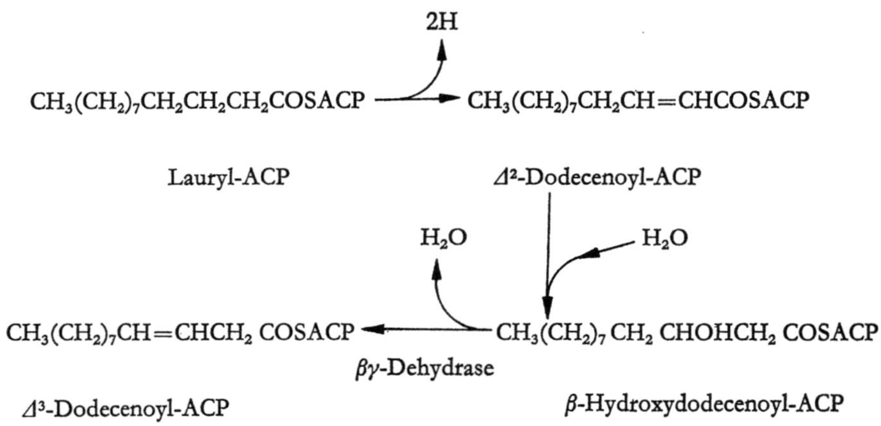

Fig. 1. Mechanism of unsaturated acid formation in *Euglena* chloroplasts by β,γ-dehydration

summarized in Fig. 2, which accounts for all the monoene acids found in *Euglena* except Δ^3-*trans*-hexadecenoic acid which is discussed later. The mechanism proposed is similar to that discovered in bacteria (Scheuerbrandt and Bloch, 1962) except that the initiating desaturation step in *Euglena* is O_2-dependent while that in bacteria is not.

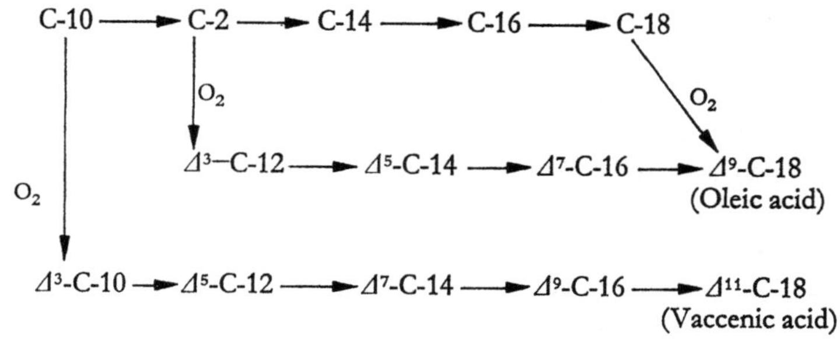

Fig. 2. The biosynthesis of mono-enes in *Euglena* chloroplasts. All the reactions take place at the ACP level

In *Chlorella* (1-^{14}C) stearate is converted into oleate and (1-^{14}C) palmitate into palmitoleic by intact cells suspended in phosphate buffer; in a rich heterotrophic medium no desaturation can be demonstrated (James *et al.*, 1965; Harris *et al.*, 1967).

In higher plants the investigations of Stumpf's group show that while intact chloroplasts synthesize mainly oleic (18:1) and palmitic acid from (^{14}C) acetate the soluble system from chloroplasts synthesize mainly stearic acid (18:0) and palmitic acid (Brooks and Stumpf, 1966). This means that the soluble system lacks an enzyme concerned with desaturation, possibly the key β,γ-dehydrase (Brock *et al.*, 1966). The results of Stumpf *et al.* (1967) suggest that in intact leaves the bulk of stearic acid is not being converted into oleic acid, but Harris *et al.* (1967) have obtained presumptive evidence that some stearate at least, can be converted into oleate in leaf tissue. If leaf discs are incubated anaerobically with (^{14}C)acetate for 6 h, considerable activity accumulates in stearate, something which does not happen under aerobic conditions. Transfer of the leaf discs from anaerobic to aerobic conditions results in the percentage of of counts in the stearic acid dropping and that in the oleic acid rising.

Trans-Δ^3-hexadecenoic acid is found in chloroplast lipids specifically attached to position 2 of phosphatidyl glycerol (Haverkate and van Deenen, 1965; van Deenen and Haverkate, 1966) and is present in no other lipid component. Although this acid was considered to be unique to chloroplasts and to phosphatidyl glycerol, it does occur as a major component of the seed oil of *Helenium bigelowii* (Hopkins and Chisholm, 1964) where it presumably occurs in a triglyceride. James *et al.* (1965) have demonstrated in *Chlorella vulgaris* that it is derived from palmitic acid by a dehydrogenation reaction which requires light and probably oxygen. They consider that the conversion may actually occur *in situ* in the phosphatidyl glycerol molecule.

(ii) *Dienoic, Trienoic and Tetraenoic acids.* As indicated earlier the presence of di-, tri- and tetra-enoic acids is very characteristic of chloroplast lipids and the presence of linoleic and α-linolenic acids has been documented for some time. However, $\Delta^{7,10,13}$ and $\Delta^{4,7,10,13}$-C-16 polyunsatured acids are now known to exist, especially in algal chloroplasts, as components of the monogalactosyl glycerides (Bloch *et al.*, 1967). With regard to these acids Bloch *et al.* (1967) found that if dark grown *Euglena* are illuminated and the changes in the C-16 fatty acid components of monogalactosyl glycerides followed with time then there is a fall in the amount of 16:0 material and an increase in 16:4 acids, which indicates a precursor-product relationship.

With regard to the C-18 acids, HARRIS *et al.* (1967) showed that chopped lettuce and castor leaf and intact cells of *Chlorella vulgaris* will desaturate oleic acid to linoleic acid and α-linolenic acid in the mandatory presence of oxygen. The mechanism of the reaction remains to be settled and a cell-free system necessary for further study has not yet been obtained from any tissue.

JAMES's group (HARRIS *et al.*, 1967; NICHOLS *et al.*, 1967) propose that some lipid components may themselves be involved in desaturation. For example, phosphatidyl choline may play an important role in α-linolenic acid formation from oleic acid. When the specific activities of oleic acid and linoleic acid from various lipids in *Chlorella* exposed to $^{14}CO_2$ for 2 to 4 h were measured, they were highest in both cases in phosphatidyl choline; furthermore the ratio of the specific activity of oleic acid to that of linoleic acid changed from 3:1 after 2 h to 1:3 after 4 h. In no other lipid was the specific activity of linoleic acid ever greater than that of oleic acid. In the same way phosphatidyl glycerol is also probably concerned with the formation of *trans* Δ^3-hexadecenoic acid (see previous section). Other experiments have indicated that monogalactosyl diglycerides may be similarly involved in the synthesis of C-14, C-16 and C-18 saturated acids and C-16 and C-18 unsaturated acids (NICHOLS *et al.*, 1967) (see also BLOCH *et al.*, 1967).

3. Complete Lipids

a) Phospholipids

(^{14}C)Acetate (STUMPF and JAMES, 1962) and ^{32}Pi (SISSAKIAN, 1958) are incorporated into phospholipids in isolated chloroplasts. No direct information is however available on the mechanism of formation of phospholipids in chloroplasts, but they probably arise in the same way as in other tissues (summarized in Fig. 3, see also p. 130). The results of FERRARI and BENSON (1961) on the uptake of $^{14}CO_2$ into the phospholipids of *Chlorella pyrenoidosa* during steady state photosynthesis support the view that phosphatidyl glycerol provides an acylated phosphatidyl residue for the synthesis of phosphatidyl-choline, -ethanolamine, and -inositol. The observations of Nichols *et al.* (1967) tend to confirm this view except in the case of phosphatidyl inositol.

b) Galactosyl Glycerides

As UDP-galactose can transfer galactose to an endogenous precursor in isolated spinach chloroplasts to form a galactolipid, which on hydrolysis yields 1-[β-D-galactosyl]-glycerol (II) (NEUFELD and HALL, 1964), the final step in the biosynthetic pathway can be visualized as in Reaction (16). The enzyme involved has now been isolated from chloroplasts of spinach (CHANG and KULKARNI, 1970).

The time curve for the incorporation of $^{14}CO_2$ into galactosyl diglycerides by *Chlorella pyrenoidosa* during steady state photosynthesis suggest that the digalactosyl diglyceride (III) is formed from the monogalactosyl diglyceride (FERRARI and BENSON, 1961). UDP Gal might reasonably be considered the galactose donor, but a problem is that the first galactoside linkage is β- and the second is α-. Furthermore, the difference in the fatty acid composition of the two components have to be ex-

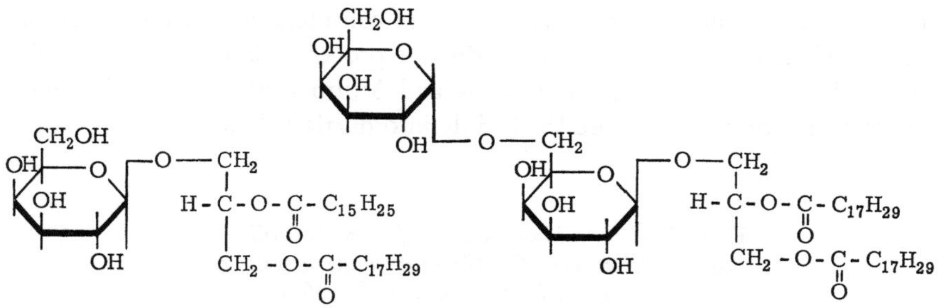

(II) (III)

UDP-Galactose + $\alpha\beta$-diglyceride $\rightleftharpoons$ UDP + monogalactosyl diglyceride (16)

plained. The difficulties inherent in this problem are well illustrated by the recent work of BLOCH and his colleagues (1967) with algae. Table 2 shows the fatty acid distribution in monogalactosyl diglycerides (MGG) and digalactosyl diglycerides (DGG) in *Euglena*. The same type of distribution is observed in *Chlamydomonas* and spinach. These results would indicate that the formation of DGG from MGG by direct

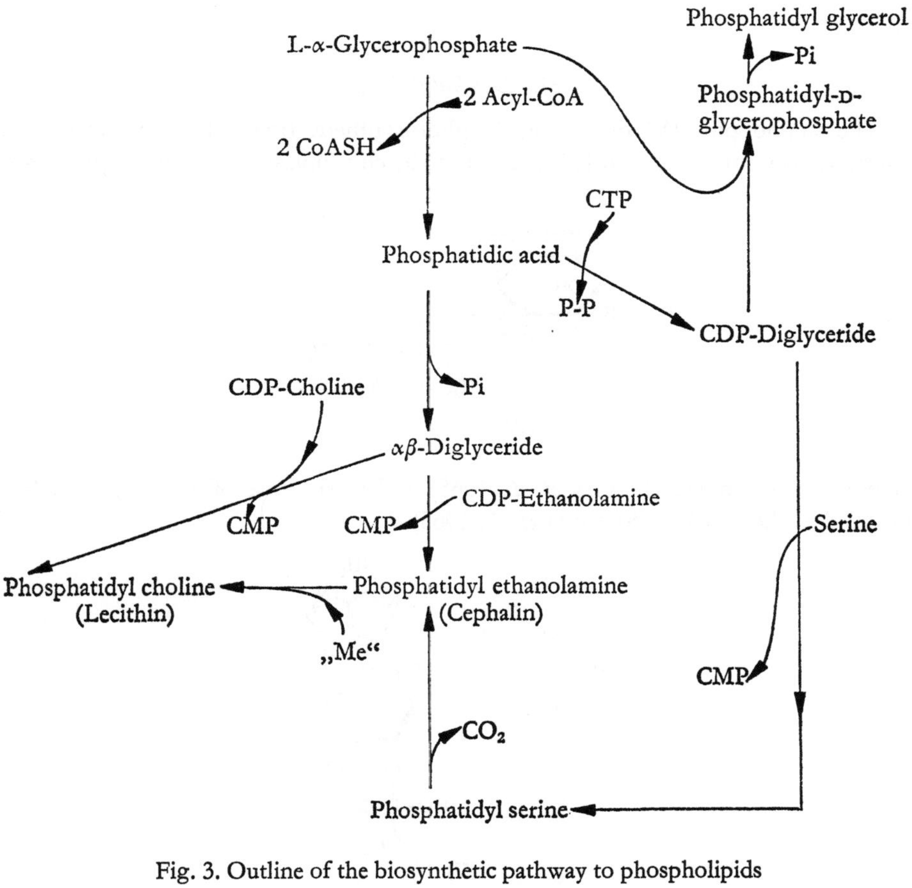

Fig. 3. Outline of the biosynthetic pathway to phospholipids

galactosylation does not occur, because if it did considerable hydrogenation of the fatty acid residue must occur. No such reductive process is known in plants. This is supported by the recent work of ONGUN and MUDD (1967) which showed that chloroplasts incorporated glucose ^{14}C mainly into digalactolipids.

Table 2. *Fatty acid composition of mono (MGG) and digalactosylglycerides (DGG) in* Euglena gracilis *cultured in the light* (BLOCH *et al.*, 1967)

	MGG	DGG
Palmitate (16:0)	6	17
Stearate (18:0)	1	0
Oleate (18:1)	9	19
Linoleate (18:2)	6	12
α-Linolenate (18:3)	41	26
$\Delta^{4,7,10,13}$-Hexadecenate (16:4)	32	7
Total saturated	7	17
Total unsaturated	73	33

c) Plant Sulpholipid

The sulpholipid (IV) discovered in photosynthetic tissues by BENSON and his colleagues (DANIEL *et al.*, 1961) is present in the chloroplasts of all higher plants and

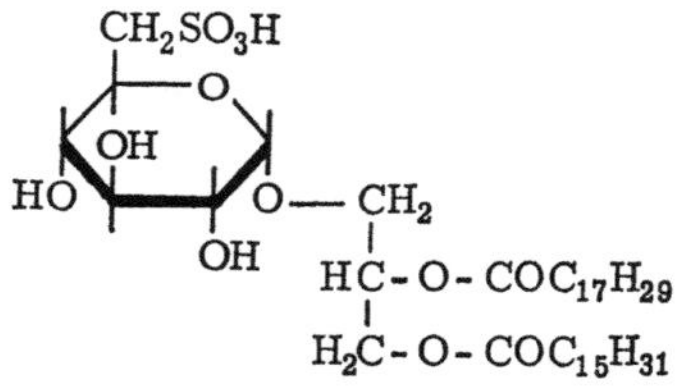

(IV)

algae so far examined (DAVIES *et al.*, 1965) and in the case of *Lemna perpusilla* is localized in the lamellae (SHIBUYA *et al.*, 1965).

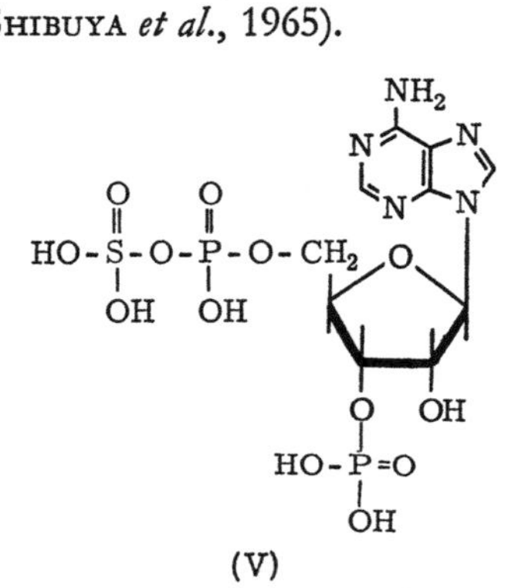

(V)

It raises a number of important biosynthetic problems not least the origin of the sulphonic acid residue in 6-sulphoquinovose. No C_5 or C_6 sugar is incorporated into this residue so the existence of sulpholactaldehyde, sulpholactic acid and sulphopropandiol in *Chlorella pyrenoidosa* (BENSON and SHIBUYA, 1961) suggests that the sulphur enters the molecule at the C_3 stage, possibly via phosphoenolpyruvate and "active sulphur", 3′-phosphoadenosine, 5′-sulphatophosphate (PAPS, V). As PAPS synthesis has been demonstrated in *Euglena* chloroplast fragments and as sulpholipid synthesis is inhibited by MoO_4^{2-} (DAVIES *et al.*, 1966) [MoO_4^{2-} inhibits formation of adenosine 5′-sulphatophosphate (APS), the precursor of PAPS (WILSON and BANDURSKI, 1958)], it is possible that PAPS takes part in sulpholipid biosynthesis by transferring sulphur to phosphoenolpyruvate. The resulting 2-phospho-3-sulpholactate could eventually be converted into 3-sulpholactaldehyde [indicated in the scheme, outlined in Fig. 4], which in turn could undergo an aldol condensation with

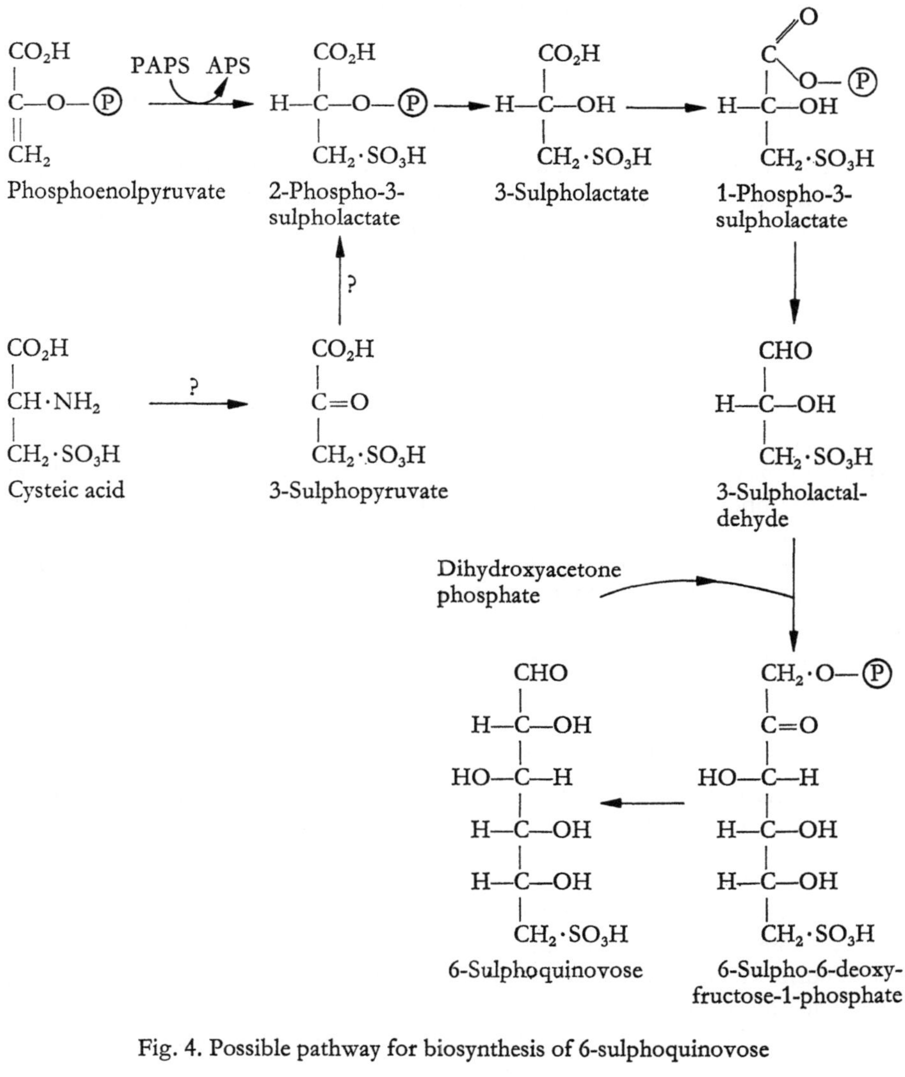

Fig. 4. Possible pathway for biosynthesis of 6-sulphoquinovose

dihydroxyacetone phosphate to yield 6-sulpho-6-deoxyfructose 1-phosphate; this could dephosphorylate and isomerize to 6-sulphoquinovose. This scheme would accommodate the conclusion from experiments with (^{35}S)cysteic acid and (3-^{14}C)cysteic acid, that this compound can provide an intact C-S unit for sulpholipid biosynthesis in *E. gracilis* (Davies *et al.*, 1966), and also for the fact that (^{14}C)cysteine is not incorporated into sulpholipid in *Chlorella* (Nissen and Benson, 1964). Presumably neither organism can oxidize the SH of cysteine to the -SO_3H level of cysteic acid.

If ^{35}S-labelled *Chlorella ellipsoidea* is transferred to a sulphur-deficient medium and allowed to photosynthesize, the label rapidly disappears from the sulpholipid. On the other hand, no loss of ^{35}S from this component occurs if the cells are transferred to a complete medium (Miyachi and Miyachi, 1966). This has led to the view that the sulpholipid is a reservoir of both sulphur and carbon in this alga.

C. Chloroplast Pigments

1. General

The ubiquitous pigments of the chloroplasts are the chlorophylls *a* and *b* (VI, a, b) and the carotenoids, mainly β-carotene (VII), lutein (VIII), violaxanthin (IX) and neoxanthin (X). Only slight variations in this carotenoid pattern are noticed when in some leaves, α-carotene (XI), cryptoxanthin (XII), zeaxanthin (XIII) and antheraxanthin (XIV) are noted. In algae the chloroplast pigments show much greater structural variations, these need not be considered here in detail (see Goodwin, 1965). The phycobilins are found only in algae.

2. Chlorophylls

The chlorophylls are metalloporphyrins and thus the early stages of their biosynthetic pathway would be expected to be similar if not identical with those leading to haem. It is now clear that the chlorophylls and heme do share a common pathway as far as protoporphyrin IX (XV). Discussion of the steps leading to protoporphyrin can be conveniently divided into three stages: (a) the steps leading to δ-aminolevulinic acid (ALA) (XVI); (b) the formation of the precursor pyrrole, porphobilinogen (PBG, XVII) from ALA; (c) the formation of protoporphyrin from PBG. All the steps have been demonstrated in animal tissues and photosynthetic bacteria and most but not all, in higher plants. It still remains to be shown whether every step can occur in chloroplasts.

a) δ-Aminolevulinic Acid (ALA) Synthesis

Succinyl-CoA synthetase (Reaction 17) has been purified from spinach leaves (Hager, 1962), tobacco (Bush, 1969) and wheat (Nandi and Waygood, 1965),

$$\text{Succinate} + \text{ATP} + \text{CoASH} \rightleftharpoons \text{Succinyl-CoA} + \text{ADP} + \text{Pi} \qquad (17)$$

but its intracellular location remains to be demonstrated. The pathway of ALA synthesis from glycine and succinyl-CoA (Reaction 18) takes place in animal mitochondria which supply succinyl-CoA from the Krebs cycle. Glycine is readily available in chloroplasts (p. 219) but there is yet no evidence that chloroplasts can make succinyl-CoA (Granick, 1967). ALA synthetase activity has not yet been

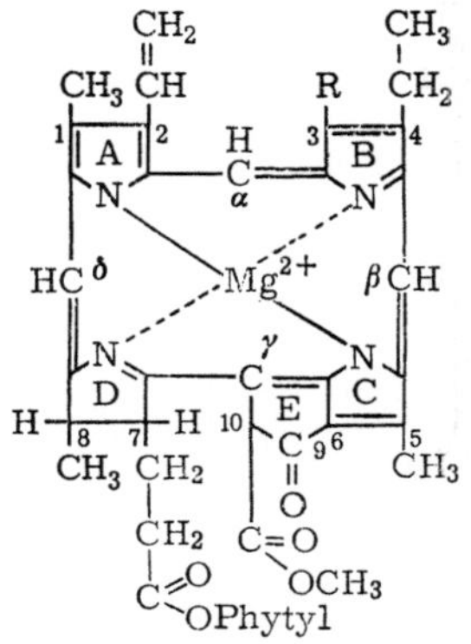

(VI)a Chlorophyll *a* R=CH_3
(VI)b Chlorophyll *b* R=CHO

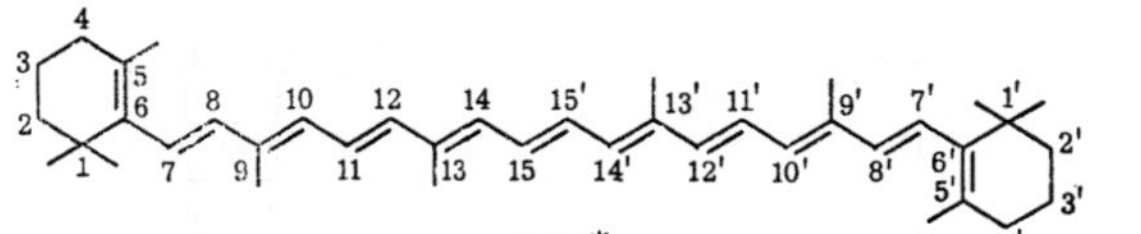

(VII)*

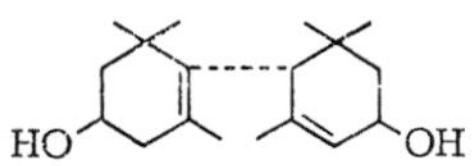

(VIII)

(IX)

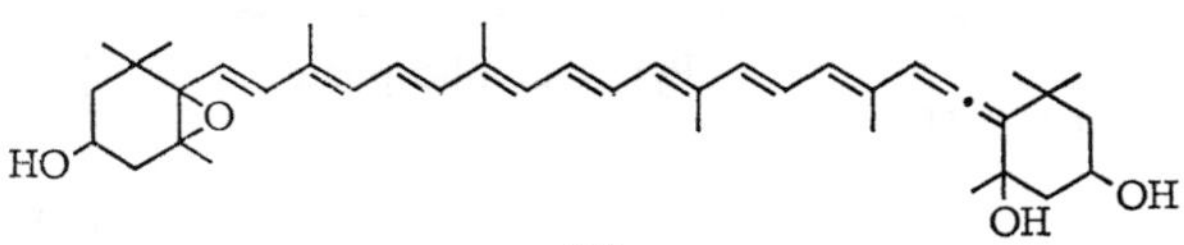

(X)

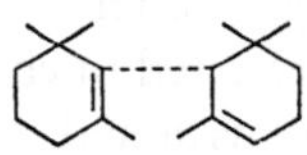

(XI)

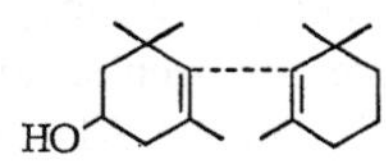

(XII)

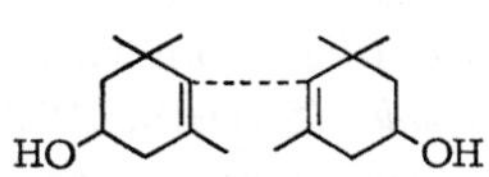

(XIII)

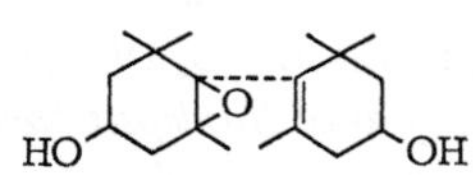

(XIV)

clearly demonstrated in higher plants but it has been obtained as a soluble enzyme from the chromatophores of *Rhodopseudomonas spheroides* (Kikuchi *et al.*, 1958). So the production of ALA in higher plants still remains to be clarified.

* Unless otherwise indicated the polyene chain (6′ – 6) in all the carotenoids discussed is as in β-carotene (VII).

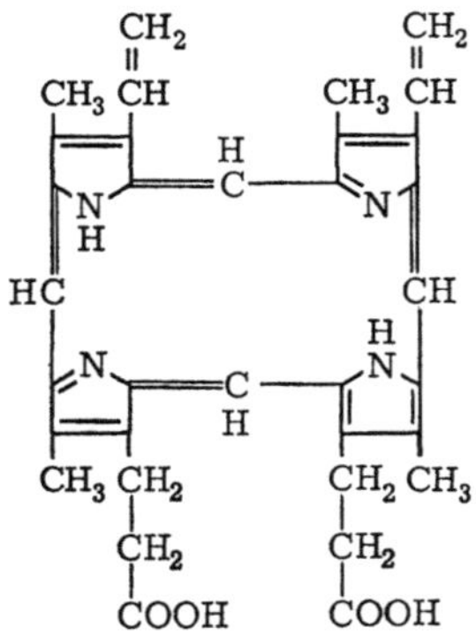

(XV)

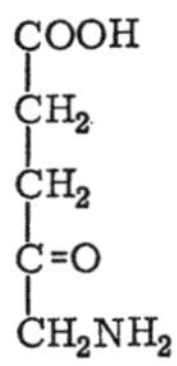

(XVI)

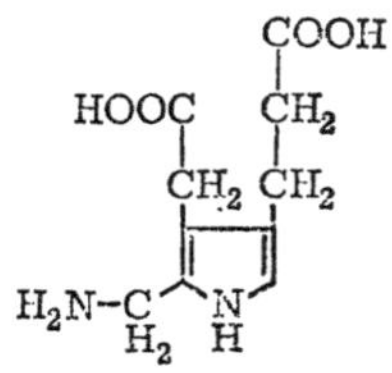

(XVII)

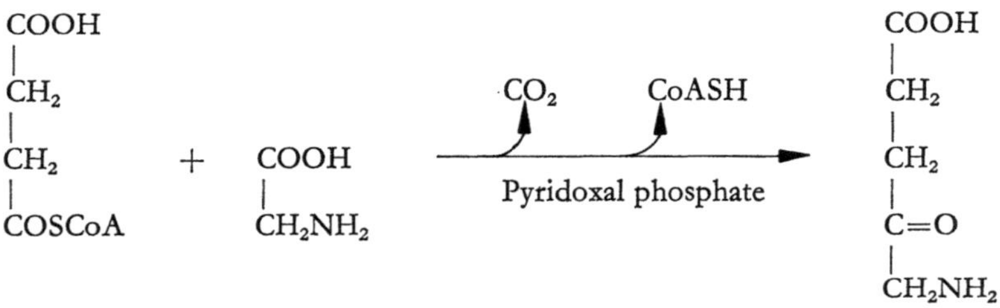

(18)

b) Porphobilinogen (PBG) Synthesis

It should be emphasized that although all the steps known to exist between ALA and protoporphyrin have not been separately demonstrated in chloroplasts, isolated *Euglena* chloroplasts will convert ALA into uroporphyrin and coproporphyrin but not protoporphyrin. However, after lysis protoporphyrin is formed from ALA (Carell and Kahn, 1964).

The enzyme ALA-dehydrase (PBG synthetase) catalyzing Reaction (19) has been detected in plants (see Bogorad, 1965) and partly purified from wheat leaves (Nandi and Waygood, 1967) and *Rhodopseudomonas spheroides* (Shemin, 1962). About 22% of the activity in *E. gracilis* is in the chloroplasts, firmly bound to the grana (Carell

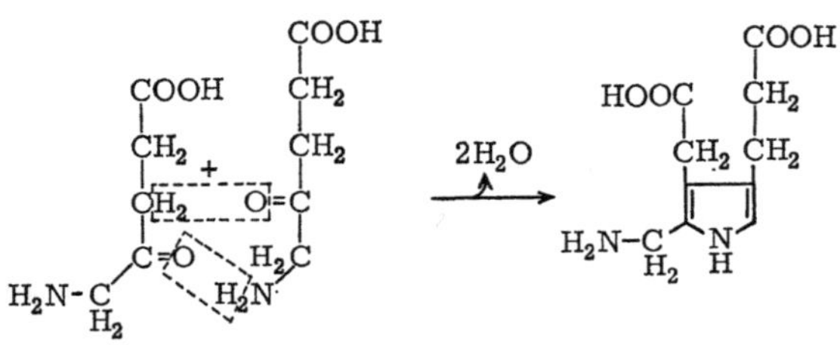

(19)

and KAHN, 1964) and it appears to be present in chloroplasts of wheat (NANDI and WAYGOOD, 1967) and callus tissue of *Kalanchoë cranata* (STOBART and THOMAS, 1968). Further, the level of the enzyme increases about three times when dark-grown *E. gracilis* are illuminated (EBBON and TAIT, 1969). Kinetic studies (GRANICK and MAUZERALL, 1958) suggest the mechanism indicated in Fig. 5. It is assumed that both molecules

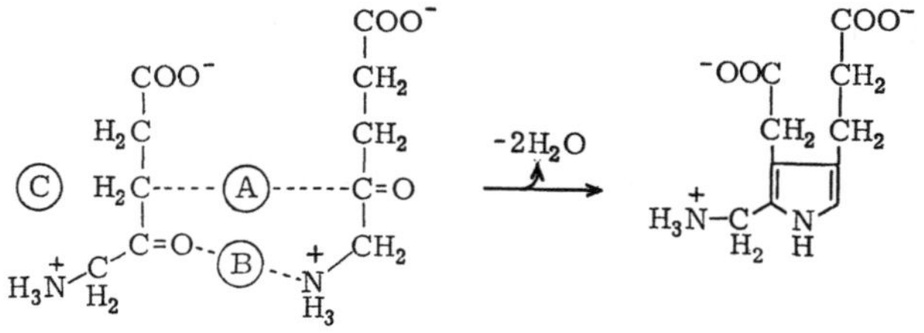

Fig. 5. Mechanism for the formation of PBG from ALA

of ALA are bound to the enzyme site, the first at least ten times more strongly than the second, the binding of which involves the formation of a ketimine (B in Fig. 5). The formation of an enolate ion at C (which could be favoured by a metal, a requirement for which was not, however, found) would allow an aldol condensation to take place by enolate ion attack on the carbonyl of the second molecule of ALA. Condensation at A with resulting hydrogen migration would result in PBG formation (GRANICK, 1967).

c) Formation of Porphyrinogens

It is now well established that the formation of the macrocyclic porphyrin molecule takes place at the oxidation level of a porphyrinogen and that oxidation to a porphyrin takes place as the last step in the process. The expected product of the linear condensation of four PBG molecules followed by cyclization would be uroporphyrinogen I [Urogen I; (XVIII)], the derivatives of which are only rarely encountered in nature. It is the derivatives of uroporphyrinogen III [Urogen III (XIX)],

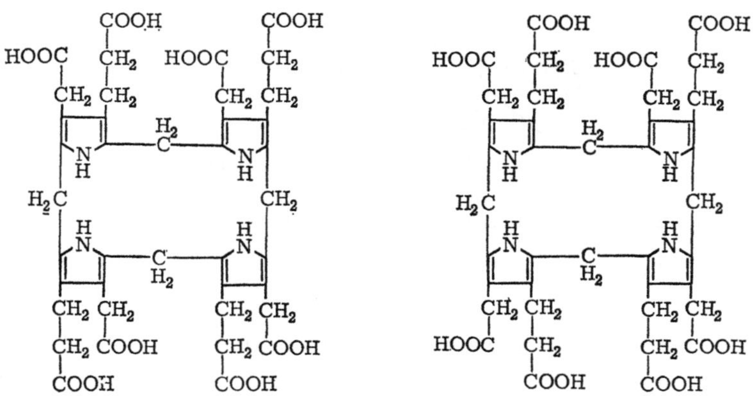

(XVIII) (XIX)

such as chlorophyll and haem, which predominate. In this porphyrinogen the PBG residue utilized as ring D appears to have been flipped over before cyclization occurs. It is in the area of urogen synthesis that studies with higher plants have played a significant part in clarifying the position.

BOGORAD (1960) showed that two enzymes working simultaneously are required for the synthesis of Urogen III in spinach and *Chlorella* preparations. Acetone-dried powders from spinach will, under anaerobic conditions, convert PBG into Urogen III and a trace of Urogen I, but after heating to 60° the preparations convert PBG quantitatively into Urogen I. The heat-stable enzyme has been called PBG deaminase, or Urogen I synthetase, and is specific for PBG, exhibiting no activity on isoporphobilinogen (XX) or opsopyrrole dicarboxylic acid (XXI).

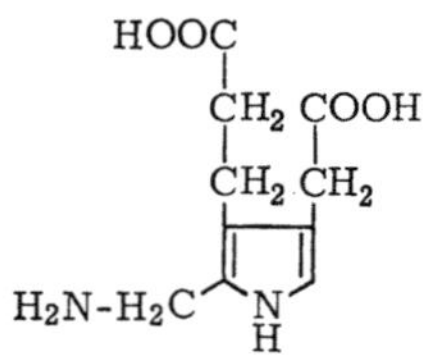

(XX)

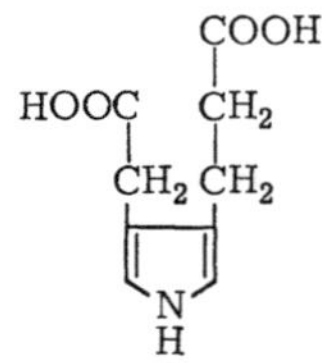

(XXI)

The second enzyme, which will not act on PBG or Urogen I, but will, in the presence of Urogen I synthetase convert PBG into Urogen III, has been isolated from wheat germ and named Urogen III co-synthetase (BOGORAD, 1962). The enzyme has recently been highly purified (STEVENS and FRYDMAN, 1968). The mechanism of the combined activity of the two enzymes has not been finally elucidated but a

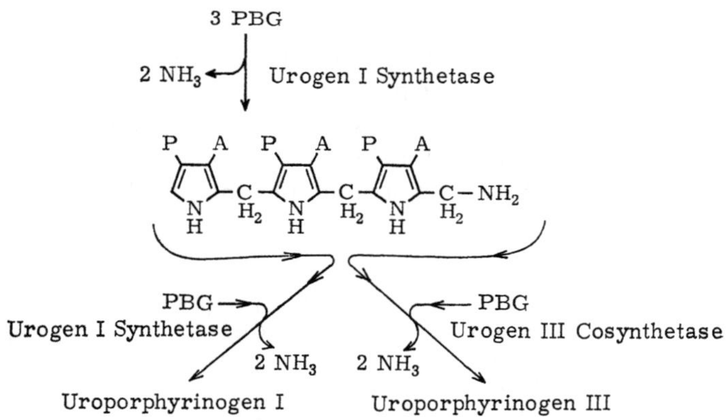

Fig. 6. Possible relationship between Urogen I synthetase and Urogen III Co-synthetase (BOGORAD, 1965)

possible scheme (Fig. 6) has been proposed by BOGORAD (1965). Isolated chloroplasts from *Euglena* will not convert PBG into porphyrins until osmotically ruptured (CARELL and KAHN, 1964).

d) Conversion of Urogen III into Protoporphyrin

The first step involves an enzyme uroporphyrinogen decarboxylase which specifically decarboxylates the acetic acid residues of Urogen III to Coprogen III [Reaction (20)] (MAUZERALL and GRANICK, 1958). The enzyme has been isolated from human red cells and has been demonstrated in tissues of higher plants and broken *Chlorella* cells (BOGORAD, 1965). No intermediates in the sequence have yet

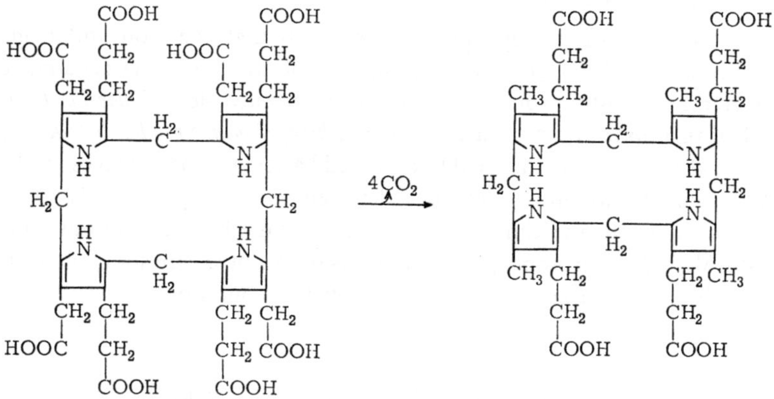

(20)

been found, and as the enzyme acts on other urogen isomers, although much less quickly (III > IV > II > I; III 7.5 times faster than I), it would appear that decarboxylations occur at random. The enzymes will not act on porphyrins *per se*, but a series of porphyrins with 7,6 and 5 carboxyl groups has been observed in chlorophyll-less mutants of *Chlorella* (see GRANICK, 1967); these were probably formed as urogens but were rapidly oxidized as they were formed.

Coprogen III is next converted into protoporphyrinogen IX (Protogen IX) by Coprogen oxidase which has been obtained from *Euglena* (SANO and GRANICK, 1961) as well as animal sources. The enzyme will only attack the III isomer at any significant speed and coproporphyrin III is not a substrate. It is, therefore, this enzyme and the urogen co-synthetase which together determine the specificity of formation of the III series. Coprogen oxidase attacks only two of the four propionic acid residues and up to now it has not been possible to replace oxygen by any other electron acceptor, although some variation must be necessary for porphyrin formation in the photosynthetic bacteria which are obligate anaerobes. GRANICK and SANO (1961) from experiments with T_2O suggest that the mechanism may involve the simultaneous removal of a hydride ion and CO_2.

Protogen IX is at the end of the line as far as porphyrinogens are concerned because they cannot chelate with metals. At this stage protogen is converted into protoporphyrin. The reaction can proceed spontaneously with great rapidity but it may also be enzymically catalyzed *in vivo* (PORRA and FALK, 1964).

e) Chelation of Protoporphyrin

A ferrochelatase has been obtained from a number of sources, mainly mitochondria (PORRA and JONES, 1963) and duck red cells (LABBE and HUBBARD, 1961), but no report on its existence in plants has appeared. As chloroplasts contain specific cytochromes (*f* and b_6 in higher plants) such an enzyme probably exists in the chloroplast. Demonstration of enzymic activity is difficult because of the ease of spontaneous chelation; necessary requirements for enzymic activity are (i) the iron must be maintained in the ferrous form by anaerobiosis or reducing agents such as ascorbic acid or glutathione, (ii) buffers which will complex with iron (e.g. phosphates or amines) must be avoided; (iii) the water-insoluble protoporphyrin must be effectively dispersed by appropriate detergents.

It is at the stage of protoporphyrin formation that the iron and magnesium pathways are assumed to bifurcate. In support of this view the accumulation of protoporphyrin and Mg protoporphyrin by chlorophyll-less *Chlorella* (GRANICK, 1948) and of protoporphyrin by a mutant of *Chlamydomonas reinhardi* (SAGER, 1955) can be cited; also, if ALA is supplied to etiolated barley and bean leaves in the dark, the same two compounds can be detected in the proplastids (GRANICK, 1961). However, up to the time of writing no enzyme has been found which will incorporate Mg^{2+} into protoporphyrin and this does not occur spontaneously as it can do with iron, copper and zinc under appropriate conditions.

f) Conversion of Mg Protoporphyrin into Chlorophyll a

The proposed steps from Mg protoporphyrin to chlorophyll *a* are outlined in Fig. 7. The evidence for each step in plants and bacteria will now be considered: Mg protoporphyrin methyl ester has been isolated from a *Chlorella* mutant and from etiolated barley treated with ALA (see GRANICK, 1967); in photosynthetic bacteria the evidence is even stronger because apart from the fact that it accumulates when *Rhodopseudomonas spheroides* is grown on an iron-deficient medium (JONES, 1963), a methylating enzyme, firmly attached to the chromatophores, esterifies Mg protoporphyrin 15 times faster than protoporphyrin in the presence of S-adenosylmethionine (TAIT and GIBSON, 1961). A similar enzyme exists in maize chloroplasts (RODMER and BOGORAD, 1967), in *Euglena* chloroplasts (EBBON and TAIT, 1969), and also in *Chlorobium thiosulfatophilum* although the major chlorophyll in this organism lacks the methyl ester group (HOLT, 1966).

The probable intermediates between Mg protoporphyrin methyl ester and the intermediate with intact fifth ring of chlorophyll have been isolated from *Chlorella* mutants [ELLSWORTH and ARONOFF, 1968, 1969]. The appropriate structures arranged in a logical biosynthetic sequence are

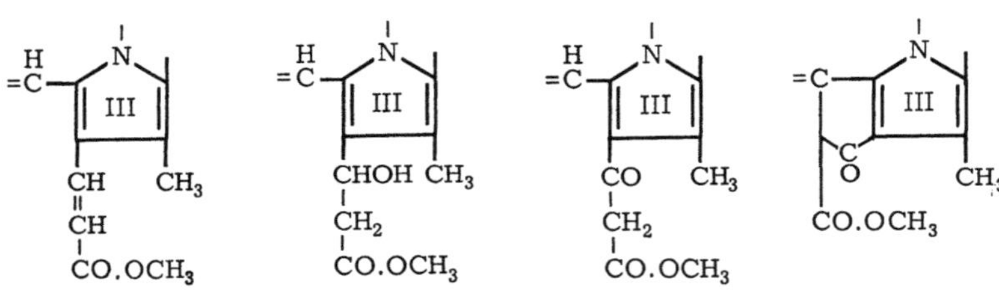

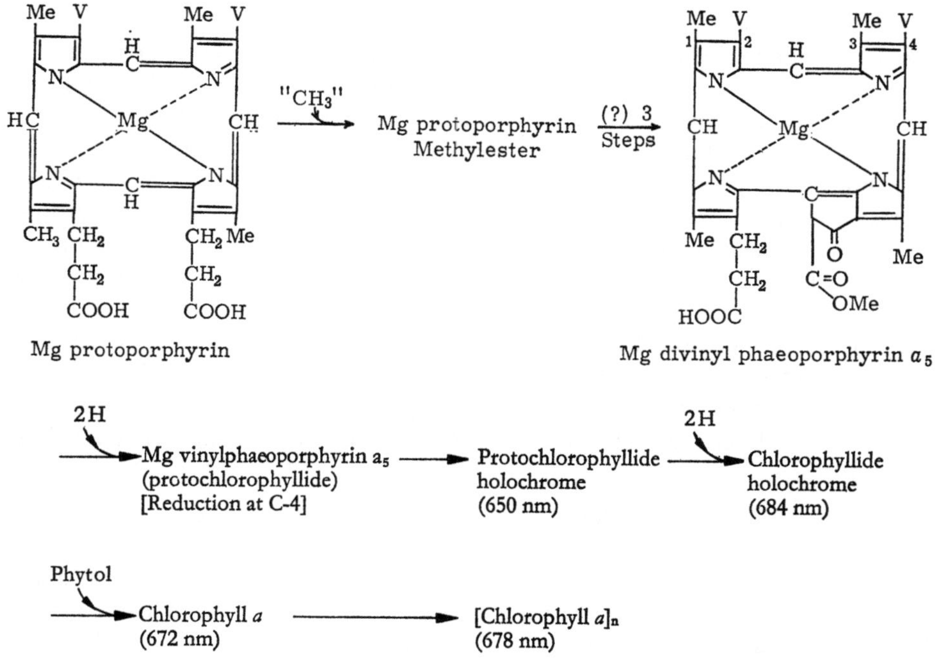

Fig. 7. Pathway of formation of chlorophyll *a* from Mg protoporphyrin

If Mg protoporphyrin is the intermediate which undergoes these changes then the last compound indicated in the above sequence is Mg divinylphaeoporphyrin a_5 which has been isolated by Jones (1963, 1966, 1968). The reduction of the vinyl group at C-4 in this compound yields Mg vinylphaeoporphyrin a_5 (protochlorophyllide) which has been known for some time to accumulate in etiolated seedlings (Smith, 1960). However, it is possible that reduction can take place at any stage in the sequence [Ellsworth and Aronoff, 1968, 1969] so it is difficult to decide which is the major pathway *in vivo*. Protochlorophyllide has been detected in chlorophyll-less *Chlorella* mutants (Granick, 1948) and can be demonstrated by fluorescence microscopy in etiolated barley leaves fed ALA in the dark (Granick, 1967). The action spectrum for the conversion of protochlorophyllide into chlorophyllide indicates that the photoreceptor is protochlorophyllide itself (Koski *et al.*, 1951).

Smith (1960) isolated from etiolated seedlings a protein-bound protochlorophyllide (holochrome) which is converted quantitatively into chlorophyllide holochrome on illumination. The observed quantum yield is about 0.6, which might indicate that the reaction is a photosentized reduction requiring two quanta for each molecule reduced; however, Smith prefers to consider it a one quantum process with an efficiency of 0.6. The source of the reducing power may be a component of the protein complex itself because although the reduction proceeds effectively at − 70° it is completely inhibited at − 195°; furthermore, Smith (1951) showed that water was not the source of the hydrogen. The holochrome from bean leaves is 100 Å in diameter with a molecular weight of 600,000 (Boardman, 1962). Trown (1965) believes that the protein component of the holochrome is fraction I protein (Kawa-

SHIMA and WILDMAN, 1970) (p. 20). Three types of protochlorophyllide exist in the proplastids of etiolated leaves; they are characterized by different absorption maxima; one (λ_{max} 631 nm) is easily bleached by light and is probably free protochlorophyllide; the second (λ_{max} 650 nm) is easily converted into chlorophyllide on illumination, this is bound in SMITH's holochrome; the third has the same absorption maximum as the holochrome complex but is not photoreduced (GRANICK, 1961). The last may represent the holochrome complex with an essential reducing component missing, because a leaf homogenate from a mutant of *Arabidopsis* which accumulates protochlorophyllide, will convert this compound into chlorophyllide in the presence of a colourless fraction from etiolated leaves of normal *Arabidopsis* (RÖBBELEN, 1956). Very recently, the protochlorophyllide absorbing in the region 634 to 638 nm has been resolved into two forms, one absorbing at 637 nm and the other at 628 nm. The former, which occurs in about equal amounts with the 650 nm form, is also photoreduced; the 628 nm form is not photoreduced (KAHN *et al.*, quoted by KIRK, 1970).

On illumination of the holochrome chlorophyllide is formed even at $-60°$ and the absorption maximum of the complex shifts to 684 nm within 1 sec of a 1 to 2 msec flash. After about 10 min in the dark this has moved to 672–673 nm (SHIBATA, 1967; GOEDHEER, 1961; SCHOPFER and SIEGELMAN 1968). Later (up to 2 days) the maximum shifts again to longer wavelengths due to the synthesis of large amounts of chlorophyll–683 and chlorophyll–673 during the major phase of greening (see KIRK, 1970).

Recently an intermediate change has been observed between the 650 nm and 678 nm forms of the protochlorophyllide; this takes about 1 sec and the absorption maximum is at 668 nm (LITVIN and BELYAEV, 1968). In sedimentation studies BOARDMAN (1967) has shown that protochlorophyllide is associated with an 18S component and concluded that after illumination either the chlorophyllide formed is split from the 18S material, or the conversion causes a conformational change in the protein which causes its aggregation. New protochlorophyllide which is rapidly formed when a seedling is returned to darkness is also associated with an 18S protein.

The reason for the 684 → 673 nm spectral shift in newly formed chlorophyll is still obscure although a number of reasonable explanations have been put forward. BUTLER and BRIGGS (1966) suggest that it is due to disaggregation of the pigment molecules on the holochrome; because a 650 → 630 nm and a 684 → 673 nm shift can be achieved with the protochlorophyllide holochrome and chlorophyllide holochrome, respectively by freezing and thawing in the dark. The 630 nm protochlorophyllide holochrome is converted into the 673 chlorophyllide holochrome on illumination. WETTSTEIN (1967) considers that the 650 → 684 nm shift corresponds to vesicle formation from the prolamellar body and that the 684 → 673 nm shift corresponds to vesicle dispersion, furthermore a mutant albino 17 will not disperse vesicles or carry out the 684 → 673 nm shift. However, the time-course of BOARDMAN's recent experiments did not appear to coincide with these morphological changes. Another view is that the 684 → 673 nm shift in barley is due to the gradual phytolation of chlorophyllide (SIRONVAL *et al.*, 1965). However, it appears that this enzymic change takes a longer time to complete than do the spectral shifts (WOLFF and PRICE, 1957; BOARD-

MAN, 1967). A recent suggestion is that the shift is due to a change in conformation of ring D (BALLSCHMITER and KATZ, 1968).

Protochlorophyll, that is phytolated protochlorophyllide, is present in small amounts in some plants, but there is doubt whether it can be photoreduced to chlorophyll *a*. VIRGIN (1960) considers it unlikely and GODNEV *et al.* (1968) have confirmed this and have calculated that phytolation would cause steric interference with the reduction at C-7 and C-8 of ring D.

The mechanism of phytolation of chlorophyll remains undecided, but by analogy with other terpenoid reactions, phytol pyrophosphate would be the active substrate. However, it is clear that under certain conditions the enzyme chlorophyllase, which is localized in the plastid lamellae (ARDAO and VENNESLAND, 1960; KLEIN and VISHNIAC, 1961) can synthesize chlorophyll rather than hydrolyze it (SHIMIZU and TAMAKI, 1963). Whatever the mechanism involved it is under the most delicate control; there is normally no reservoir of phytol in etiolated seedlings and when, on illumination, chlorophyll begins to be synthesized the amount of phytol formed exactly parallels the amount of chlorophyllide available for esterification (FISHER and BOHN, 1958). It has been claimed, however, that a store of phytol exists in young pine needles.

Experiments with stereospecifically labelled mevalonic acid (discussed in detail on p. 252) indicate that during the biosynthesis of phytol the component isoprenoid units are all biosynthetically *trans*, which strongly suggests that it is formed by the hydrogenation of all-*trans* geranylgeranyl pyrophosphate (WELLBURN *et al.*, 1966).

The comparatively small amount of protochlorophyllide which accumulates in dark-grown seedlings has sometimes been taken to indicate that it is not a precursor of chlorophyllide and thus chlorophyll. The true situation appears to be that there is a very tight regulatory control of protochlorophyllide levels in chloroplasts and only the amount which can be accommodated on the holochrome accumulates. In support of this view it is found that if etiolated leaves are given a series of very short exposures (milliseconds) to light of appropriate intensity interspersed with longer dark periods (10 to 15 min) then considerable levels of chlorophyll can be achieved (MADSEN, 1963). The small amount of protochlorophyllide present is converted into chlorophyllide by the light flash and during the recovery period in the dark this is removed from the holochrome which becomes charged with further protochlorophyllide ready for conversion during the next light flash.

The chlorophyll so formed is chlorophyll *a*, but chlorophyll *b* (methyl at C-3 replaced by formyl) is also always formed in higher plants, although under certain conditions the *a*:*b* ratio can vary considerably. Although blue-green and brown algae contain no chlorophyll *b*, no plant is known which does not contain chlorophyll *a*. Furthermore, a number of mutants of higher plants are known in which chlorophyll *b* is lacking although chlorophyll *a* levels are normal, but none are known in which the reverse situation exists [see KIRK and TILNEY-BASSETT (1967) for full details]. It is true to say that although from mutant studies and on chemical grounds one might expect chlorophyll *a* to be a precursor of chlorophyll *b*, the evidence so far available is not completely compelling (see BOGORAD, 1965; GRANICK, 1967). The experiments of SHLYK [an outspoken champion of the origin of chlorophyll *b* from *a* (BOGORAD, 1965)] are based on specific activity measurements of the two compounds under various conditions after exposure to $^{14}CO_2$, (^{14}C) glycine, (^{14}C) ALA or $^{28}Mg^{2+}$. These results and those of other workers who have carried out similar experiments are open to various interpretations [see BOGORAD (1965), for critical discussion] and more direct experiments are now required, such as extensions of GODNEV and AKULOVICH's experiments in 1960 when they added (^{14}C) chlorophyll *a* to the excised tips of onion leaves and found labelled chlorophyll *b* after 48 h. Similar experiments with (^{14}C) chlorophyll *b* did not lead to labelled chlorophyll *a*. Furthermore, the experiments of KUPKE and HUNTINGTON (1963) and WIECKOWSKI (1963) still require explanation. These investigators found that chlorophyll *b* disappeared as chlorophyll *a* appeared when young, green leaves of *Phaseolus vulgaris* were placed in the dark for 12 h.

g) Chlorophyll Synthesis in Relation to Plastid Development

The pattern of chlorophyll synthesis in leaves parallels the differentiation of proplastids into functional chloroplasts. In most cases transference of seedlings grown in the dark directly into light results in a lag phase before chlorophyll synthesis begins; the length depends on a number of factors such as species, variety and age of leaf, light intensity and other similar environmental conditions (VIRGIN, 1961; SISLER and KLEIN, 1963). One reason for this would be to allow for the formation of the enzymes concerned when their synthesis is derepressed in some way by light. In support of this view is the fact that protein inhibitors prevent the greening of illuminated etiolated tissues; bean leaves are inhibited by chloramphenicol (MARGULIES, 1962); bean and maize leaves by actinomycin D (BOGORAD and JACOBSON, 1964) and *Euglena* by 5-fluorouracil (SMILLIE, 1963) and actidione (KIRK and ALLEN, 1965). The derepression is a phytochrome-mediated effect. If etiolated seedlings are exposed for a very short period to red (690 nm) light and returned to darkness, then during the next few hours plastids expand to the size of chloroplasts and protein (enzymes) synthesis rapidly takes place. If these seedlings are then taken into continuous light the lag period, noted above, is absent, and chlorophyll synthesis begins immediately (MEGO and JAGENDORF, 1961). If etiolated seedlings are illuminated with a short exposure to red light and this is followed by a shot of far-red (740 nm) light then the development of the plastid in the dark does not occur. This is characteristic of phytochrome-mediated reactions. The decrease in lag period is inhibited by chloramphenicol (BERIDZET *et al.*, 1966), which also inhibits greening when applied early in the greening process (KLEIN, 1966) this suggests that inhibition of protein synthesis is due to inhibition of synthesis of new messenger RNA and that phytochrome acts by activating a gene controlling synthesis of new messenger RNA (MARGULIES, 1967).

BOGORAD [1967 (2)] interprets the situation not as derepression but the result of the need constantly to synthesize ALA synthetase; this is basically similar to MARGULIES' view. If leaves are treated with chloramphenicol or puromycin prior to initial illumination then on return to darkness there is little, if any synthesis of protochlorophyll (GASSMAN and BOGORAD, 1967). This means that the enzyme ALA synthetase has disappeared and not been resynthesized. Similar results with actinomycin D indicate that messenger RNA for ALA synthetase is also labile. ALA synthetase is also rapidly degraded in *Rhodopseudomonas spheroides* when protein synthesis stops (BULL and LASCELLES, 1963).

Where is the specific site of action of this derepression effect in relation to chlorophyll synthesis? As etiolated leaves accumulate about ten times their normal amount of protochlorophyllide in their plastids when fed ALA, none of the steps from ALA to protochlorophyllide would appear to be the limiting factor (GRANICK, 1967). As the excess protochlorophyllide is not converted into chlorophyll when the seedlings are illuminated, it may be that the amount of holochrome is limiting or the structural protein to which the chlorophyllide is attached is in short supply (KIRK and ALLEN, 1965). However it is generally agreed (GRANICK, 1967) that under normal conditions the formation of ALA is the controlling factor, and ALA is the obvious point at which to exert control because it is the first specific step in the biosynthetic sequence to porphyrins. In liver, it has been shown that haem represses the synthesis of ALA synthetase and there is no evidence of allosteric inhibition (GRANICK, 1966). This same type of mechanism probably functions in plants although no direct evidence is yet available. The chlorophyll pathway could be regulated by feed-back control of synthesis of Mg chelates, which would tend to channel the precursor protoporphyrin into haem synthesis which would then inhibit ALA synthetase (LASCELLES, 1965).

Synthesis of ALA synthetase in liver is also regulated by certain steroids (KAPPAS and GRANICK, 1968). The view is that an apo-repressor and a co-repressor, haem (end product), combine to form an active repressor and that certain steroids inactivate the repressor by displacing haem from it. Similar low molecular weight repressors and activators function in *Rhodopseudomonas spheroides* (MARRIOTT, 1968), but there are no sterols in these organisms.

In the photosynthetic bacterium *Rhodopseudomonas spheroides* haem inhibits ALA synthetase by a repressor mechanism and by an allosteric effect (LASCELLES, 1964).

In mature chloroplasts of monocotyledons the repression of chlorophyll synthesis is almost complete; very little, if any turn-over, as evidenced by the incorporation of $^{14}CO_2$ into the chlorophylls is apparent. A slight turn-over occurs in dicotyledons (PERKINS and ROBERTS, 1963).

h) Timing of Chlorophyll Biosynthesis

Genes exist which affect the timing rather than the mechanism of synthesis of chlorophylls. Seeds of a "virescent" mutant of barley reach their maximum chlorophyll content some 2 to 3 days later than seeds of normal plants. Development of normal chloroplasts is also delayed from about 4 to 12 days (MACLACHLAN and ZALIK, 1963). A maize mutant, pale yellow-1, synthesizes no chlorophylls for about 8 days after germination, then the mutant block is broken and synthesis becomes normal. The same pattern is also seen in carotenoid synthesis and, in the dark, synthesis of protochlorophyll and xanthophylls is similarly affected (KAY and PHINNEY, 1956).

Other genes are known which cause a failure in the switching off of chlorophyll synthesis in ripening fruit. Such mutants are known in tomatoes ("green flesh" and "dirty red") (RAMIREZ and TOMES, 1964) and *Capsicum* (peppers) (SMITH, 1950).

So little is known of the details of regulation of chlorophyll synthesis that it is difficult to say more about these interesting mutants except that perhaps they involve the mutation of a regulator gene (see also KIRK and TILNEY-BASSETT, 1967).

i) Effect of Iron on Chlorophyll Synthesis

It has long been known that lack of iron in the nutrition of plants results in chlorosis owing to the failure to synthesize chlorophyll. There appear to be a number of reasons for this: inability to produce sufficient succinyl-CoA for ALA synthesis (p. 232); lower levels of ferredoxin which result in reduced photophosphorylation activity and thus insufficient production of ATP required for the endergonic steps in chlorophyll synthesis, and inhibition of the step between coproporphyrinogen and protoporphyrin; MARSH *et al.* (1963) showed that iron-deficient young leaves of the cow pea had impaired ability to incorporate labelled citrate, α-ketoglutarate and succinate into chlorophyll although the levels of ALA synthetase were normal. Ferredoxin levels were also lowered. Coproporphyrinogen and to a lesser extent protoporphyrin accumulates in darkness in leaf discs on a Fe-deficient medium, the former but not the latter disappears as soon as iron is added (HSU and MILLER, 1965). In some higher plants chlorophyll-deficient mutants exist in which the fundamental lesion is a failure in iron metabolism. For example, in a yellow stripe (ys_1) mutant of maize, the lesion is overcome by growing the plants in a medium containing ferrous iron. The mutant is apparently unable to take up ferric iron via the roots because ferric iron applied to the leaf surface is effective (BELL *et al.*, 1958).

In the photosynthetic bacteria large amounts of coproporphyrin are excreted into the culture medium in iron-deficiency. This is apparently due to the fact that lack of iron results in haem being synthesized in amounts insufficient to repress activity and synthesis of ALA synthetase (LASCELLES, 1964). Thus as ALA synthetase is normally the rate-limiting enzyme the removal of inhibition and repression results in additional synthesis of ALA which is channelled into porphyrin synthesis. The accumulation of coproporphyrin is probably due to coproporphyrinogen oxidase (p. 237) becoming the rate-limiting step under these conditions.

j) Chlorophyll Synthesis in the Dark

Although most higher plants cannot make leaf chlorophyll in the dark, some conifers produce green cotyledons in the dark as do some members of the order Gnetales. Many ferns make chlorophyll in the dark but neither members of the Equisetaceae nor the Bryophytes do so (see KIRK and TILNEY-BASSETT, 1967 for complete references). Many higher plants (e.g. maize) lose chlorophyll if the green plants are returned to darkness (FRANK and KENNEY, 1955). On the other hand a number of algae, for example, certain strains of *Chlorella* and *Scenedesmus* and certain tissues of plants, for example, the inner tissues of tomato fruit and conifer cotyledons, can synthesize chlorophylls in the dark. This means that a non-photosensitized enzymic reduction of protochlorophyllide must occur in these cases. There also exists the possibility that both photo- and non-photoreduction of protochlorophyllide can occur side by side in such algae as *Chlorella*, because mutants of *Chlorella* and *Chlamydomonas* which have lost the ability to synthesize chlorophyll in the dark continue to do so in the light (GRANICK, 1967). Nothing is known about how regulation of synthesis occurs in the dark.

3. Phycobilins

The phycobilins, which are linear tetrapyrroles, normally occur in the lamellae of the procaryotic blue-green algae and in the chloroplasts of the Rhodophyta and some Cryptophyta. The structure of the two main pigments phycoerythrobilin (XXII a) and phycocyanobilin (XXII b) have only recently been elucidated (SIEGELMAN *et al.*, 1968) so very little is known about their biosynthesis (see e.g. TROXLER and BOGORAD, 1967). OHÉOCHA (1968) has indicated that they can be formed by

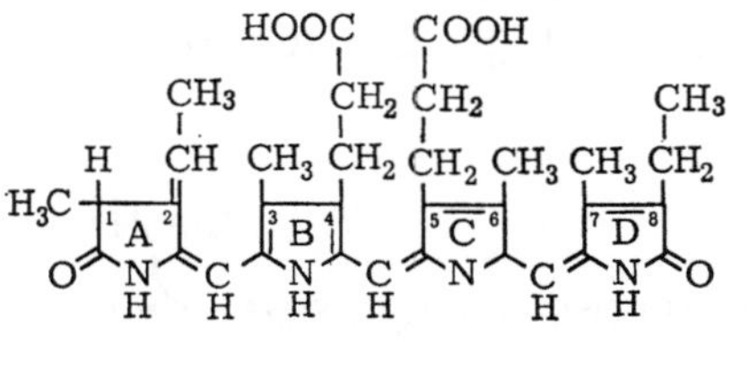

(XXIIa)

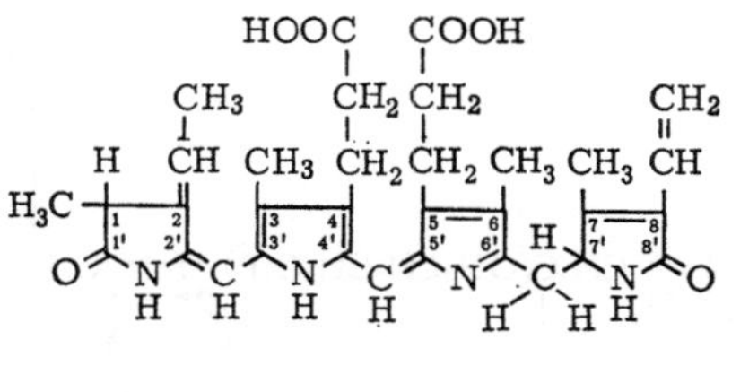

(XXIIb)

addition of two moles of hydrogen to biliverdin, but attempts to demonstrate reduction of biliverdin by extracts of red algae have so far been inconclusive. The mechanism of bilverdin formation from porphyrins is still undecided, but it is clear from experiments with ALA that they arise conventionally from this compound, and specific activity studies suggest that porphobilinogen and a porphyrin (? Urogen III, Coprogen III, protoporphyrin IX and, possibly, haem) are intermediates (TROXLER and LESTER, 1967). However, in *Cyanidium caldarium* actively synthesizing phycocyanin none of the possible precursors could be detected (TROXLER and LESTER, 1967). In *Anacystis nidulans*, on the other hand, added protoporphyrin or haem did increase phycobilin production under certain circumstances and also diluted out the incorporation of (^{14}C) acetate into the pigments (GODNEV *et al.*, 1966).

MURPHY *et al.* (1967) in experiments designed to throw light on the opening of the pyrrole ring showed that an extract of red algae contained a reducing compound which by a coupled oxidation reaction would oxidize pyridine haemochrome to verdihaemochrome. The nature of the reducing compound is not known but the amount present in the algal extracts varies with the season. Similar changes which can be brought about by heat-stable systems have been found in mammalian liver (MURPHY *et al.*, 1967; LEVIN, 1967) and in fish (TSUCHIYA and NOMURA, 1968). Earlier work had suggested that the reaction was enzymic (NAKAJIMA *et al.*, 1963).

The action spectrum for formation of C-phycocyanin in *C. caldarium* indicates that a haem-type pigment is the photoreceptor and that this pigment may also be the precursor of the prosthetic group of the pigment (NICHOLS and BOGORAD, 1962).

An additional complication is the matching of the biosynthesis of the protein carrier with that of the pigment prosthetic group. Various interesting speculations about phycobilin biosynthesis have been summarized by BOGORAD and TROXLER (1967).

4. Carotenoids

a) Formation of Carotenes

The carotenoids accompany chlorophylls in all photosynthetic tissues, and in the higher plants the chloroplast carotenoids are qualitatively and quantitatively very similar in all the genera examined (p. 232) (GOODWIN, 1965; STRAIN, 1966).

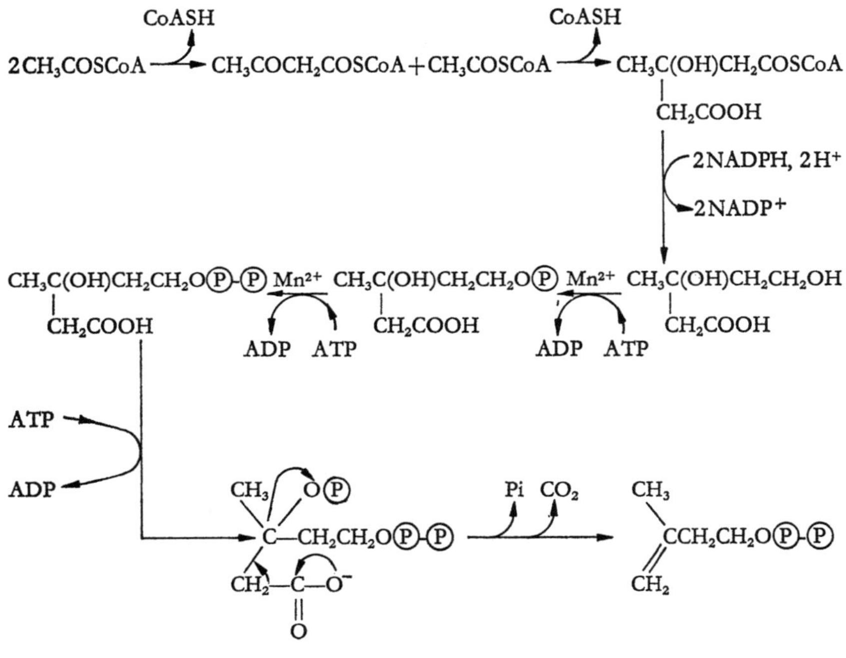

Fig. 8. The conversion of acetyl-CoA into isopentenyl pyrophosphate

As already indicated, the qualitative distribution in algae (GOODWIN, 1965) and photosynthetic bacteria (JENSEN, 1965) is much more complex.

Carotenoids are synthesized from the universal biological isoprenoid precursor, isopentenyl pyrophosphate, IPP, which in turn is synthesized from acetyl-CoA by a well established pathway (Fig. 8) via mevalonic acid (MVA).

Although most of the steps to IPP have been demonstrated in plants, e.g. carrot and tomato slices and fungi (GOODWIN, 1965), their occurrence in chloroplasts has only been clearly demonstrated for mevalonic kinase and mevalonic 5-phosphokinase. ROGERS *et al.* [1966 (1, 2)] reported the presence of these enzymes in chloroplast fragments obtained by ultra sonication of chloroplasts isolated by non-aqueous techniques; they are also present in chloroplast fractions of *Phaseolus vulgaris* (Grey and Kekwick, 1969) and in the chloroplasts of plant tissue cultures (Thomas and Stobart, 1970). In experiments with illuminated excised etiolated leaves exogeneous MVA is only slightly incorporated into β-carotene and other chloroplast terpenoids (Goodwin, 1967); this is due mainly to the comparative impermeability of the chloroplast outer membrane to MVA. The significance of this regulation of terpenoid metabolism is discussed on p. 261.

IPP isomerizes to dimethylallyl pyrophosphate (Fig. 9) (Lynen *et al.*, 1959) which acts as a starter for chain elongation. The isomerase has been demonstrated in *Hevea* latex (Chesterton and Kekwick, 1966); it is an enzyme which is inhibited by iodo-

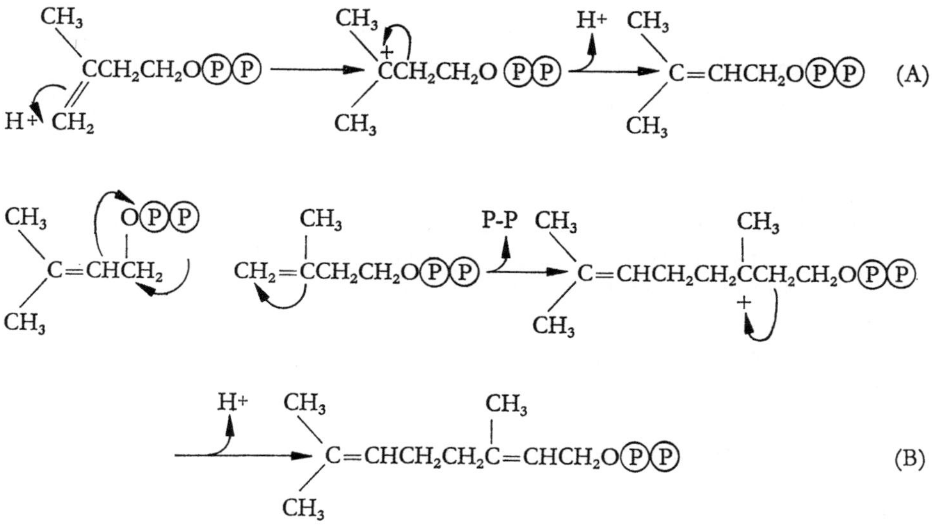

Fig. 9. The conversion of IPP into DMAPP (A) and the formation of geranyl pyrophosphate from DMAPP and IPP (B)

acetamide and as this compound inhibits the incorporation of MVA into carotenoids in the fungus *Phycomyces blakesleeanus* the isomerase has been implicated in carotenoid biosynthesis (Yokoyama *et al.*, 1962). Dimethylallyl pyrophosphate then condenses with a molecule of IPP to form geranyl pyrophosphate (C_{10}) which condenses with a further molecule of IPP to form farnesyl pyrophosphate (C_{15}) which itself undergoes a similar elongation to produce geranylgeranyl pyrophosphate (C_{20}) (Fig. 9). It is not clear whether the chain elongations are carried out by the same or different enzymes; the consensus appears to be that one enzyme is involved (Lynen *et al.*, 1958; Popják and Cornforth, 1960).

Geranyl pyrophosphate (C_{10}) has not yet been experimentally demonstrated as a carotenoid precursor but farnesyl pyrophosphate is incorporated into carotenoids by a cell-free system from *P. blakesleeanus* (Yokoyama *et al.*, 1961) by carrot plastids and rather inefficiently by chloroplasts supplemented with a soluble factor from carrots (Anderson and Porter, 1962). As the incorporation of farnesyl pyrophosphate is stimulated by the simultaneous addition of MVA (source of additional C_5 unit) (YOKOYAMA *et al.*, 1961) it is reasonable to assume that geranylgeranyl pyrophosphate, which is formed by isolated carrot plastids is a carotenoid precursor (WELLS *et al.*, 1964), and this has recently been demonstrated in tomato plastids (SHAH *et al.*, 1968) and extracts from *Phycomyces blakesleeanus* (LEE and CHICHESTER, 1969).

The first C-40 compound formed is phytoene (XXIII)[1] and the mechanisms of conversion of geranylgeranyl pyrophosphate into phytoene is considered later.

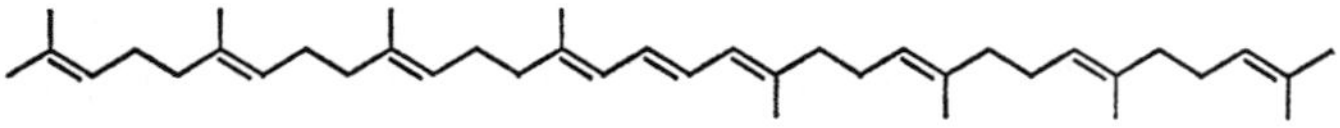

(XXIII)

Synthesis of phytoene from MVA by fragments of chloroplasts prepared by the non-aqueous technique has recently been demonstrated (CHARLTON *et al.*, 1967). The co-factors required are Mg^{2+} and ATP; NAD^+ or $NADP^+$ is not required (compare the formation of squalene. Tomato plastids will also incorporate isopentenyl pyrophosphate into phytoene (JUNGALWALA and PORTER, 1965). Phytoene is then stepwise dehydrogenated to phytofluene, ζ-carotene and neurosporene (Fig. 10). The first step has been demonstrated in tomato plastids (BEELER and PORTER, 1962), but the next two steps have not been directly demonstrated although circumstantial evidence is strong. For example, together with phytoene and phytofluene these compounds accumulate when carotenoid synthesis is specifically blocked in various organisms either by mutations, e.g. in higher plants such as maize (W-1) (ANDERSON and ROBERTSON, 1960) or by selective inhibitors such as diphenylamine (see GOODWIN, 1965). In the latter case when the inhibitor is removed the kinetics of the disappearance of these compounds and the appearance of the fully unsaturated pigments indicate a precursor/product relationship (see JENSEN, 1965).

The carotenoids of green leaves are all bicyclic compounds and some evidence points to cyclization beginning at the neurosporene level (Fig. 11). β-Zeacarotene, a key intermediate (Fig. 11), has been isolated from maize (PETZOLD *et al.*, 1959). It also accumulated in DPA-inhibited cultures of *Phycomyces* (DAVIES *et al.*, 1963), under certain cultural conditions in *Rhodotorula* (SIMPSON *et al.*, 1964) and is probably compound 'X' in Claes's *Chlorella* mutant 5/520. This mutant, when grown in the dark synthesizes phytoene and the other partly saturated carotenoids; on illumination anaerobically β-carotene and 'X' are synthesized at the expense of the phytoene group (CLAES, 1958).

The other possible precursor of the cyclic carotenes is lycopene (Fig. 12). DECKER and UEHLEKE (1961) first reported that isolated chloroplasts can convert (^3H) lycopene into β-carotene, and support has been provided by GODNEV and ROTFARB (1962), WELLS *et al.* (1964), HILL and ROGERS (1969), KUSHWAHA *et al.* (1969). On the other hand, DAVIES (1963) showed that in *Rhizophlictis rosea* lycopene is not converted into γ-carotene; it is also established that while lycopene synthesis in commercial tomatoes is inhibited by ripening temperatures above 30°, β-carotene synthesis is unaffected (GOODWIN and JAMIKORN, 1952). There may, be a second mechanism in certain to-

[1] Naturally occurring phytoene has a *cis* configuration at the central double bond; it is indicated here as *trans* for simplification in discussing biosynthetic transformations.

mato crosses in which the usual high levels of lycopene are replaced by an equivalent amount of β-carotene; in these tomatoes the increased β-carotene synthesis is temperature-sensitive (TOMES, 1963).

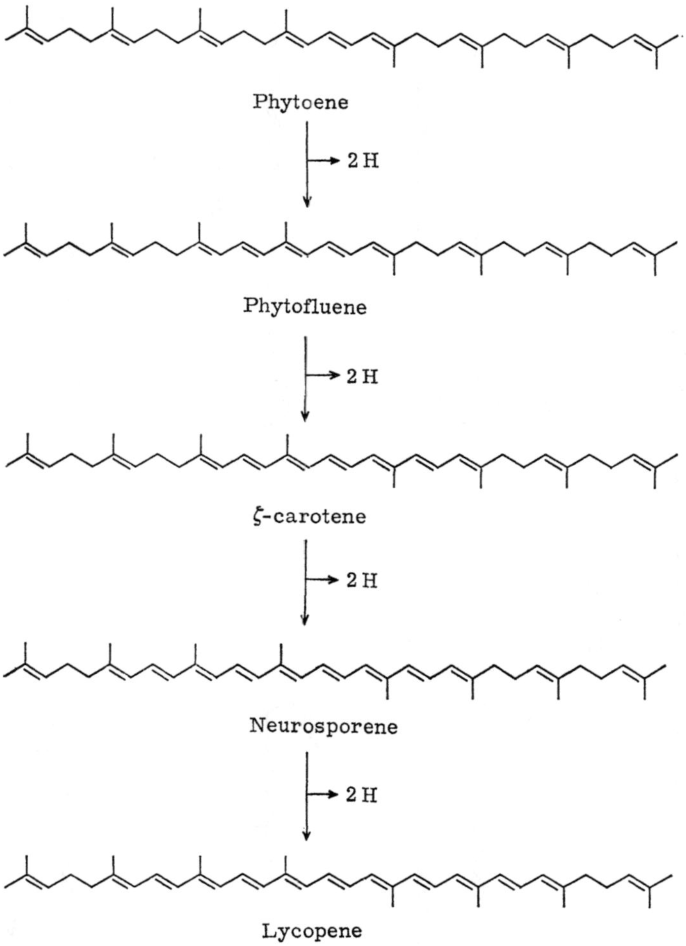

Fig. 10. The stepwise dehydrogenation of phytoene to lycopene

Whatever the nature of the immediate acyclic precursor of the cyclic carotenoids, it is now clear that in seedlings α-carotene is not formed from β-carotene or *vice versa*. This is discussed in detail later (see p. 253).

b) Formation of Xanthophylls

The insertion of oxygen into carotenoids occurs late on in the biosynthetic sequence, usually at the fully dehydrogenated level. The *Chlorella* mutant 5/520,

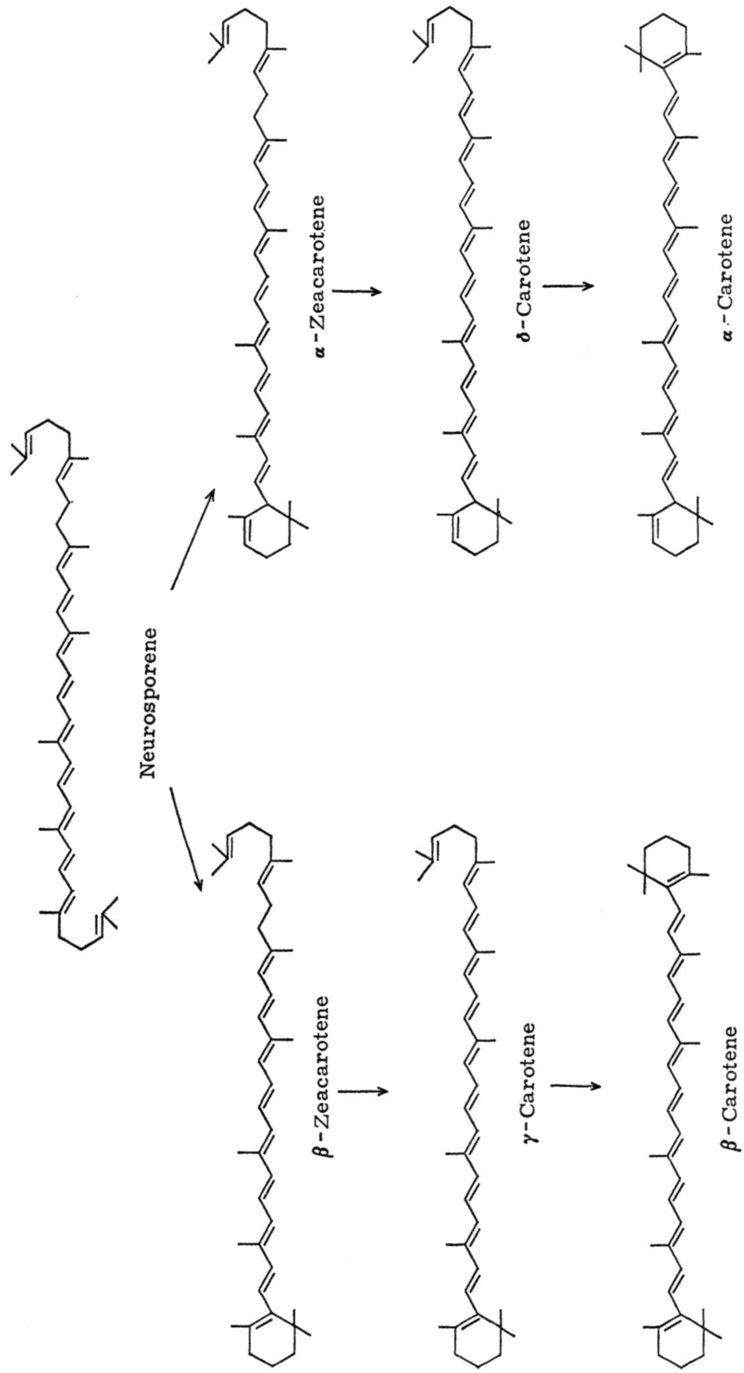

Fig. 11. The conversion of neurosporene into α-carotene and β-carotene

mentioned above has been used to demonstrate this. If this organism is grown heterotrophically in the dark, subsequently illuminated anaerobically and then returned to darkness and allowed access to oxygen, then the carotenes formed in the light are converted into xanthophylls as indicated by a precursor-product relationship between the rate of disappearance of the hydrocarbons and rate of appearance of the oxygenated pigments (CLAES, 1957, 1959). A similar demonstration of an oxidation of this type

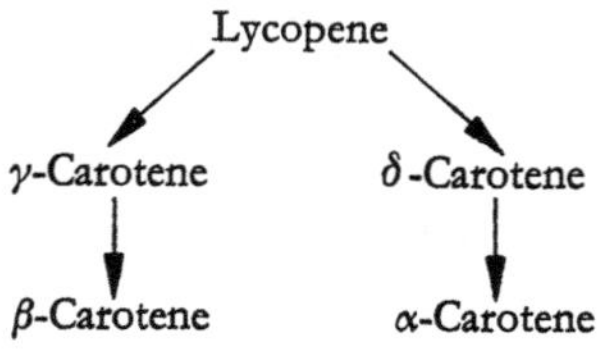

Fig. 12. The conversion of lycopene into α-carotene and β-carotene

comes from studies on the photosynthetic bacterium *Rhodopseudomonas spheroides;* under anaerobic conditions the cells are yellowish-brown and the major carotenoid is the hydrocarbon, spheroidene (Fig. 13); on exposure to oxygen the cells turn purple and the yellow spheroidene is quantitatively converted into the reddish purple ketocarotenoids, spheroidenone (Fig. 13) (VAN NIEL, 1947; GOODWIN *et al.*, 1956;

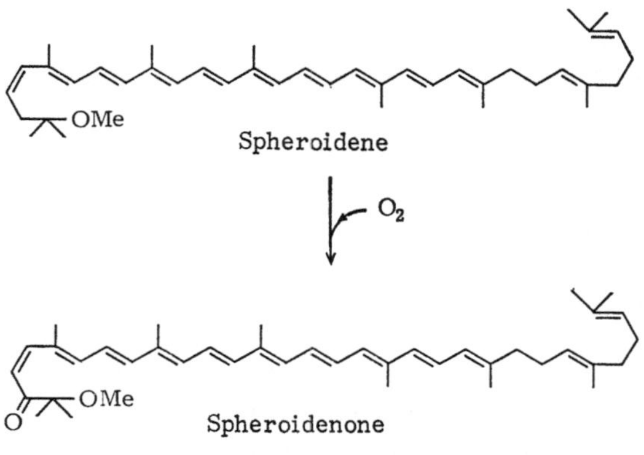

Fig. 13. The conversion of spheroidene into spheroidenone

DAVIES *et al.*, 1961). The oxygen of spheroidenone arises from oxygen and not water (SHNEOUR, 1962).

The oxygen in the hydroxyl functions of lutein (VIII) also arises from oxygen (YAMAMOTO *et al.*, 1962), but the mechanism involved is not known. Similarly the epoxide functions in antheraxanthin, (XIV) and violaxanthin (IX) are O_2-derived and are inserted stepwise in spinach leaves (YAMAMOTO *et al.*, 1962) and in *Euglena* chloroplasts (KRINSKY, 1966). BAMJI and KRINSKY (1965) have shown that in *Euglena* preparations the reverse reaction (de-epoxidation) can take place provided either NADPH or $FMNH_2$ is present as a reducing agent (Fig. 14). According to these workers the epoxidation in *Euglena* is a light-catalyzed step with a mandatory requirement for oxygen and which *in vivo* may not be an enzyme reaction. In leaves, however, YAMAMOTO *et al.* (1967) considers that both epoxidation and de-epoxidation are light-induced dark reactions.

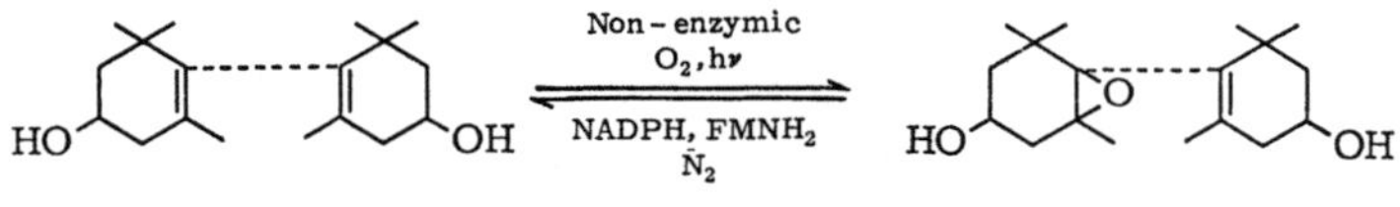

Zeaxanthin Antheraxanthin de-epoxidase Antheraxanthin

Fig. 14. The reversible epoxidation of zeaxanthin to antheraxanthin in *Euglena*

c) Stereospecificity of Carotenoid Biosynthesis

In the isomerization of IPP to dimethylallyl pyrophosphate a proton is lost from the carbon which originated from C-4 of MVA (Fig. 9). A loss of a similar proton occurs at each addition of an IPP molecule (Fig. 9, p. 247) so that in phytoene synthesis 8 protons are lost from 8 molecules of MVA as indicated by asterisks in Formula (XXIV). As C-4 of MVA carries two hydrogen atoms the probability existed that the enzymes concerned with the removal of protons from C-4 act stereospecif-

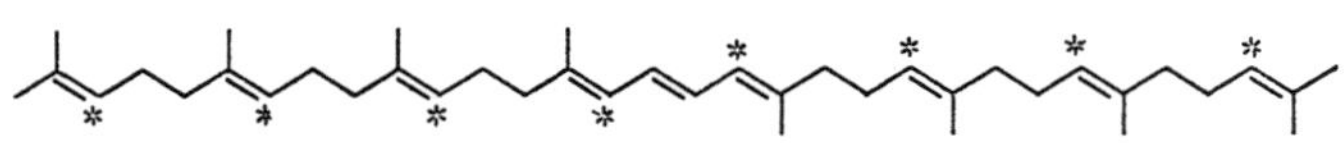

(XXIV)

ically, although chemically the two hydrogens cannot be distinguished. To test this concept it was necessary to be able to differentiate experimentally between the two hydrogen atoms. This was achieved by CORNFORTH and POPJÁK and their colleagues (see CORNFORTH and POPJÁK, 1966), who synthesized two species of ($2\text{-}^{14}C$) MVA one with tritium in the position which confers the *R* configuration on C-4 (XXV) and the other with tritium in the position which confers the *S* configuration on C-4 (XXVI), which in both cases becomes an asymmetric carbon atom. In the conversion

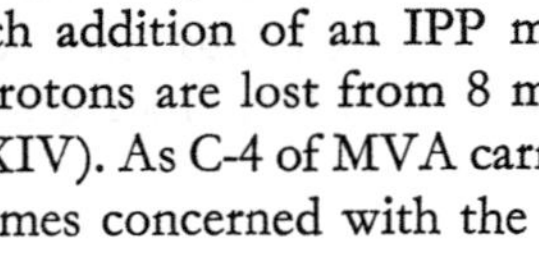

(XXV)

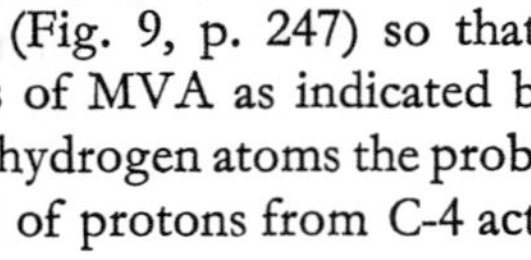

(XXVI)

of these substrates into phytoene four possibilities exist: (a) if the 4 *R* tritium is stereospecifically removed at each step then with ($2\text{-}^{14}C$)-4 $R(4\text{-}^3H_1)$ MVA no tritium would be retained in the phytoene while with ($2\text{-}^{14}C$)-4 $S(4\text{-}^3H_1)$ MVA no tritium would be lost and the $^{14}C{:}^{3}H$ ratio in the phytoene would be the same as in the original MVA; (b) the situation could be exactly reversed if the 4*S* tritium were stereospecifically removed; (c) if a non-stereospecific removal occurred then the $^{14}C{:}^{3}H$ ratio would be 1:0.5 with each substrate; (d) if different stereospecific removals occurred at each condensation then the ratios would be either 1:0.25, 1:0.5 or 1:0.75 with one species of MVA and correspondingly 1:0.75, 1:0.5 and 1:0.25 with the other species. Experiments with these species of MVA have shown that in carrot slices, tomato slices, the fungus *Phycomyces blakesleeanus* and spinach chloroplasts the incorporation of MVA into phytoene involves the stereospecific removal of the 4-*pro-S* hydrogen[2] at each step [Goodwin and Williams, 1966; Williams *et al.*, 1967 (1)]. This represents the same stereospecificity as in squalene biosynthesis in animals (Cornforth *et al.*, 1966) and in plants (Goodwin and Williams, 1966).

When phytoene is dehydrogenated to the more saturated carotenes, then a hydrogen atom is lost from what was originally C-5 of six of the 8 molecules of MVA incorporated into one molecule. (Two are also lost from the center of the molecule during dimerization of geranyl pyrophosphate to phytoene, but this case is considered below.) The possibility of stereospecific removal of hydrogen also exists in this situation because, like C-4, C-5 of MVA also carries two hydrogen atoms. Recent investigations with tomato mutants and (2-^{14}C)-5 *R*(5-$^{3}H_1$) MVA and (2-^{14}C-5-$^{3}H_2$) MVA have shown that the dehydrogenation of phytoene does occur stereospecifically and that at each step the *pro-R* hydrogen atom is lost from the carbons originating from C-5 of MVA [Williams *et al.*, 1967 (2)].

d) Mechanism of Phytoene Formation

The exact mechanism by which two molecules of geranylgeranyl pyrophosphate condense to yield phytoene cannot be stated definitely at the moment, but experiments with (2-^{14}C)-5 *R*(5-$^{3}H_1$) MVA and (2-^{14}C-5-$^{3}H_2$) MVA have resulted in the possibilities being narrowed down. In the first place it has been shown that the dimerization involves the elimination of the *pro-S* hydrogen from C-1 of each participating geranylgeranyl pyrophosphate [Williams *et al.*, 1967 (2)]. A number of possibilities can be formulated which are based on those proposed by Cornforth and Popják (1966) for the formation of squalene from two molecules of farnesyl pyrophosphate. In the case of phytoene biosynthesis Fig. 15 is perhaps the most attractive; instead of a nucleophilic displacement of the sulphur-enzyme by H^- from NADPH as in squalene formation, the incipient charge on the C from the sulphur enzyme is neutralized by loss of a proton from the adjacent methylene group, thus introducing the central double bond characteristic of phytoene.

e) Mechanism of Ring Formation

The substrate (2-^{14}C)-4 *R*(4-$^{3}H_1$) MVA has been used to elucidate the mechanism of the formation of the α- and β-ionone rings in the carotenes. The reaction (Fig. 16) can formally be considered to be initiated by proton attack at C-2 (carotenoid number-

[2] The nomenclature of Cahn and Ingold (1951) modified by Hanson (1966) to name paired ligands g,g at a tetrahedron atom Xggij.

ing); the carbonium ion so formed is stabilized by loss of a proton from either C-4 or C-6 to form an α-ionone or β-ionone ring respectively. If the acyclic precursor, be it lycopene or neurosporene, is synthesized from (2-^{14}C)-4 R(4-$^{3}H_1$) MVA it will

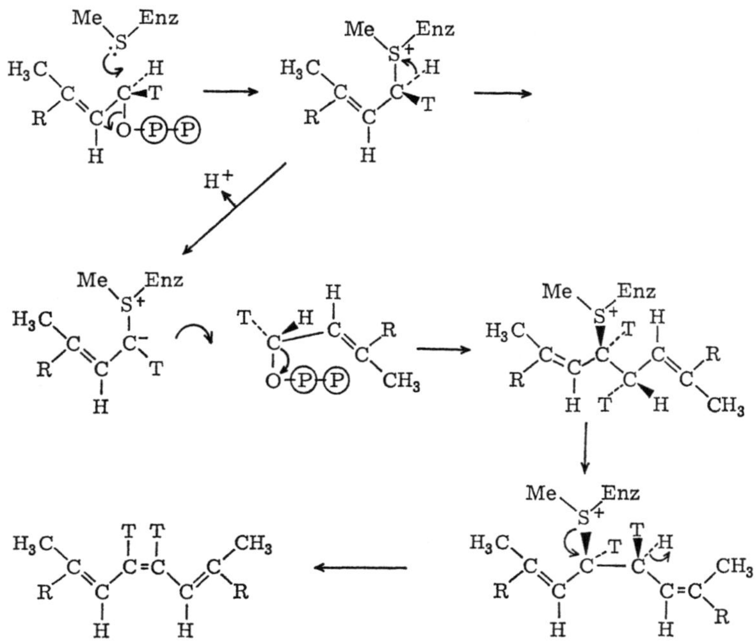

Fig. 15. A possible mechanism for phytoene formation

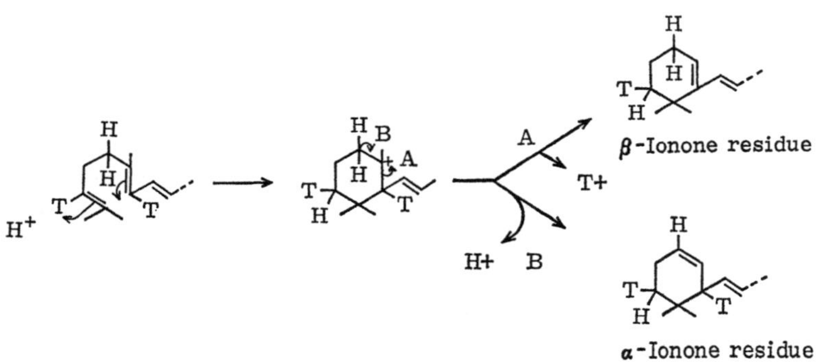

Fig. 16. The mechanism of formation of α- and β-ionone residues in carotenoids

have the same ^{14}C:^{3}H ratio as the starting material [Williams *et al.*, 1967 (1)]; as 8 MVA molecules make up the C-40 precursor then, calling the ^{14}C:^{3}H ratio 1:1 this represents a C:H atomic ratio of 8:8 (all labelled C and H atoms conserved in the molecule). When an α-ionone residue is formed no tritium atom would be lost whilst one tritium atom would be lost for every β-ionone residue formed. As no carbon atom is lost during these transformations the ratio of 8:8 should change to 8:7 in the case of α-carotene (one α- and one β-ionone residue) and to 8:6 in the case of β-carotene (two β-ionone residues). These ratios have been obtained experimentally in carrot and tomato slices and in *Phycomyces blakesleeanus* [GOODWIN and WILLIAMS, 1965 (1, 2); WILLIAMS *et al.*, 1967 (1)]. Similarly the expected ratios have been observed with δ-carotene (one α-ionone end and open end (8:8), and γ-carotene (one β-ionone end one open end) (8:7). The mechanism indicated in Fig. 16 clearly indicates that α-carotene does not arise from β-carotene, but does not eliminate the reverse process. This has now been eliminated by experiments with ($2\text{-}^{14}C\text{-}2\text{-}^{3}H_2$) MVA in which it was found that α-carotene contains one tritium atom fewer than β-carotene because C-4 of the α-ionone ring arises from C-2 of MVA one hydrogen of which is lost in reaction B (Fig. 16). This could not occur if β-carotene were formed from α-carotene (WILLIAMS *et al.*, 1967).

f) Mechanism of Xanthophyll Formation

Experiments with the stereospecifically labelled MVA's mentioned previously have shown that the hydroxylation reactions do not involve a keto intermediate but probably involve the direct insertion of OH without change in configuration (WALTON *et al.*, 1969). Furthermore, experiments with ($2\text{-}^{14}C$)-$5R$($5\text{-}^{3}H_1$) MVA indicate that the carotenes contain more tritium than do the xanthophylls (GOODWIN *et al.*, 1968), so that there is little possibility in the systems examined that the carotenes arise from xanthophylls as was recently suggested (COSTES, 1965; SAAKOV, 1965).

g) Effect of Light

As with chlorophyll (p. 242) if etiolated seedlings are brought immediately into the light there is a short lag before synthesis of carotenoids begins; this lag can be abolished by a short treatment with red light before full illumination. The red light effect can be nullified by far-red light which indicates that the effect is phytochrome-mediated (HENSHALL and GOODWIN, 1965).

Small amounts of carotenoids are found in etiolated seedlings and these consist mainly of highly oxidized forms usually associated with non-photosynthetic tissues such as flowers and fruit (GOODWIN, 1952), bean cotyledons (GOODWIN and PHAGPOLNGARM, 1960) and plant tissue cultures, which do not contain chloroplasts (WILLIAMS and GOODWIN, 1965).

h) Genetic Control of Carotenoid Synthesis in Chloroplasts

Although there is little doubt that chloroplasts can synthesize carotenoids, there are indications that synthesis of the enzymes involved are under nuclear control rather than cytoplasmic control. An albino mutant of maize (W-3) is blocked between phytoene and the coloured carotenoids and the lesion is located on maize chromosome 2 (ROBERTSON, 1958). Two other maize mutants have been reported which are blocked at the carotene and lycopene stage and both show Mendelian type inheritance (FALUDI-DANIEL *et al.*, 1956). Furthermore a white mutant of *Helianthus annuus* contains no coloured carotenoids and this character is inherited in the Mendelian manner, being controlled by a single recessive factor (Wallace and Habermann, 1959).

An odd mutation results in a yellow mutant of *Helianthus annuus* accumulating xanthophylls but no carotenes (Wallace and Schwarting, 1954; Wallace and Habermann, 1960). It has been suggested that this is due to mutation in a regulator gene (Kirk and Tilney-Bassett, 1967).

i) Evidence for Formation of Carotenoids in Chloroplasts

There is now good evidence that isolated chloroplasts can synthesize carotenoids from mevalonic acid. This has been achieved with non-aqueously prepared chloroplasts (Charlton *et al.*, 1967) and isolated purified preparations effectively covert IPP into phytofluene and lycopene in the presence of $NADP^+$, FAD and light (Subbarayan *et al.*, 1970), although the non-aqueous chloroplasts will not desaturate phytoene (Charlton *et al.*, 1967). Isolated chloroplasts can also cyclize lycopene to the cyclic carotenes (Suzue and Porter, 1969).

It has not yet been demonstrated that isolated chloroplasts can form mevalonic acid or even the basic building unit acetyl-CoA. Various possibilities have been considered (see e.g. Kirk, 1970) but all require further investigation before a decision on the problem can be made.

D. Chloroplast Quinones

The main quinones localized in chloroplasts are the plastoquinones, with plastoquinone A (PQA, XXVII) as the major component, phylloquinone (vitamin K_1) (XXVIII) and the tocopherolquinones (XXIX). The derivatives, plastochromanol (XXX) and tocopherols (XXXI) are also present in the chloroplasts (see e.g. Wallwork and Pennock, 1968; Crane *et al.*, 1966; Dunphy *et al.*, 1966). Some 95% of the quinones are probably in the lamellae (Bucke *et al.*, 1966), although recent work suggests that they are mainly in the plastoglobuli (Tevini and Lichtenthaler, 1970).

The general pattern of synthesis in developing chloroplasts is indicated in Table 3 for illuminated dark grown maize seedlings and *Euglena* cultures [Threlfall and Griffiths, 1967; Threlfall *et al.*, 1967 (1); Threlfall and Goodwin, 1967]. There is a massive increase in all the quinones which correlates well with chlorophyll synthesis[3]. There is little, if any change in tocopherol levels (Threlfall and Griffiths, 1967; Hall and Laidman, 1966) or in ubiquinone, which is located in mitochondria and not in chloroplasts, and thus would not be expected to increase under these circumstances. Tocopherol levels do increase in whole leaves of *Ficus elastica* (Lichtenthaler, 1969).

[3] In normal seedlings chlorophyll is a good guide to chloroplast development. There are, however, mutants (e.g. *xantha*-3 barley) which do synthesize significant amounts of chlorophyll although their chloroplast structures are atypical (von Wettstein, 1961).

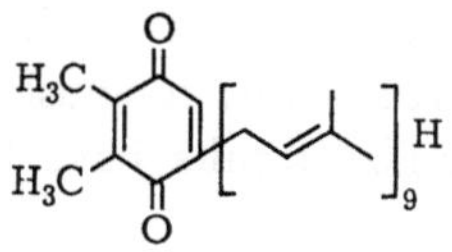

(XXVII)

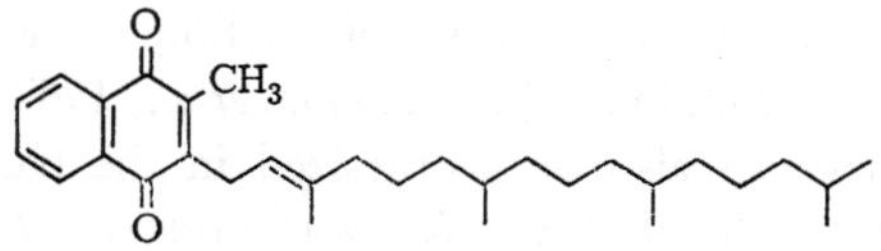

(XXVIII)

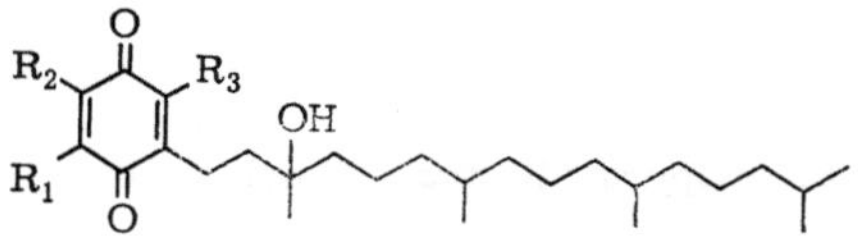

α-Tocopherolquinone $R_1=R_2=R_3=CH_3$
β-Tocopherolquinone $R_1=R_3=CH_3$; R=H
γ-Tocopherolquinone $R_1=R_2=CH_3$; R_3=H

(XXIX)

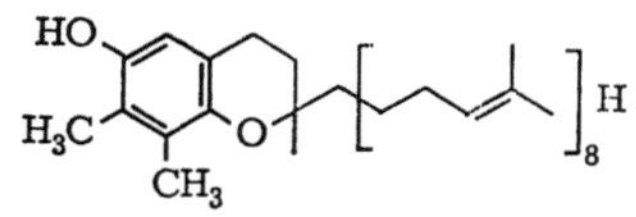

(XXX)

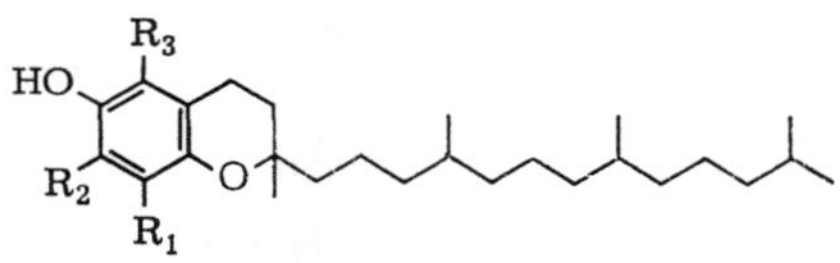

(XXXI)

(see (XXIX), for designations of R_1, R_2 and R_3)

Plastoquinone is synthesized *in toto* in the chloroplast. The pattern of labelling from (2-^{14}C) MVA and CO_2 in greening etiolated *Euglena* cells and maize seedlings is the same as with other terpenoids synthesized in the chloroplast (see p. 261); MVA is poorly incorporated while CO_2 is specifically and effectively incorporated. Furthermore, when plastoquinone isolated from the $^{14}CO_2$ experiment was degraded, the specific activity of the carbon atoms of the ring system was the same as that of the carbon atoms of the side chain [THRELFALL *et al.*, 1967 (2)].

Table 3. *The synthesis of chloroplast quinones and ubiquinone when etiolated tissues are illuminated for 24 h*

	Maize 6 days shoots		*Euglena gracilis Z* 7 days cultures	
	before illumination	after illumination	before illumination	after illumination
	(μg/100 shoots)		(μg/g dry wt.)	
Chlorophyll	0	23000	0	1900
Plastoquinone	241	942	17	103
α-Tocopherolquinone	0	413	0	44
Ubiquinone	220	275	234	261

The pattern of synthesis of phylloquinone and tocopherolquinone is similar to that of plastoquinone [THRELFALL *et al.*, 1967 (1)] and the reasonable conclusion is that they can also be synthesized in the chloroplasts. Similar results have been obtained with *Nicotiana tabacum* (GRIFFITHS *et al.*, 1968).

(^{14}C) Shikimate is effectively incorporated into these quinones and into the tocopherols [WHISTANCE *et al.*, 1966 (1)] and so is (^{14}C) tyrosine. Experiments with (β-^{14}C) tyrosine and *p*-hydroxyphenyl-(3-^{14}C) pyruvate has revealed that label is

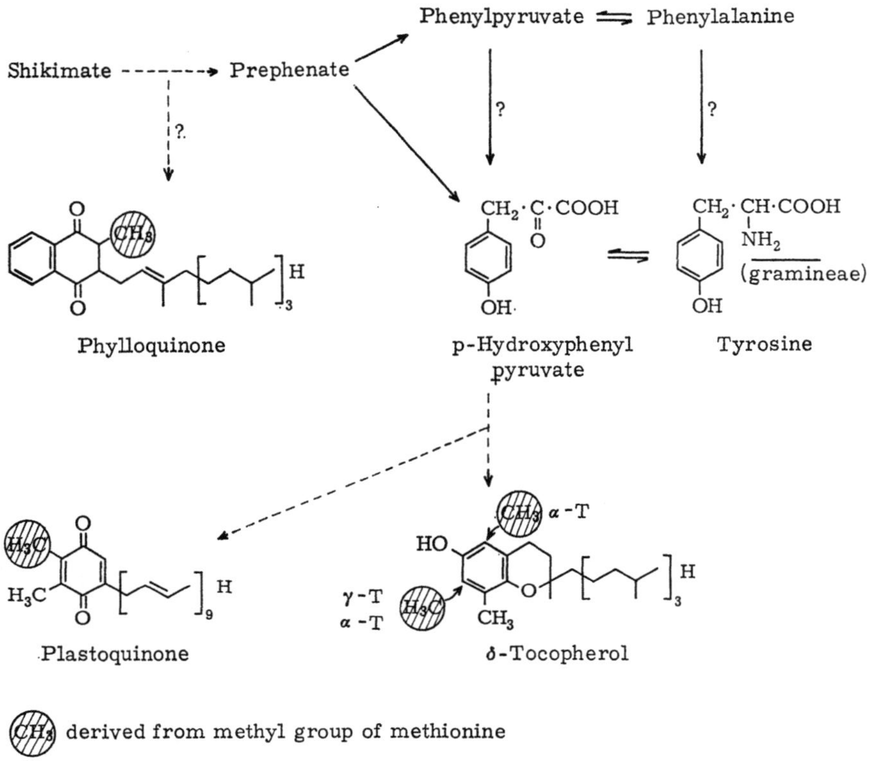

Fig. 17. The probable biosynthetic route for plastoquinones and tocopherols

incorporated specifically into one of the three ring methyl groups of plastoquinone and the single ring methyl group of phylloquinone and δ-tocopherol [THRELFALL *et al.*, 1967 (3); THRELFALL and WHISTANCE, 1968]. The other ring methyl groups in plastoquinone and in α-tocopherol arise from methionine (WHISTANCE and THRELFALL, 1967) and experiments with (^{14}C-$^{3}H_2$-methyl) methionine has led to the view that methylation arises by direct transfer of the intact methyl group (WHISTANCE *et al.*, 1967). All these experiments indicate that an intramolecular rearrangement takes place when *p*-hydroxyphenylpyruvate (from tyrosine or shikimate) is converted into the plastid quinones (Fig. 17). The rearrangement would be analogous to that involved in homogentisic acid in the conversion of tyrosine into melanin.

Indeed homogentisic acid has now been shown to be a precursor of plastoquinones, tocopherols, and α-tocopherol quinones (WHISTANCE and THRELFALL, 1968, 1970).

The failure to incorporate *p*-hydroxybenzoic acid into plastoquinone in plants although it is readily incorporated into ubiquinone [THRELFALL *et al.*, 1966 (1, 2), 1967 (3)] as previously reported for microorganisms (PARSON and RUDNEY, 1964), and animals (OLSON *et al.*, 1963), combined with the results with tyrosine, clearly

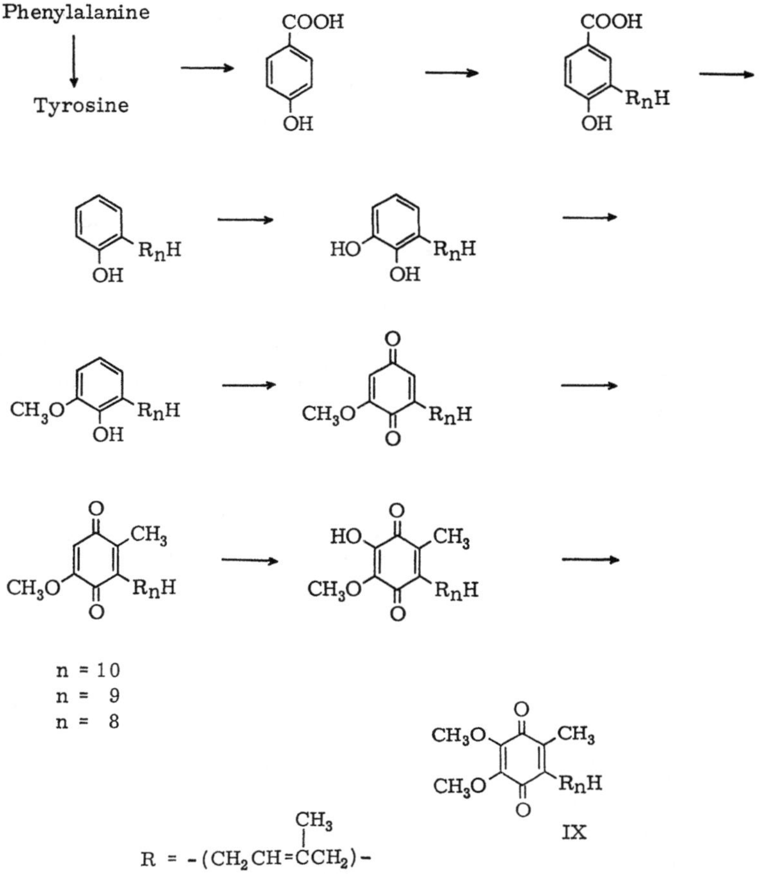

Fig. 18. The probable route for the biosynthesis of ubiquinone from phenylalanine

differentiate the biosynthetic pathways of the two compounds. A possible pathway for ubiquinone synthesis has recently been proposed (Fig. 18) (FRIIS *et al.*, 1966; OLSON *et al.*, 1966).

E. Polyprenols

The first chloroplast polyprenol to be fully characterized was solanesol, an all *trans* C-45 alcohol, which is abundant in tobacco leaf chloroplasts (STEVENSON *et al.*,

1963). Further investigations revealed that a series of polyprenols with 9 to 13 isoprene residues are present in the chloroplasts of a number of plants; they have been named castaprenols and ficaprenols (WELLBURN and HEMMING, 1966; WELLBURN *et al.*, 1967; STONE *et al.*, 1967) and contain both *cis-* and *trans*-double bonds and structures such as (XXXIII) (castaprenol-12) have been suggested. Similar, if not idential, prenols are present in the wood of *Betula verrucosa* (LINDGREN, 1964).

There is an increase in the content of castaprenols in horse chestnut leaves during ageing. This can be accounted for by the increase in number of osmiophilic globules in the chloroplast with age, because the castaprenols are located mainly in the chloroplast globules although some are found in the lamellae.

(2-^{14}C) MVA is incorporated into castaprenols in 12 week-old horse-chestnut leaves to a slight extent (about 1% of the incorporation into sterols) but to a somewhat greater extent in older leaves. This is an opposite trend to that observed with chlorophylls (see p. 263).

The pattern of incorporation of (2-^{14}C)-4 R(4-$^3H_1$) MVA into these compounds indicates that both *cis* and *trans* double bonds are present in the molecule. As structural studies suggest that the first four isoprenoid units in the castaprenols are *trans*- and the remainder probably *cis*, it may be that the first four units of these compounds arise from *trans*-geranylgeranyl pyrophosphate in the usual way but that further isoprenoid units are added in the *cis* configuration by unique chain-elongating enzymes (see HEMMING, 1970).

CH_3—C(CH_3)=CHCH_2—[—CH_2—C(CH_3)=CH—CH_2]$_{10}$—CH_2—C(CH_3)=CHCH_2OH

cis, trans — 6 *cis-4 trans* — *trans*

(XXXIII)

F. Sterols

Sterols are found both inside and outside the chloroplast, although the two fractions may be qualitatively different. For example, the sterol which is tightly bound to the chloroplasts of beans is cholesterol. This sterol represents some 15% of the total chloroplast sterol which in turn represents about 0.5% of the dry weight of the chloroplast (MERCER and TREHARNE, 1966).

Although there may be limited synthesis of sterol in chloroplasts (see later in the section) the main site of synthesis is outside the chloroplast. Evidence for this statement (see GOODWIN, 1966; KNAPP *et al.*, 1969) includes (i) synthesis of sterols is active in etiolated seedlings in the dark and illumination of the seedlings actually reduces the rate of synthesis; (ii) incorporation of (2-^{14}C)—MVA into greening seedlings is very marked whilst that of CO_2 is very limited—exactly the opposite situation to that observed with the plastid terpenoids (see next section); (iii) chloroplast-free systems will rapidly incorporate (2-^{14}C) MVA into the sterol precursor although little is incorporated into phytoene; on the other hand purified chloroplasts are feeble in synthesizing squalene, although very active in synthesizing phytoene (CHARLTON *et al.*, 1967).

A small fraction (4 to 5%) of sterols in plants is esterified and this is always very rapidly labelled from (2-^{14}C) MVA and is clearly in a state of rapid turnover. It is possible that this fraction may be "transfer sterol" moving newly synthesized material from microsomes to other cell organelles including the chloroplast (GOODWIN, 1967).

As just indicated, a very limited synthesis of squalene occurs in chloroplasts isolated by the non-aqueous technique (CHARLTON *et al.*, 1967). This means that either the chloroplast preparations were slightly contaminated with a cytoplasmic squalene synthesizing system or that chloroplasts can carry out limited synthesis of sterols. With regard to the first possibility HEBER (1960) indicates that about 6% of the total protein of non-aqueously isolated chloroplasts is cytoplasmic in origin. If chloroplasts can synthesize sterols, it is tempting to suggest that they synthesize cholesterol whilst the major, cytoplasmic system synthesizes the characteristic phytosterols which carry a supernumerary side chain at C-24.

G. Regulation of Terpenoid Synthesis

1. In Developing Chloroplasts

Evidence which has accumulated over the past few years on the biosynthesis of terpenoids in developing chloroplasts has led to the elaboration of a hypothesis to explain the control of terpenoid synthesis which is based on the simultaneous functioning of two possible regulatory mechanisms: (i) enzyme segregation and (ii) specific membrane permeability (GOODWIN and MERCER, 1963). In the first case it is assumed that the enzymes concerned with chain elongation of terpenoids are present both inside and outside the chloroplast, whilst those concerned with specific steps in the synthesis of plastid terpenoids (e.g. carotenoids) are found only inside the chloroplast. Secondly, it is assumed that the limiting membrane of the chloroplast is relatively impermeable either way to MVA, the first specific intermediate in terpenoid biosynthesis. The evidence for these views are now summarized.

Enzyme segregation. As indicated on p. 248, ROGERS *et al.* [1966 1, 2)] have shown that mevalonic kinase, the first enzyme concerned in the utilization of MVA, is present both inside and outside the chloroplasts. The crude enzymes appear to have different pH optima (7.5 for chloroplasts and 5.5 for cytoplasm) but this is apparently not so with purified enzyme (GREY and KEKWICK, 1969). The experiments of CHARLTON *et al.* (1967) indicate that phytoene synthetase activity is concentrated in the chloroplast while squalene synthetase is largely outside the chloroplast. The fact that the terpenoid side chain of various plastoquinones is very actively labelled from $^{14}CO_2$, while that of ubiquinone, a mitochondrial terpenoid quinone, is not (p. 258) indicates two different sites of synthesis of the C-45 and C-50 prenols. Furthermore the incorporation of $^{14}CO_2$ into terpenoids (mainly chloroplastidic) in greening etiolated seedlings depends on the stage of development of the chloroplasts, while incorporation of (^{14}C) MVA (mainly into extra-chloroplastidic terpenoids) is independent of chloroplast development (TREHARNE *et al.*, 1966). The situation is summarized in Fig. 19.

Membrane permeability. The differential incorporation of $^{14}CO_2$ and (2-^{14}C) MVA into chloroplastic and extra-chloroplastic terpenoids discussed above is circumstantial evidence for a permeability barrier at the limiting membrane of the chloroplast. Some evidence for this was obtained by ROGERS *et al.* (1965) who showed that isolated chloroplasts obtained by both aqueous and non-aqueous techniques would not phosphorylate MVA significantly until they were broken by ultrasonication; similarly MVA is incorporated into phytoene by non-aqueously prepared chloroplasts more efficiently when they are disrupted (CHARLTON *et al.*, 1967).

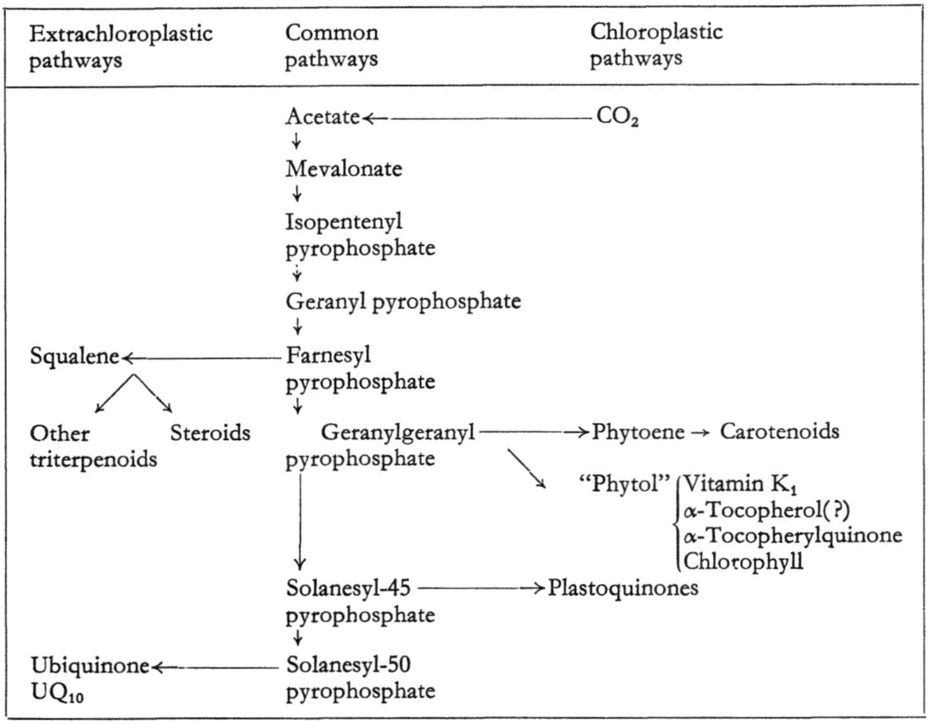

Fig. 19. Proposed scheme of segregation of enzymes concerned with terpenoid biosynthesis in developing seedlings. Centre column: enzymes present inside and outside the chloroplast; Right: enzymes present only or predominantly inside the chloroplast; Left: enzymes present only or predominantly outside the chloroplast

On this view of regulation, the situation during germination of seeds in the dark would be that synthesis of sterols required for membrane production occurs at the expense of endogenous food supplies from which MVA is synthesized via the conventional route; the MVA formed cannot significantly penetrate the immature plastid to synthesize pigments which are not required at this stage of development. As soon as a seedling emerges from soil and is exposed to light, the plastids rapidly become functional and fix CO_2 into MVA which is incorporated into the chloroplast terpenoids. Because of the barrier of the chloroplast membrane, MVA cannot pass into the cytoplasm in significant amounts to be channelled into the synthesis of triterpenoids including sterols which are not required at that time. Later, sucrose, formed in the chloroplast, can move out and provide the extraplastidic MVA needed to form sterols required for further growth of the plant. The problem in algae, such as *Scenedesmus* and *Chlorella*, which can produce fully functional chloroplasts in the dark remains to be examined. However, in pine seedling cotyledons which are green when the seeds are germinated in the dark, the MVA/CO_2 pattern of incorporation into plastid terpenoids in the light is the same as in other seedlings (WIECKOWSKI and GOODWIN, 1967).

2. In Mature Chloroplasts

The extent of incorporation of $^{14}CO_2$ into chloroplast terpenoids depends on the stage of development of the chloroplast; this means that when sufficient terpenoid

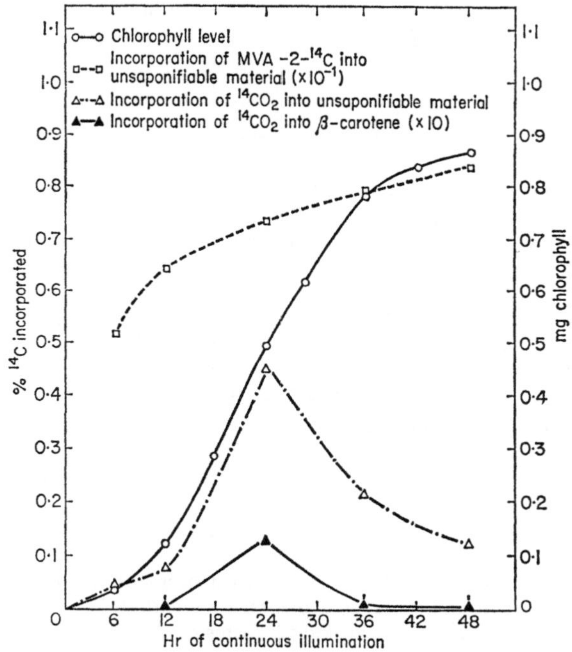

Fig. 20. Relationship between CO_2 uptake and MVA incorporation into terpenoids and the stage of chloroplast developmen in maize seedlings

material has been synthesized, a mechanism as yet unknown, comes into play which not only prevents synthesis of further terpenoids, but also almost completely eliminates "turnover" of CO_2 in the terpenoids. The pattern of incorporation into developing chloroplasts is indicated in Fig. 20. If maize seedlings are allowed to develop chloroplasts for different times before being exposed to $^{14}CO_2$ for 3 h or to (2-^{14}C) MVA for 6 h, then during the early stages of illumination, chloroplast development, as indicated by chlorophyll synthesis, parallels the amount of $^{14}CO_2$ fixed into the plastid terpenoids, while that from MVA is independent of chlorophyll concentration. As chloroplasts reach maturity (36 to 48 h illumination) chlorophyll synthesis slows down and CO_2 fixation drops precipitately (TREHARNE *et al.*, 1966). The turn-over of CO_2 in pigments of mature chloroplasts is generally a little greater in dicotyledons than in monocotyledons (PERKINS and ROBERTS, 1963).

3. Loss of Regulation during Changes of Chloroplasts into Chromoplasts

In some fruit, for example tomatoes and red peppers, the ripening process results in the conversion of chloroplasts into chromoplasts (FREY-WYSSLING and KREUTZER, 1958; KIRK and JUNIPER, 1967) with many well known associated changes such as loss of chlorophyll and accumulation of carotenoids. The means by which these massive biochemical changes are triggered off to coincide with the equally far-reaching morphological changes remain entirely unknown and offer a fascinating new field for investigation. Useful biochemical material would be the mutants with genes which retard the disappearance of chlorophyll (p. 243) and those with the lutescent gene which delays the onset of lycopene synthesis in tomatoes by up to 2 weeks (MACKINNEY, 1952).

H. Literature

ALBERTS, A. W., GOLDMAN, P., VAGELOS, P. R.: The condensation reaction of fatty acid synthesis. J. biol. Chem. **238**, 557—565 (1963).

ALLEN, C. F., HIRAYAMA, O., GOOD, P.: Lipid composition of photosynthetic systems. In: Biochemistry of chloroplasts, Vol. I, pp. 195—202. (T. W. GOODWIN, Ed.). London: Academic Press 1966.

ANDERSON, D. G., PORTER, J. W.: The biosynthesis of phytoene and other carotenes by enzymes of isolated higher plant plastids. Arch. Biochem. **97**, 509—528 (1962).

ANDERSON, I. C., ROBERTSON, D. W.: Role of carotenoids in protecting chlorophyll from photodestruction. Plant Physiol. **35**, 531—534 (1960).

APP, A. A., JAGENDORF, A. T.: (1) Incorporation of labelled amino acids by chloroplast ribosomes. Biochim. biophys. Acta (Amst.) **76**, 286—292 (1963).

— — (2) Repression of chloroplast development in *Euglena gracilis* by substrates. Protozool. **10**, 340—343 (1963).

ARDAO, C., VENNESLAND, B.: Chlorophyllase activity of spinach chloroplastin. Plant Physiol. **35**, 368—371 (1960).

BACON, J. S. D., PALMER, M. J., DE KOCK, P. C.: Measurement of aconitase activity in the leaves of various normal and variegated plants. Biochem. J. **78**, 198—203 (1963).

BALLSCHMITER, K., KATZ, J. J.: Long wavelength forms of chlorophyll. Nature (Lond.) **220**, 1231—1233 (1968).

BALTUS, E., QUERTIER, J.: A method for extraction and characterization of RNA from subcellular fractions of *Acetabularia*. Biochim. biophys. Acta (Amst.) **118**, 192 (1966).

BAMJI, M. S., KRINSKY, N. I.: Carotenoid de-epoxidations in algae II. Enzymatic conversion of antheraxanthin to zeaxanthin. J. biol. Chem. **240**, 467—470 (1965).

— JAGENDORF, A. T.: Amino acid incorporation by wheat chloroplasts. Plant Physiol. **41**, 764—770 (1966).

BASSHAM, J. A.: Photosynthesis: The path of carbon. In: Plant biochemistry, pp. 875—903. (J. BONNER, J. E. VARNER, Eds.). New York: Academic Press 1966.

BEELER, D., PORTER, J. W.: The enzymatic conversion of phytoene to phytofluene. Biochem. biophys. Res. Commun. **8**, 367—371 (1962).

BELL, W. D., BOGORAD, L., MCILRATH, W. J.: Response of the yellow-stripe maize mutant (ys_1) to ferrous and ferric iron. Botan. Gaz. **120**, 36—39 (1958).

BENSON, A. A., SHIBUYA, I.: Sulfocarbohydrate metabolism. Fed. Proc. **20**, 79 (1961).

BERIDZE, G., ODINTSOVA, M. S., CHERKASHINA, N. A., SISSAKIAN, N. M.: The effect of nucleic acid synthesis inhibitors on the chlorophyll formation by etiolated bean leaves. Biochem. biophys. Res. Commun. **23**, 683—689 (1966).

BEZINGER, E. N., MOLCHANOV, M. I., SISSAKIAN, N. M.: Significance of lipid compounds in the biosynthesis of chloroplast proteins. Biochymika **29**, 749—768 (1964).

BLOCH, K., CONSTANTOPOULOS, G., KENYON, C., NAGAI, J.: Lipid metabolism of algae in the light and in the dark. In: Biochemistry of chloroplasts, Vol. II, pp. 197—212 (T. W. GOODWIN, Ed.). London: Academic Press 1967.

BOARDMAN, N. K.: Studies on a protochlorophyll-protein complex II. The photo-conversion of protochlorophyll to chlorophyll *a* in the isolated complex. Biochim. biophys. Acta (Amst.) **64**, 279—293 (1962).

— Ribosome composition and chloroplast development in *Phaseolus vulgaris*. Exp. Cell Res. **43**, 474—482 (1966).

— FRANCKI, R. I. B., WILDMAN, S. G.: Protein synthesis by cell-free extracts of tobacco leaves III. Comparison of the physical properties and protein synthesizing activities of 70s chloroplast and 80s cytoplasmic ribosomes. J. molec. Biol. **17**, 470—484 (1966).

— Chloroplast structure and development. In: Harvesting the sun, pp. 211—230 (A. SAN PIETRO, F. A. GREER, T. J. ARMY, Eds.). New York: Academic Press 1967.

BOGORAD, L.: The biosynthesis of protochlorophyll. In: Comparative biochemistry of photoreactive systems, pp. 227—256 (M. B. ALLEN, Ed.). New York: Academic Press 1960.

— Porphyrin synthesis. In: Methods in Enzymology, Vol. V, pp. 885—895. (S. P. COLOWICK, N. O. KAPLAN, Eds.). New York-London: 1962.

— JACOBSON, A.: Inhibition of greening of etiolated leaves by actinomycin D. Biochem. biophys. Res. Commun. **14**, 113—117 (1964).

— Chlorophyll biosynthesis. In: Chemistry and biochemistry of plant pigments, pp. 29—74 (T. W. GOODWIN, Ed.). London: Academic Press 1965.

— (1) Biosynthesis and morphogenesis in plastids. In: Biochemistry of chloroplasts, Vol. II, pp. 615—632 (T. W. GOODWIN, Ed.). London: Academic Press 1967.

— (2) Chloroplast structure and development. In: Harvesting the sun, pp. 191—210 (A. SAN PIETRO, F. A. GREER, T. J. ARMY, Eds.). New York: Academic Press 1967.

BOVÉ, J., RAACKE, I. D.: Amino acid-activating enzymes in isolated chloroplasts from spinach leaves. Arch. Biochem. **85**, 521—531 (1958).

BRACHET, J.: Protein synthesis in the absence of the nucleus. Nature (Lond.) **213**, 650—655 (1961).

BRAWERMAN, G., EISENSTADT, J. M.: Template and ribosomal ribonucleic acids associated with the chloroplast and the cytoplasm of *Euglena gracilis*. J. molec. Biol. **10**, 403—411 (1964).

— Nucleic acid associated with the chloroplasts of *Euglena gracilis*. In: Biochemistry of chloroplasts, Vol. I, pp. 301—318 (T. W. GOODWIN, Ed.). London: Academic Press 1966.

BREIDENBACH, R. W., KAHN, A., BEEVERS, H.: Characterization of glyoxysomes from castor bean endosperm. Plant Physiol. **43**, 705—713 (1968).

BROCK, D. J. H., KASS, L. R., BLOCH, K.: The role of β-hydroxydecanol thiol ester dehydrase in fatty acid biosynthesis in *E. coli*. Fed. Proc. **25**, 340 (1966).

BROOKS, J. L., STUMPF, P. K.: Fat metabolism in higher plants XXXIX. Properties of a soluble fatty acid synthesizing system from lettuce chloroplasts. Arch. Biochem. **116**, 108—116 (1966).

BUCKE, C., LEECH, R. M., HALLAWAY, M., MORTON, R. A.: The taxonomic distribution of plastoquinone and tocopherolquinone and their intracellular distribution in leaves of *Vicia faba* L. Biochim. biophys. Acta (Amst.) **112**, 19—34 (1966).

BULL, M. J., LASCELLES, J.: The association of protein synthesis with the formation of pigments in some photosynthetic bacteria. Biochem. J. **87**, 15—28 (1963).

BURKARD, G., ECHLANCHAR, B., WEIL, J. M.: Presence of N-formyl-methionyl-t RNA in bean chloroplasts. FEBs Letters **4**, 571—574 (1967).

BUSH, L. P.: Influence of certain cations on activity of succinyl CoA synthetase from tobacco Plant. Physiol. **44**, 347—350 (1969).

BUTLER, W. L., BRIGGS, W. R.: The relation between structure and pigments during the first stages of proplastid greening. Biochim. biophys. Acta (Amst.) **112**, 45 (1966).

CAHN, R. S., INGOLD, C. K.: Specification of configuration about quadricovalent asymmetric atoms. J. chem. Soc. **1951**, 612—622.

Carell, E. F., Kahn, J. S., Synthesis of porphyrins by isolated chloroplasts of *Euglena*. Arch. Biochem. **108**, 1—6 (1964).
Chang S. B., Kulkarni, N. D.: Enzymatic reactions for galactolipid synthesis with a soluble, sub-chloroplast fraction from *Spinacia oleracea*. Phytochemistry **9**, 927—934 (1970).
Chang, W. H., Tolbert, N. E.: Distribution of C^{14} in serine and glycine after $C^{14}O_2$ photosynthesis by isolated chloroplasts. Modification of serine-C^{14} degradation. Plant Physiol. **40**, 1048—1052 (1965).
Charlton, J. M., Treharne, K. J., Goodwin, T. W.: Incorporation of (2-^{14}C) mevalonate into terpenoids by chloroplast preparations. Biochem. J. **101**, 7P—8P (1966).
Chen, J. L., Wildman, S. G.: Functional chloroplast polyribosomes from tobacco leaves. Science **155**, 1271—1273 (1967).
Chesterton, C. J., Kekwick, R. G. O.: Evidence for the presence of Δ^3-isopentenyl pyrophosphate isomerase activity from the formation of squalene in *Hevea brasiliensis* latex. Biochem. J. **100**, 56P-57P (1966).
Chiang, K. S., Sueoka, N.: Replication of chloroplast DNA in *Chlamydomonas reinhardi* during vegetative cell cycle; its mode and regulation. Proc. nat. Acad. Sci. (Wash.) **57**, 1506—1513 (1967).
Claes, H.: Biosynthese von Carotinoiden bei *Chlorella*. Z. Naturforsch. **12b**, 401—407 (1957).
— Biosynthese von Carotenoiden bei *Chlorella*. Z. Naturforsch. **13b**, 222—224 (1958).
— Biosynthese von Carotenoiden bei *Chlorella*. Z. Naturforsch. **14b**, 4—7 (1959).
Clark, M. F., Matthews, R. E. F., Ralph, R. K.: Ribosomes and polyribosomes in *Brassica pekinensis*. Biochim. biophys. Acta (Amst.) **91**, 289—304 (1964).
Cornforth, J. W., Popják, G.: Substrate stereochemistry in squalene biosynthesis. Biochem. J. **191**, 553—568 (1966).
— Cornforth, R. H., Donninger, C., Popják, G.: Studies on the biosynthesis of cholesterol XIX. Steric course of hydrogen eliminations and of C-C bond formation in squalene biosynthesis. Proc. roy. Soc. B **163**, 492—514 (1966).
Costes, C.: Metabolism et role physiologique des carotenoids dans les feuilles vertes. Ann. Physiol. véget. **7**, 105—142 (1965).
Crane, F. L., Henninger, M. D., Wood, P. M., Barr, R.: Quinones in chloroplasts. In: Biochemistry of chloroplasts, Vol. I, pp. 133—152 (T. W. Goodwin, Ed.). London: Academic Press 1966.
Daniel, H., Miyana, M., Mumma, R. O., Yagi, T., Lepage, M., Shibuya, I., Benson, A. A.: The plant sulfolipid. Identification of 6-sulfo-quinovose. J. Amer. chem. Soc. **83**, 175—176 (1961).
Davies, B. H.: Biosynthesis of carotenoids. Biochem. J. **85**, 2P (1963).
— Villoutreix, J., Williams, R. J. H., Goodwin, T. W.: The possible role of β-zeacarotene in carotenoid cyclization. Biochem. J. **89**, 96P (1963).
Davies, J. B., Jackman, L. M., Siddons, P. T., Weedon, B. C. L.: The structures of phytoene, phytofluene, β-carotene, and neurosporene. Proc. Chem. Soc. **1961**, 261—263.
Davis, W. H., Mercer, E. I., Goodwin, T. W.: The occurrence and intracellular distribution of the plant sulfolipid in maize, runner beans, plant tissue cultures, and *Euglena gracilis*. Phytochemistry **4**, 741—749 (1965).
— — Some observations on the biosynthesis of the plant sulfolipid by *Euglena gracilis*. Biochem. J. **98**, 369—373 (1966).
Decker, K., Ueheleke, H.: Eine enzymatische Isomerisierung von Licopin und β-Carotin. Hoppe-Seylers Z. physiol. Chem., **323**, 61—76 (1961).
Dunphy, P. J., Whittle, K. J., Pennock, J. F.: Plastochomanol. In: Biochemistry of chloroplasts, Vol. I, pp. 165—172 (T. W. Goodwin, Ed.). London: Academic Press 1966.
Ebbon, J. G., Tait, C. H.: Studies on S-adenosyl-magnesium protoporphyrin methyltransferase in *Euglena gracilis*. Biochem. J. **111**, 573—582 (1969).
Edelman, M., Schiff, J. A., Epstein, H. T.: Studies of chloroplast development in *Euglena*. XII. Two types of satellite DNA. J. molec. Biol. **11**, 769—774 (1965).
Eisenstadt, J. M., Brawerman, G.: The incorporation of amino acids into the protein of chloroplasts and chloroplast ribosomes of *Euglena gracilis*. Biochim. biophys. Acta (Amst.) **76**, 319—321 (1963).

— — The protein-synthesizing systems from the cytoplasm and the chloroplast of *Euglena gracilis*. J. molec. Biol. **10**, 392—402 (1964).
— Protein synthesis in chloroplasts and chloroplast ribosomes. In: Biochemistry of chloroplasts, Vol. II, pp. 341—350 (T. W. Goodwin, Ed.). London: Academic Press 1967.
Ellis, R. J.: Chloroplast ribosomes: Stereospecificity of inhibition by chloramphenicol. Science **163**, 477—478 (1969).
Ellsworth, R. K., Aronoff, S.: Investigations on the biogenesis of chlorophyll *a*. III. Biosynthesis of Mg-vinylpheoporphine *a*, methylester from Mg-protoporphine IX monomethylester as observed in *Chlorella* mutants. Arch. Biochem. **125**, 269—277 (1968).
— — Investigations on the biogenesis of chlorophyll *a*. IV. Isolation and partial characterization of some biosynthe ticintermediates obtained from *Chlorella* mutants. Arch. Biochem. **130**, 374—383 (1969).
Faludi-Daniel, A., Lang, F., Fràdkin, L. I.: The state of chlorophyll *a* in leaves of carotenoid mutant maize. In: Biochemistry of chloroplasts, Vol. I, pp. 269—274 (T. W. Goodwin, Ed.). London: Academic Press 1966.
Ferrari, R. A., Benson, A. A.: The paths of carbon in photosynthesis of the lipids. Arch. Biochem. **93**, 185—192 (1961).
Fischer, F. G., Bohn, H.: Über die Bildung und das Vorkommen von Phytol. Ann. Chem. **611**, 224—235 (1958).
Frank, S., Kenney, A. L.: Chlorophyll and carotenoid destruction in the absence of light in seedlings of *Zea mays* L. Plant Physiol. **30**, 413—418 (1955).
Frey-Wyssling, A., Kreutzer, E.: The submicroscope development of chromoplasts in the fruit of *Capsicum annum* L. J. Ultrastruct. Res. **1**, 397—411 (1958).
Friis, P., Davies, G. D., Folkers, K.: Complete sequence of biosynthesis from p-hydroxybenzoic acid to ubiquinone. J. Amer. chem. Soc. **88**, 4754—4756 (1966).
Gassman, M., Bogorad, L.: Control of chlorophyll production in rapidly greening bean leaves. Plant Physiol. **42**, 774—780 (1967).
Gibbons, G. C., Raison, J., Smillie, R. M.: Proc. Aust. Biochem. Soc. **43**. (Quoted by Kirk, J. T. O., 1970 (1969).
Gibor, A.: Chloroplast heredity and nucleic acids. Amer. Nat. **99**, 229—239 (1965).
— DNA synthesis in chloroplasts. In: Biochemistry of chloroplasts, Vol. II, pp. 321—328 (T. W. Goodwin, Ed.). London: Academic Press 1967.
— Granick, S.: Plastids and mitochondria: inheritable systems. Science **145**, 890—897 (1964).
Godnev, T. N., Akulovich, N. K.: Nature of pigments of the protochlorophyll group. Dokl. Akad. Nauk SSSR. (transl.) **134**, 710—712 (1960).
— — Khodasevich, E. V.: Participation of esterified and nonesterified forms of protochlorophyll of etiolated sprouts in formation of chlorophyll *a*. Dokl. Akad. Nauk SSSR. (transl.) **150**, 590—593 (1963).
Godnev, T. N., Galaktionov, S. G., Raskin, V. I.: Dokl Akad. Nauk SSSR. **181**, 167 (1968). (Quoted by Kirk, J. T. O., 1970).
— Kondrateva, E. N., Usbenkaya, V. E.: Possible pathways of the biosynthesis of bacterioviridin (*Chlorobium*-chlorophyll). Izv. Akad. Nauk SSSR. Ser. Biol. **1966**, 525—531.
— Rotfarb, R. M.: Lycopene as the probable precursor of other carotenoids. Dokl. Akad. Nauk SSSR. **147**, 962—963 (1962).
Goedheer, J. C.: Effect of changes in chlorophyll concentration on photosynthetic properties. I. Fluorescence and absorption of greening bean leaves. Biochim. biophys. Acta (Amst.) **51**, 494—504 (1961).
Goldman, P., Alberts, A. W., Vagelos, P. R.: The condensation reaction of fatty acid synthesis. J. biol. Chem. **238**, 3579—3583 (1963).
Goodwin, T. W.: The comparative biochemistry of carotenoids. London: Chapman & Hall 1952.
— Jamikorn, M.: Biosynthesis of carotenes in ripening tomatoes. Nature (Lond.) **170**, 104—105 (1952).
— Land, D. G., Sissons, M. E.: Studies in carotenogenesis. Biochem. J. **64**, 486—491 (1956).

— Phagpolngarm, S.: Studies in carotenogenesis. 28. The effect of illumination on carotenoid synthesis in french bean *(Phaseolus vulgaris)* seedlings. Biochem. J. **76** 197—199 (1960).
— Mercer, E. I.: The regulation of sterol and carotenoid metabolism in germinating seedlings. In: Control of lipid metabolism. Biochem. Soc. Symp. No. 24 (J. K. Grant, Ed.). London: Academic Press 1963.
— (Ed.): Biochemistry of plant pigments. London: Academic Press 1965.
— Williams, R. J. H.: (1) A mechanism for the cyclization of an acyclic precursor to form β-carotene. Biochem. J. **94**, 5C—7C (1965).
— (2) A mechanism for the biosynthesis of α-carotene. Biochem. J. **97**, 28C (1965).
— The biosynthesis of carotenoids (pp. 36—57). Regulation of tepenoid synthesis in higher plants (pp. 57—72). In: Biosynthetic pathways in higher plants (J. B. Pridham, T .Swain, Eds.). London: Academic Press 1966.
— Williams, R. J. H.: The stereochemistry of phytoene biosynthesis. Proc. roy. Soc. B **163**, 515—518 (1966).
— Terpenoids and chloroplast development. In: Biochemistry of chloroplasts, Vol. II, pp. 721—734 (T. W. Goodwin, Ed.). London: Academic Press 1967.
— Britton, G., Walton, T. J.: Xanthophyll biosynthesis in maize seedlings. Plant Physiol. **43**, S-46 (1968).
Granick, S.: The structural and functional relationships between heme and chlorophyll. Harvey Lectures **44**, 220—245 (1948).
— Magnesium protoporphyrin monoester and protoporphyrin monomethylester in chlorophyll biosynthesis. J. biol. Chem. **236**, 1168—1172 (1961).
— The induction *in vitro* of the synthesis of δ-aminolevulinic acid synthetase in chemical porphyria: A response to certain drugs, sex hormones, and foreign chemicals. J. biol. Chem. **241**, 1359—1375 (1966).
— The heme and chlorophyll biosynthetic chain. In: Biochemistry of chloroplasts, Vol. II, pp. 373—410 (T. W. Goodwin, Ed.). London: Academic Press 1967.
— Mauzerall, D.: Porphyrin biosynthesis in erythrocytes II. Enzymes converting δ-aminolevulinic acid to coproporphyrinogen. J. biol. Chem. **232**, 1119—1140 (1958).
— Sano, S.: Mitochondrial coproporphyrinogen oxidase and the formation of protoporphyrin. Fed. Proc. **20**, 376 (1961).
Grey, J. C., Kekwick, R. G. O.: Mevalonate kinase from etiolated cotyledons of french beans. Biochem. J. **113**, 37-P (1969).
Griffiths, W. T., Threlfall, D. R., Goodwin, T. W.: Observations on the nature and biosynthesis of terpenoid quinones and related compounds in tobacco shoots. Europ. J. Biochem. **5**, 124 (1968).
— — — Nature, intracellular distribution and formation of terpenoid quinones in maize and barley shoots. Biochem. J. **103**, 589—600 (1967).
Gunning, B. E. S.: The fine structure of chloroplast stroma following aldehyde osmium tetroxide fixation. J. Cell Biol. **24**, 79—93 (1965).
Gyldenholm, A. O.: Unpublished work quoted by D. Wettstein, (q.v.).
Habermann, H. M.: Spectra of normal and pigment-deficient mutant leaves of *Helianthus annuus* L. Physiol. Plantarum **13**, 718—725 (1960).
Hager, L. P.: Succinyl CoA synthetase. In: The enzymes, Vol. VI, pp. 387—400 (P. D. Boyer, H. A. Lardy, K. Myrbäck, Eds.). New York: Academic Press 1962.
Hall, G. S., Laidman, D. L.: Tocopherols and ubiquinone in the germinating wheat grain. Biochem. J. **101**, 5P (1966).
Hanson, K. R.: Applications of the sequence rule. I. Naming the paired ligands g,g at a tetrahedral atom Xggij. II. Naming the two faces of a trigonal atom Yghi. J. Amer. chem. Soc. **88**, 2731—2742 (1966).
Harris, R. V., James, A. T., Harris, P.: Synthesis of unsaturated fatty acids by green algae and plant leaves. In: Biochemistry of chloroplasts, Vol. II, pp. 241—254 (T. W. Goodwin, Ed.). London: Academic Press 1967.
Haverkate, F., Van Deenen, L. L. M.: Isolation and chemical characterization of phosphotidyl glycerate from spinach leaves. Biochim. biophys. Acta (Amst.) **106**, 78—92 (1956).

Heber, U.: Vergleichende Untersuchungen an Chloroplasten, die durch Isolierungs-Operationen in nicht-wäßrigem und in wäßrigem Milieu erhalten wurden. Z. Naturforsch. **156,** 95—99, 100—109 (1960).
Hemming, F. W.: Polyprenols. In: Natural substances formed biologically from mevalonic acid (T. W. Goodwin, Ed.). Biochem. Soc. Symp. **29,** 105—117 (1970).
Henshall, J. D., Goodwin, T. W.: Amino acid-activating enzymes in germinating pea seedlings. Phytochemistry **3,** 677—691 (1964).
— — The effect of red and far red light on carotenoid and chlorophyll formation in pea seedlings. Photochem. Photobiol. **3,** 234—247 (1965).
Hiatt, A. J.: Condensing enzyme from higher plants. Plant Physiol. **37,** 85—89 (1962).
Hill, M., Rogers, L. J.: Conversion of lycopene to β-carotene by chloroplasts of higher plants. Biochem. J. **113,** 31 P (1969).
Hiller, R. G.: Studies on the incorporation of $^{14}CO_2$ into glutamic acid in *Chlorella pyrenoidosa*. Phytochemistry **3,** 569—579 (1964).
Holt, A. A.: Recently characterized chlorophylls. In: The chlorophylls, Vol. III, pp. 1115—1118 (L. P. Vernon, G. R. Seely, Eds.). New York: Academic Press 1966.
Hoober, J. K., Siekevitz, P.: Effects of chloramphenicol and cyclaheximide on chloroplast membrane formation in *Chlamydomonas reinhardi*. J. Cell Biol. **39,** 12 A (1968).
Hopkins, C. Y., Chisholm, J.: Occurrence of trans-3-hexadecanoic acid in a seed oil. Canad. J. Chem. **42,** 2224—2227 (1964).
Hsu, W. P., Miller, G. W.: Chlorophyll and porphyrin synthesis in relation to iron in *Nicotiana tabacum*, L. Biochim. biophys. Acta (Amst.) **111,** 393—402 (1965).
Hudock, G. A., Fuller, R. C.: Control of triosephosphate dehydrogenase in photosynthesis. Plant Physiol. **40,** 1205—1211 (1965).
Huffaker, R. C., Oberdorf, R. B., Keller, C. J., Kleinkoff, G. E.: *In vivo* light stimulation of carboxylative phase enzyme activities in greening barley. Plant Physiol. **39,** S-14—S-15 (1964).
James, A. T., Harris, R. V., Harris, P.: The biosynthesis of oleic acid in a green alga. Biochem. J. **95,** 6 P (1965).
Jensen, S. L.: Biosynthesis and function of carotenoid pigments in microorganisms. Ann. Rev. Microbiol. **19,** 167—182 (1965).
Johnson, U. E., Togasaki, R. K., Levine, R. P.: A mutant strain of *Chlamydomonas* defective in chloroplast ribosome production. Plant Physiol. **43,** S-6 (1968).
Jones, O. T. G.: Magnesium 2,4-divinylphaeoporphyrin a_5 monomethyl ester, a protochlorophyll-like pigment produced by *Rhodopseudomonas spheroides*. Biochem. J. **89,** 182—188 (1963).
— A protein-protochlorophyll complex obtained from inner seed coats of *Cucurbita pepo*. Biochem. J. **101,** 153—160 (1966).
— Biosynthesis of chlorophylls. In: Porphyrins and related compounds (T. W. Goodwin, Ed.). Biochem. Soc. Symp. No. 28, p. 131. London: Academic Press 1968.
Jungalwala, F. B., Porter, J. W.: The biosynthesis of phytoene from isopentenyl and farnesyl pyrophosphate by a partially purified tomato enzyme system. Plant. Physiol. **40,** S-18 (1965).
Kappas, A., Granick, S.: Steroid induction of porphyrin synthesis in liver cell culture II. The effects of heme, uridine diphosphate glucuronic acid, and inhibitors of nucleic acid and protein synthesis on the induction process. J. biol. Chem. **243,** 346—351 (1968).
Karlsson, U., Miller, A., Boardman, N. K.: Electron microscopy of ribosomes isolated from tobacco leaves. J. molec. Biol. **17,** 487—489 (1966).
Kawashima, N., Wildman, S. G.: Fraction I protein. Ann. Rev. Plant Physiol. **21,** 325—358 (1970).
Kay, R. E., Phinney, B. O.: The control of plastid pigment formation by a virescent gene, pale-yellow-1 of maize. Plant Physiol. **31,** 415 (1956).
Kikuchi, G., Kumin, A., Talmage, P., Shemin, D.: The enzymatic synthesis of δ-aminolevulinic acid. J. biol. Chem. **233,** 1214—1219 (1958).
Kirk, J. T. O.: Studies on RNA synthesis in chloroplast preparations. Biochem. biophys. Res. Commun. **16,** 233—238 (1964).
— Studies on the dependence of chlorophyll synthesis on protein synthesis in *Euglena gracilis*. Planta **78,** 200—207 (1968).

— Biochemical aspects of chloroplast development. Ann. Rev. Plant Physiol. **21**, 1—42 (1970).
— Allen, R. L.: Dependence of chloroplast pigment synthesis on protein synthesis. Biochem. biophys. Res. Commun. **21**, 523—530 (1965).
— Juniper, B. E.: The ultrastructure of the chromoplasts of different colour varieties of *Capsicum*. In: Biochemistry of chloroplasts, Vol. II, pp. 691—702 (T. W. Goodwin, Ed.). London: Academic Press 1967.
— Tilney-Bassett, R. A. E.: The plastids. London: W. H. Freeman & Co. 1967.
Klein, A. O.: Metabolism during light-induced expansion of etiolated leaves. Plant Physiol. **41**, lvi (1966).
— Vishniac, W.: Activity and partial purification of chlorophyllase in aqueous systems. J. biol. Chem. **236**, 2544—2547 (1961).
Knapp, F. F., Aexel, R. T., Nicholas, H. J.: Sterol biosynthesis in sub-cellular particles of higher plants. Plant Physiol. **44**, 442 (1969).
Koski, V. M., French, C. S., Smith, J. H. C.: The action spectrum for the transformation of protochlorophyll *a* in normal and albino corn seedlings. Arch. Biochem. **31**, 1—17 (1951).
Krinsky, N. I.: The role of carotenoid pigments as protective agents against photosensitized oxidations in chloroplasts. In: Biochemistry of chloroplasts, Vol. I, pp. 423—430 (T. W. Goodwin, Ed.). London: Academic Press 1966.
Kupke, D. W., Huntington, J. L.: Chlorophyll *a* appearance in the dark in higher plants: Analytical notes. Science **140**, 49—51 (1963).
Kushwaha, S. C., Subbarayan, C., Beeler, D. A., Porter, J. W.: The conversion of lycopene-15,15′-3H to cyclic carotenes by soluble extract of higher plant plastids. J. biol. Chem. **244**, 3635—3642 (1969).
Labbe, R. F., Hubbard, N.: Metal specificity of the iron-protoporphyrin chelating enzyme from rat liver. Biochim. biophys. Acta (Amst.) **52**, 130—135 (1961).
Lascelles, J.: Tetrapyrrole biosynthesis and its regulation. New York: W. A. Benjamin 1964.
— In: Biosynthetic pathways in higher plants, pp. 163—178 (J. B. Pridham, T. Swain, Eds.). London: Academic Press 1965.
Lee, T. G., Chichester, C. O.: Geranylgeranyl pyrophosphate as the condensing unit enzymate synthesis of carotenes. Phytochemistry **8**, 603—609 (1969).
Lennarz, W. J., Light, R. J., Bloch, K.: A fatty acid synthetase from *E. coli*. Proc. nat. Acad. Sci. (Wash.) **48**, 840—846 (1962).
Levin, E. Y.: The conversion of photohemochrome to verdohemochrome with liver homogenates. Biochim. biophys. Acta (Amst.) **136**, 155—158 (1967).
Lichtenthaler, H. K.: Die Bildung überschüssiger Plastidenchinone in den Blättern von *Ficus elastica*. Roxb., Z. Naturforsch. **24b**, 1461 (1970).
Lindgren, B. O.: Isolation from birch wood of an aliphatic oligoterpene alcohol, a methyl ester of a hydroxy-triterpene carboxylic acid and a triterpene alcohol. Acta chem. scand. **18**, 836—837 (1964).
Litvin, F. F., Belyaev, O. B.: Biokhimiya **33**, 754 (1968). (Quoted by Kirk, J. T. O., 1970.)
Loening, U. E.: RNA structure and metabolism. Ann. Rev. Plant Physiol. **19**, 37 (1968).
— Ingle, J.: Diversity of RNA components in green plant tissues. Nature (Lond.) **215**, 363—367 (1967).
Lynen, F.: The role of biotin-dependent carboxylations in biosynthetic reactions. Biochem. J. **102**, 381 (1967).
— Eggerer, H., Henning, U., Kessel, I.: Zur Biosynthese der Terpene. III. Farnesyl-pyrophosphat und 3-Methyl-Δ^3-butenyl-1-pyrophosphat, die biologischen Vorstufen des Squalens. Angew. Chem. **70**, 738—742 (1958).
— Agranoff, B. W., Eggerer, H., Henning, U., Moslein, E. M.: Zur Biosynthese der Terpene, IV. γ,γ-Dimethyl-allyl-pyrophosphat und Geranyl-pyrophosphat, biologische Vorstufen des Squalens. Angew. Chem. **71**, 657—684 (1959).
Mackinney, G.: Carotenoids. Ann. Rev. Biochem. **21**, 473—492 (1952).
Maclachlan, D., Zalick, S.: Plastid structure, chlorophyll concentration, and free amino acid composition of a chlorophyll mutant of barley. Canad. J. Bot. **41**, 1053—1063 (1963).

MADSEN, A.: On the formation of chlorophyll and initiation of photosynthesis in etiolated plants. Photochem. Photobiol. **2**, 93—100 (1963).
MAJERUS, P. W., ALBERTS, A. W., VAGELOS, P. R.: Acyl carrier protein IV. The identification of 4′-phosphopantetheine as the prosthetic group of the acyl carriei protein. Proc. nat. Acad. Sci. (Wash.) **53**, 410—417 (1965).
MARCUS, A.: Amino acid dependent exchange between pyrophosphate and adenosine triphosphate in spinach preparations. J. biol. Chem. **234**, 1238—1240 (1959).
MARGULIES, M. M.: Effect of chloramphenicol on light dependent development of seedlings of *Phaseolus vulgaris* var. black valentine, with particular reference to development of photosynthetic activity. Plant Physiol. **37**, 473—480 (1962).
— Effect of chloramphenicol on light-dependent synthesis of proteins and enzymes of leaves and chloroplasts of *Phaseolus vulgaris*. Plant Physiol. **39**, 579—585 (1964).
— Effect of chloramphenicol on chlorophyll synthesis of bean leaves. Plant Physiol. **42**, 218—220 (1967).
— PARENTI, F.: Synthesis of ribulose diphosphate carboxylase by chloroplasts *in vitro*. Plant Physiol. **43**, S-18 (1968).
MARSH, H. V., EVANS, H. J., MATRONE, G.: Investigation of the role of iron in chlorophyll metabolism I. Effect of iron deficiency on chlorophyll and heme content and on the activities of certain enzymes in leaves. Plant Physiol. **38**, 632—637 (1963).
MARRIOTT, J.: Regulation of porphyrin synthesis. In: Porphyrins and related compounds (T. W. GOODWIN, Ed.). Biochem. Soc. Symp. No. 28, pp. 61—74. London: Academic Press 1968.
MAUZERALL, D., GRANICK, S.: Porphyrin biosynthesis in erythrocytes. III. Uroporphyrinogen and its decarboxylase. J. biol. Chem. **232**, 1141—1162 (1958).
MEGO, J. L., JAGENDORF, A. T.: Effect of light on growth of black valentine bean plastids. Biochim. biophys. Acta (Amst.) **53**, 237—254 (1961).
MERCER, E. I., THOMAS, G.: The occurrence of ATP-adenosylsulphate-3′-phosphotransferase in the chloroplasts of higher plants. Phytochemistry **8**, 2281—2285 (1969).
— TREHARNE, K. J.: Occurrence of sterols in chloroplasts. In: Biochemistry of chloroplasts, Vol. I, pp. 181—186 (T. W. GOODWIN, Ed.). London: Academic Press 1966.
MIYACHI, S., MIYACHI, S.: Sulfolipid metabolism in *Chlorella*. Plant Physiol. **41**, 479—486 (1966).
MUDD, J. B., MCMANUS, T. T.: Metabolism of acetate by cell-free preparations from spinach leaves. J. biol. Chem. **237**, 2057—2063 (1962).
— — Participation of sulfhydryl groups in fatty acid synthesis by chloroplast preparations. Plant Physiol. **39**, 115—119 (1964).
— — Relationship of the syntheses of lipid and water-soluble acids by chloroplast preparations. Plant Physiol. **40**, 340—344 (1965).
MURPHY, R. F., OH'EOCHA, C., O'CARRA, P.: The formation of verdohaemochrome from pyridine protohaemochrome by extracts of red algae and of liver. Biochem. J. **104**, 6C—8C (1967).
NAGAI, J., BLOCH, K.: Synthesis of oleic acid by *Euglena gracilis*. J. biol. Chem. **240**, PC3702—PC3703 (1965).
NAKAJIMA, H., TAKAMURA, T., NAKAJIMA, O., YAMNOKO, K.: Studies on heme α-methenyl oxygenase I. The enzymatic conversion of pyridine-hemichromogen and hemoglobin-haptoglobin into a possible precursor of biliverdin. Biol. Chem. **238**, 3784—3796 (1963).
NEUFELD, E. F., HALL, C. W.: Formation of galactolipids by chloroplasts. Biochem. biophys. Res. Commun. **14**, 503—508 (1964).
NICHOLS, K. E., BOGORAD, L.: Action spectra studies of phycocyanin formation in a mutant of *Cyanidium caldarium*. Botan. Gaz. **124**, 85—93 (1962).
— HARRIS, P., JAMES, A. T.: The biosynthesis of Trans-Δ^3-hexadecanoic acid by *Chlorella vulgaris*. Biochem. biophys. Res. Commun. **21**, 473—479 (1965).
NISSEN, P., BENSON, A. A.: Absence of selenate ester and "selenolipid" in plants. Biochim. biophys. Acta (Amst.) **82**, 400—402 (1964).
OH'EOCHA, C.: The formation of bile pigments. In: Porphyrins and related compounds (T. W. GOODWIN, Ed.). Biochem. Soc. Symp. No. 28, pp. 91—105. London: Academic Press 1968.

Olsen, R. K., Bentley, R., Aiyar, A. S., Djalameh, G. H., Gold, P. H., Ramsey, V. G., Spunger, C. M.: Benzoate derivatives as intermediates in the biosynthesis of coenzyme Q in the rat. J. biol. Chem. **238**, PC3146—PC3148 (1963).

— Davis, G. D., Jr., Moore, H. W., Folkers, K., Parson, W. W., Rudney, H.: 2-multiprenylphenols and 2-decaprenyl-6-methoxyphenol, biosynthetic precursors of ubiquinones. J. Amer. chem. Soc. **1966**, 5919—5923.

Ongun, A., Mudd, J. B.: Biosynthesis of galactolipids in plants. J. biol. Chem. **243**, 1558—1566 (1968).

Overath, P., Stumpf, P. K.: Fat metabolism in higher plants. XXIII. Properties of a soluble fatty acid synthetase from avocado mesocarp. J. biol. Chem. **239**, 4103—4110 (1964).

Parson, W. W., Rudney, H.: The biosynthesis of the benzoquinone ring of ubiquinone from p-hydroxybenzaldehyde and p-hydroxybenzoic acid in rat kidney, *Azotobacter vinelandii* and baker's yeast. Proc. nat. Acad. Sci. (Wash.) **51**, 444—450 (1964).

Perkins, H. J., Roberts, D. W. A.: On chlorophyll turnover in monocotyledons and dicotyledons. Canad. J. Bot. **41**, 221—226 (1963).

Petzold, E. N., Quackenbush, F. W., McQuistan, M.: Zeacarotenes, new provitamins A from corn. Arch. Biochem. **82**, 117—124 (1959).

Pierpoint, W. S.: The distribution of succinate dehydrogenase and malate dehydrogenase among components of tobacco-leaf extracts. Biochem. J. **88**, 120—125 (1963).

Pollack, R. W., Davies, P. J.: Kinetic studies on chloroplastic ribosomal RNA and chlorophyll development during greening of pea buds. Phytochemistry **9**, 471—476 (1970).

Pollard, C. J., Stemler, A., Blaydes, D. F.: Ribosomal ribonucleic acids of chloroplastic and mitochondrial preparations. Plant Physiol. **41**, 1323—1329 (1966).

Popják, G., Cornforth, J. W.: The biosynthesis of cholesterol. Advanc. Enzymol. **22**, 281—335 (1960).

Porra, R. J., Falk, J. F.: The enzymic conversion of coproporphyrinogen III into protoporphyrin IX. Biochem. J. **90**, 69—75 (1964).

— Jones, O. T. G.: Studies on ferrochelatase. 2. An investigation of the role of ferrochelatase in the biosynthesis of various haem prosthetic groups. Biochem. J. **87**, 186—192 (1963).

Porter, J. W., Anderson, D. G.: Biosynthesis of carotenes. Ann. Rev. Plant Physiol. **18**, 197—228 (1967).

Pritchard, G. G., Whittingham, C. P., Griffin, W. J.: Effect of isonicotinyl hydrazide on the path of carbon in photosynthesis. Nature (Lond.) **190**, 553—554 (1961).

Pugh, E. L., Sauer, F., Wakil, S. J.: On the mechanism of fatty acid synthesis in *E. coli*. Fed. Proc. **23**, 166 (1964).

Radmer, R. J., Bogorad, L.: (-)S-adenosyl-L-methionine-magnesium protoporphyrin methyltransferase, an enzyme in the biosynthetic pathway of chlorophyll in *Zea mays*. Plant Physiol. **42**, 463—465 (1964).

Ramirez, D. A., Tomes, M. L.: Relationship between chlorophyll and carotenoid biosynthesis in dirty-red (green-flesh) mutant in tomato. Botan. Gaz. **125**, 221—226 (1964).

Ramirez, J. M., de Campo, F. F., Arnon, D. I.: Photosynthetic phosphorylation as energy source for protein synthesis and CO_2 assimilation. Proc. nat. Acad. Sci. (Wash.) **606**, 59 (1968).

Ranaletti, M. L., Gnanam, A., Jagendorf, A. T.: Amino acid incorporation by isolated chloroplasts. Biochim. biophys. Acta (Amst.) **186**, 192—204 (1969).

Rawson, J. R., Stutz, E.: Sedimentation characteristics in sucrose gradients of *Euglena* cytoplasmic and chloroplast ribosomal RNA. Plant Physiol. **43**, S-18 (1968).

Ray, D. S., Hanawalt, P. C.: Satellite DNA components in *Euglena gracilis* cells lacking chloroplasts. J. molec. Biol. **11**, 760—768 (1965).

Robertson, D. S.: Maize genetic coop. Newsletter **32**, 90 (1958).

Röbbelen, G.: Über die Protochlorophyll-Reduktion in einer Mutante von *Arabidopsis thaliana* (L) Heynh. Planta **47**, 532—546 (1956).

Rogers, L. J., Shah, S. P. J., Goodwin, T. W.: (1) Intracellular localization of mevalonate-activating enzymes in plant cells. Biochem. J. **99**, 381—388 (1966).

— — — (2) Mevalonate-kinase isoenzymes in plant cells. Biochem. J. **100**, 14C (1966).

ROSENBERG, L. L., CAPINDALE, J. B., WHATLEY, F. R.: Formation of oxalacetate and aspartate from phospho-enol-pyruvate in spinach leaf chloroplast extract. Nature (Lond.) **181**, 632—633 (1958).

SAAKOV, V.: Metabolism of violaxanthin-C^{14} in the leaf and its role in the photosynthesis reactions. Dokl. Akad. Nauk SSSR. **165**, 230—233 (1965).

SAGAN, L., BEN-SHAUL, Y., EPSTEIN, H. T., SCHIFF, J. A.: Studies of chloroplast in *Euglena*. XI. Radioautographic localization of chloroplast DNA. Plant Physiol. **40**, 1257—1260 (1965).

SAGER, R.: Inheritance in the green alga *Chlamydomonas reinhardi*. Genetics **40**, 476—489 (1955).

SANO, S., GRANICK, S.: Mitochondrial coproporphyrinogen oxidase and protoporphyrin formation. J. biol. Chem. **236**, 1173—1180 (1961).

SAUER, F., PUGH, F. L., WAKIL, S. J., DELANYE, R., HILL, R. L.: 2-mercoptoethylamine and β-alanine as components of acyl carrier protein. Proc. nat. Acad. Sci. (Wash.) **52**, 1360—1366 (1964).

SCHEUERBBRANDT, G., BLOCH, K.: Unstaturated fatty acids in microorganisms. J. biol. Chem. **237**, 2064—2068 (1962).

SCHIFF, J., EPSTEIN, H. T.: The replicative aspect of chloroplast continuity in *Euglena*. In: Biochemistry of chloroplasts, Vol. I, pp. 341—354 (T. W. GOODWIN, Ed.). London: Academic Press 1966.

SCHOPFER, P., SIEGELMAN, H. W.: Purification of protochlorophyllide holochrome. Plant Physiol. **43**, 990—996 (1968).

SCHWEIGER, H. G., BERGER, S.: DNA-dependent RNA synthesis in chloroplasts of *Acetabularia*. Biochim. biophys. Acta (Amst.) **87**, 533—535 (1964).

— DILLARD, W. L., GIBOR, A., BERGER, S.: RNA-synthesis in *Acetabularia*. I. RNA-synthesis in enucleated cells. Protoplasma (Wien) **64**, 1—12 (1967).

SCOTT, N. S., SHAH, V. C., SMILLIE, R. M.: Synthesis of chloroplast DNA in isolated chloroplasts. J. Cell Biol. **38**, 151—157 (1968).

SHAH, S. P. J., COSSINS, E. A.: The biosynthesis of glycine and serine by isolated chloroplasts. Phytochemistry **9**, 1545—1551 (1970).

SHEMIN, D.: δ-Aminolevulinic acid dehydrase from *Rhodopseudomonas spheroides*. Methods in Enzymology, Vol. V (S. P. COLOWICK, N. O. KAPLAN, Eds.). New York: Academic Press 1962.

SHNEOUR, E. A.: The source of oxygen in *Rhodopseudomonas spheroides* carotenoid pigment conversion. Biochim. biophys. Acta (Amst.) **65**, 510—511 (1962).

SHEPHARD, D. C.: An autoradiographic comparison of the effects of enucleation and actinomycin D on the incorporation of nucleic acid and protein precursors by *Acetabularia* chloroplasts. Biochim. biophys. Acta (Amst.) **108**, 635—643 (1965).

SHIBATA, K.: Spectroscopic studies on chlorophyll formation in intact leaves. J. Biochem. (Tokyo) **44**, 147—173 (1957).

SHIBUYA, I., MARUO, B., BENSON, A. A.: Sulfolipid localization in lamellar lipoprotein. Plant Physiol. **40**, 1251—1256 (1965).

SHIMIZU, S., TAMAKI, E.: Chlorophyllase of tobacco plants II. Enzymic phytylation of chlorophyllide and pheophorbide *in vitro*. Arch. Biochem. **102**, 152—158 (1963).

SIEGELMAN, H. W., CHAPMAN, D. J., COLE, W. J.: The bile pigments of plants. In: Porphyrins and related compounds (T. W. GOODWIN, Ed.). Biochem. Soc. Symp. No. 28, pp. 107—120. London: Academic Press 1968.

SIMONI, R. D., CRIDDLE, R. S., STUMPF, P. K.: Fat metabolism in higher plants XXXI. Purification and properties of plant and bacterial acyl carrier protein. J. biol. Chem. **242**, 573—581 (1967).

SIMPSON, K. L., NAKAYAMA, T. O. M., CHICHESTER, C. O.: Biosynthesis of yeast carotenoids. J. Bact. **88**, 1688—1694 (1964).

SIRONVAL, C., MICHEL-WOLWERTZ, M. R., MADSEN, A.: On the nature and possible functions of the 673- and 684-mμ forms *in vivo* of chlorophyll. Biochim. biophys. Acta (Amst.) **94**, 344—354 (1965).

SISLER, E. G., KLEIN, W. H.: The effect of age and various chemicals on the lag phase of chlorophyll synthesis in dark grown bean seedlings. Physiol. Plant **16**, 315—322 (1963).

SISSAKIAN, N. M.: Enzymology of the plastids. Advanc. Enzymol. **20**, 201 (1958).

— FILIPPOVICH, I. I., SVETAILO, E. N., ALIYEU, K. A.: On the protein-synthesizing system of chloroplasts. Biochim. biophys. Acta (Amst.) **95**, 474—485 (1965).

SMILLIE, R. M.: Formation and function of soluble proteins in chloroplasts. Canad. J. Bot. **41**, 123—154 (1963).
SMIRNOV, B. P.: The biosynthesis of higher acids from acetate in isolated chloroplasts of *Spinacea oleracea* leaves. Biokhimiya **25**, 419—425 (1960).
— Paper chromatography of bile acids in the form of methyl esters (R-COO-$C^{14}H_3$). Biokhimiya **27**, 197—201 (1962).
SMITH, J. H. C.: Protochlorophyll transformations. In: Comparative biochemistry of photoreactive systems, pp. 257—278 (M. B. ALLEN, Ed.). New York: Academic Press 1960.
SMITH, P.: Inheritance of brown and green mature fruit color in peppers. Heredity **41**, 138—140 (1950).
SPENCER, D.: Protein synthesis by isolated spinach chloroplasts. Arch. Biochem. **111**, 381—390 (1965).
— WILDMAN, S. G.: The incorporation of amino acids into protein by cell-free extracts from tobacco leaves. Biochemisty **3**, 954—959 (1964).
— WHITFELD, P. R.: The nature of the ribonucleic acid of isolated chloroplasts. Arch. Biochem. **117**, 337—346 (1966).
STEFFENSON, D. M., SHERIDAN, W. F.: Incorporation of H^3-thymidine into chloroplast DNA of marine algae. J. Cell Biol. **25**, 619—626 (1965).
STEVENS, E., FRYDMAN, B.: Isolation and properties of wheat germ uroporphyrinogen III cosynthetase. Biochim. biophys. Acta (Amst.) **151**, 429—437 (1968).
STEVENSON, J., HEMMING, F. W., MORTON, R. A.: The intracellular distribution of solanesol and plastoquinone in green leaves of the tobacco plant. Biochem. J. **88**, 52—56 (1963).
STOBART, A. K., THOMAS, D. R.: Aminolaevulinic acid dehydratase in tissue cultures of *Kalanchoë crenata*. Phytochemistry **7**, 1313—1316 (1968).
STONE, K. J., WELLBURN, A. R., HEMMING, F. W., PENNOCK, J. F.: The characterization of ficaprenol-10, -11 and -12 from the leaves of *Ficus elastica* (decorative rubber plant). Biochem. J. **102**, 325—330 (1967).
STRAIN, H. H.: Fat-soluble chloroplast pigments: Their identification and distribution in various Australian plants. In: Biochemistry of chloroplasts, Vol. I, pp. 387—406 (T. W. GOODWIN, Ed.). London: Academic Press 1966.
STUMPF, P. K., JAMES, A. T.: Light-stimulated enzymic synthesis of oleic and palmitic acids by lettuce-chloroplast preparations. Biochim. biophys. Acta (Amst.) **57**, 400 (1962).
— — The biosynthesis of long-chain fatty acids by lettuce chloroplasts preparations. Biochim. biophys. Acta (Amst.) **70**, 20—32 (1963).
— BOVÉ, J. M., GOFFEAU, A.: Fat metabolism in higher plants XX. Relation of fatty acid synthesis and photophosphorylation in lettuce chloroplasts. Biochim. biophys. Acta (Amst.) **70**, 260—270 (1963).
— BROOKS, J., GALLIARD, T., HAWKE, J. C., SIMONI, R.: Biosynthesis of fatty acids by photosynthetic tissues of higher plants. In: Biochemistry of chloroplasts, Vol. II, pp. 213—240 (T. W. GOODWIN, Ed.). London: Academic Press 1967.
STUTZ, E., NOLL, H.: Characterization of cytoplasmic and chloroplast polysomes in plants: evidence for three classes of ribosomal RNA in nature. Proc. nat. Acad. Sci. (Wash.) **57**, 774—781 (1967).
SUBBARAYAN, C., KUSHWAHA, S. C., SUZUE, G., PORTER, J. W.: Enzymatic conversion of isopentenyl pyrophosphate-4-^{14}C and phytoene-^{14}C to acyclic carotenes by an ammonium sulfate-precipitated spinach enzyme system. Arch Biochem. **137**, 547 (1970).
SUZUE, G., PORTER, J. W.: Enzymatic synthesis of lycopene from [4-^{14}C] isopentenyl pyrophosphate. Biochim. biophys. Acta (Amst.) **176**, 653 (1969).
TAIT, G. H., GIBSON, K. D.: The enzymic formation of magnesium protoporphyrin monomethyl ester. Biochim. biophys. Acta (Amst.) **52**, 614—616 (1961).
TEVINI, M., LICHTENTHALER, H. K.: Untersuchungen über die Pigment- und Lipochinonausstattung der zwei photosynthetischen Pigmentsysteme. Z. Pflanzenphysiol.**62**, 17—32 (1970).
TEWARI, K. K., WILDMAN, S. G.: DNA-dependent RNA polymerase of tobacco chloroplast. Fed. Proc. **26**, 869 (1967).
THOMAS, D. R., STOBART, A. K.: Mevalonate-activating enzymes in greening tissue cultures Phytochemistry **9**, 1443—1449 (1970).

THRELFALL, D. R., GRIFFITHS, W. T.: Biosynthesis of terpenoid quinones. In: Biochemistry of chloroplasts, Vol. II (T. W. GOODWIN, Ed.). London: Academic Press 1967.
— — GOODWIN, T. W.: (1) Biosynthesis of the prenyl side chains of plastoquinone and related compounds in maize and barley shoots. Biochem. J. **103**, 831—851 (1967).
— — — (2) Biosynthese of the prenyl side chain of plastoquinone and related compounds in maize and barley shoots. Biochem. J. **103**, 831 (1967).
— GOODWIN, T. W.: Nature, Intracellular distribution and formation of terpenoid quinones in *Euglena gracilis*. Biochem. J. **103**, 573—588 (1967).
— WHISTANCE, G. R., GOODWIN, T. W.: (1) Incorporation of (methyl-^{14}C, T) methionine into terpenoid quinones and related compounds by maize shoot. Biochem. J. **102**, 49 P (1967).
— — Utilization of D-tyrosine and p-hydroxyphenylpyruvate for the biosynthesis of plastoquinone and related compounds by maize shoots. Biochem. J. **108**, 24 P—25 P (1968).
TOLBERT, N. E.: Secretion of glycolic acid by chloroplasts. Brookhaven Symposia in Biol. **11**, 271—275 (1959).
— YAMAZAKI, R. K.: Peroxisomes. Ann N.Y. Acad. Sci. **168**, 325 (1969).
TOMES, M. L.: Temperature inhibition of carotene synthesis in tomato. Botan. Gaz. **124**, 180—185 (1963).
TREHARNE, K. J., MERCER, E. I., GOODWIN, T. W.: Incorporation of (^{14}C) carbon dioxide and (2^{14}C) mevalonic acid into terpenoids of higher plants during chloroplast development. Biochem. J. **99**, 239—245 (1966).
TROWN, P. W.: An improved method for the isolation of carboxydismutase. Probable identity with fraction I protein and the protein moiety of protochlorophyll holochrome. Biochemistry **4**, 908—918 (1965).
TROXLER, R. F., BOGORAD, L.: Production of bile pigment from δ-aminolevulinic acid by *Cyanidium caldarium*. In: The biochemistry of chloroplasts, Vol. II, pp. 421—425 (T. W. GOODWIN, Ed.). London: Academic Press 1967.
— LESTER, R.: Biosynthesis of phycocyanobilin. Biochemistry **6**, 3840—3846 (1967).
TSUCHIYA, Y., NOMURA, T.: Formation du verdohaemochrome par les poissons: II. Repartition du degre de formation dans divers tissus animaux et dans les cellules hepatiques chez *Cololabis saira*. Bull. Japan. Soc. Sci. Fisheries **34**, 205—209 (1968).
VAN DEENEN, L. L. M., HAVERKATE, F.: Chemical characterization of phosphatidylglycerol from photosynthetic tissues. In: Biochemistry of chloroplasts, Vol. I, pp. 117—132 (T. W. GOODWIN, Ed.). London: Academic Press 1966.
VAN NIEL, C. B.: Studies on the pigments of the purple bacteria. III. The yellow and red pigments of *Rhodopseudomonas spheroides*. Leeuvenhoed ned. Tijdschr. **12**, 156—166 (1947).
VANDOR, S. L., TOLBERT, N. E.: Glycolate biosynthesis by isolated chloroplasts. Plant Physiol. **43**, S-12 (1968).
VIRGIN, H. L.: Pigment transformation in leaves of wheat after irradiation. Physiol. Plantarum **13**, 155—164 (1960).
— Action spectrum for the elimination of the lag phase in chlorophyll formation in previously dark grown leaves of wheat. Physiol. Plantarum **14**, 439—452 (1961).
WALLACE, R. H., HABERMAN, H. M.: Genetic history and general comparison of two albino mutations of *Helianthus annuus*. Amer. J. Bot. **46**, 157—162 (1959).
WALTON, T. J., BRITTON, G., GOODWIN, T. W.: Biosynthesis of xanthophylls in higher plants: stereochemistry of hydroxylation at C-3. Biochem. J. 383—385 (1969).
WALLWORK, J. C., PENNOCK, J. F.: Some studies on the chemistry and interrelationships of plastquinones. In: Progress in photosynthesis research, Vol. 1, pp. 315—324 (H. METZNER, Ed.). (1968).
WELLBURN, A. R., STONE, K. J., HEMMING, F. W.: The stereochemistry of phytol biosynthesis in the leaves of *Ficus elasticus* and *Aesculus hippocastanum*. Biochem. J. **100**, 23 C—25 C (1966).
— — — Long chain polyisoprenoid alcohols in chloroplasts. In: Biochemistry of chloroplasts, Vol. I, pp. 173—180 (T. W. GOODWIN, Ed.). London: Academic Press 1966.
— STEVENSON, J., HEMMING, F. W., MORTON, R. A.: The characterization and properties of castaprenol-11, -12 and -13 from the leaves of *Aesculus hippocastanum* (horse chestnut). Biochem. J. **102**, 313—324 (1967).

WELLS, L. W., SCHELBE, W. J., PORTER, J. W.: The enzymatic synthesis of carotenes by isolated tomato fruit plastids and spinach leaf chloroplasts. Fed. Proc. **23**, 426 (1964).

WETTSTEIN, D. VON: On the physiology of chloroplast structures. In: Biochemistry of chloroplasts, Vol. I, pp. 14—22 (T. W. GOODWIN, Ed.). London: Academic Press 1965.

— Nuclear and cytoplasmic factors in development of chloroplast structure and function. Canad. J. Bot. **39**, 1537—1561 (1961).

— Chloroplast structure and genetics. In: Harvesting the sun, pp. 153—190 (A. SAN PIETRO, F. A. GREER, T. J. ARMY, Eds.). New York: Academic Press 1967.

WHISTANCE, G. R., THRELFALL, D. R.: Homogentisic acid: a precursor of plastoquinones, tocopherols and α-tocopherolquinone in higher plants, green algae and blue-green algae. Biochem. J. **117**, 593 (1970).

— — Biosynthesis of phytoquinones. Utilization of homogentisic acid by maize shoots for the biosynthesis of plastoquinone. Biochem. J. **110**, 482 (1968).

— — GOODWIN, T. W.: Observations on the biosynthesis of phytoterpenoid quinone and chromanol nuclei. Biochem. J. **105**, 145—154 (1967).

— — Biosynthesis of phytoquinones: an outline of the biosynthesis sequences involved in terpenoid quinone and chromanol formation by higher plants. Biochem. biophys. Res. Commun. **28**, 295—301 (1967).

— — Biosynthesis of phytoquinones: utilization of homogenetisic acid by maize shoots for the biosynthesis of plastoquinone. Biochem. J. **109**, 482—483 (1968).

— — Cinnamic acid and p-coumaric acid, precursors of ubiquinone in higher plants, green algae and fungi. Biochem. J. **118**, 55P (1970).

WIECKOWSKI, S.: The relation between the growth of the leaf and the synthesis of chlorophylls in *Phaseolus vulgaris*. Photochem. Photobiol. **2**, 199—206 (1963).

— GOODWIN, T. W.: Studies on the metabolism of the assimilatory pigments in cotyledons of four species of pine seedlings grown in darkness and in light. Biochemistry of chloroplasts, Vol. II, pp. 445—452 (T. W. GOODWIN, Ed.). London: Academic Press 1967.

WILDMAN, S.: The organization of grana-containing chloroplasts in relation to location of some enzymatic systems concerned with photosynthesis, protein synthesis, and ribonucleic acid synthesis. In: Biochemistry of chloroplasts, Vol. II, pp. 295—320 (T. W. GOODWIN, Ed.). London: Academic Press 1967.

WILLIAM, B. L., GOODWIN, T. W.: The terpenoids of tissue culture of Paul's scarlet rose. Phytochemistry **4**, 81—88 (1965).

WILLIAMS, G. R., NOVELLI, G. D.: Stimulation of an *in vitro* amino acid incorporating system by illumination of dark-grown plants. Biochem. biophys. Res. Commun. **17**, 23—27 (1964).

WILLIAMS, R. J. H., BRITTON, G., CHARLTON, J. M., GOODWIN, T. W.: (1) The stereospecific biosynthesis of phytoene and polyunsaturated carotenes. Biochem. J. **104**, 676—777 (1967).

— — GOODWIN, T. W.: (2) The biosynthesis of cyclic carotene Biochem. J. **105**, 99—105 (1967).

WILSON, L. G., BANDURSKI, R. S.: Enzymatic reactions involving sulfate, sulfite, selenate, and molybdate. J. biol. Chem. **233**, 975—981 (1958).

WOLFF, J. B., PRICE, L.: Terminal steps of chlorophyll *a* biosynthesis in higher plants. Arch. Biochem. **72**, 293—301 (1957).

WOLLGIEHN, R., MOTHES, K.: Über die Incorporation von H^3-thymidin in die Chloroplasten-DNS von *Nicotiana rustica*. Exp. Cell Res. **35**, 52—57 (1964).

YAMAMOTO, H., CHANG, J. L., AIHARA, M. S.: Light induced interconversion of violaxanthin and zeaxanthin in New Zealand spinach-leaf segments. Biochim. biophys. Acta (Amst.) **141**, 342—347 (1967).

— NAKAYAMA, T. O. M., CHICHESTER, C. O.: Studies on the light and dark interconversions of leaf xanthophylls. Arch. Biochem. **97**, 168—173 (1962).

YAKOYAMA, H., YAMAMOTO, H., NAKAYAMA, T. O. M., SIMPSON, K., CHICHESTER, C. O.: Incorporation of 5, 10, 15 ^{14}C-farnesol pyrophosphate into *Phycomyces* carotenoids. Nature (Lond.) **191**, 1299—1300 (1961).

— NAKAYAMA, T. O. M., CHICHESTER, C. O.: Biosynthesis of β-carotene by cell-free extracts of *Phycomyces blakesleeanus*. J. biol. Chem. **137**, 681—686 (1962).

ZELDIN, M. H., SCHIFF, J. A.: RNA metabolism during light-induced chloroplast development in *Euglena*. Plant Physiol. **42**, 922—932 (1967).

Subject Index